NONLINEAR
MATHEMATICS

NONLINEAR MATHEMATICS

Thomas L. Saaty

Office of Naval Research

Joseph Bram

Center for Naval Analyses

DOVER PUBLICATIONS, INC.
NEW YORK

Published in Canada by General Publishing Company, Ltd., 30 Lesmill Road, Don Mills, Toronto, Ontario.

Published in the United Kingdom by Constable and Company, Ltd., 10 Orange Street, London WC2H 7EG.

This Dover edition, first published in 1981, is an unabridged and unaltered republication of the work originally published in 1964 by the McGraw-Hill Book Company, N.Y.

International Standard Book Number: 0-486-64233-X
Library of Congress Catalog Card Number: 81-68482

Manufactured in the United States of America
Dover Publications, Inc.
180 Varick Street
New York, N.Y. 10014

to **LINDA** and **ALAN**

PREFACE

Nonlinear mathematics is the mathematics of today. Modern science demands it. So far, in many areas of applied mathematics and in the physical sciences, we have had to be content with simplifications of nonlinear problems which result in linearized versions. However, no longer may we approximate so simply. We must face up to nonlinear problems with their special demands and peculiar difficulties.

Although we are impressed with the power and diversity of linear methods, we must respond to the challenge to find, in the widely scattered literature, effective ways of viewing nonlinear problems. One such method is the iterative procedure based on the fixed-point property of contraction mappings, which has provided us with a versatile technique made effective by modern high-speed computing devices that can rapidly perform step-by-step procedures. However, in special cases an abundance of other ingenious techniques can be found.

This volume is intended for the teacher who feels a need to unify the prolific subject matter of nonlinear mathematics. We offer teachers access to a large number of related topics and include selected references to readily available modern literature on the subject.

The intent of the exercises is twofold: in some cases to train the student in the mechanics of a method, in others to guide him through the proof of a theorem.

Besides mathematicians, we have written this book with scientists, engineers, economists, operational analysts, and other interested readers in mind. For this reason we frequently include background material not generally well known by nonmathematicians.

Our purpose is to discover for the reader some essential qualities of mathematical systems and models unrestricted by linearity assumptions and to present to him, in each case, a variety of ideas, hoping that he may acquire from this variety the courage to apply his own creative talents to nonlinear mathematics.

We have attempted to construct the presentation of the ideas according to existence and uniqueness theorems, characterization (e.g., stability and asymptotic behavior), construction of solutions, convergence, approximation and errors.

This book is the product of several courses which we have given in linear and nonlinear mathematics. Our students' eagerness to discover what is known in nonlinear cases encouraged us to expand the material to as rich a context as possible while still confining the direction of the text toward senior and graduate classes in mathematics and science. The heterogeneity of our students, from economists hungering for information to mathematicians seeking new methods and new problems, enabled us to vary the level of presentation according to the needs. This volume reflects the compromise we made as teachers; we believe that its usefulness is thereby enhanced.

We have not hesitated to use methods, where they are applicable, from such disciplines as functional analysis and the calculus of variations. In so doing, it has also been our purpose to promote a greater familiarity with such methods within the chosen context. It goes without saying that a book, to be effective, must be comprehensible. Having tested the material on our own students, we feel that we have achieved a measure of success.

In selecting our material for teaching and for the text, it was natural for us to choose those ideas, theorems, and methods which appeared to be of current interest. We regret that considerable material had to be rejected because of the limited space available and because of our initial decision to be accessible to as wide an audience as possible. However, as the topic of nonlinear mathematics in its many facets is a dynamic one in which many individuals are conducting research, we attempted to cover as wide a base for academic purposes as possible. This book is not intended to be the last word on any of the subjects included. However, it is hoped that the material contained in this volume and the manner in which it is presented will be suitable for a semester course in selected topics in nonlinear mathematics.

We have postponed for a future volume nonlinear partial differential, integral, and difference equations and the subject of numerical methods, etc., which can only be treated after the subjects to which it is applied have been accounted for.

This is the first book attempting to penetrate such a large area of nonlinear mathematics. At present there is no simple unifying theory in nonlinear mathematics analogous to vector spaces and operators for linear mathematics. We have tried to give more than a compilation of known techniques, to convey a spirit. In this spirit we integrate different topics of mathematics to show their interrelations, to stimulate inventiveness through variety, and to emphasize their origins from interaction with natural phenomena in physics, engineering, and economics. This must be first if there is ever to be a unifying theory. And that unifying theory will hardly be a simple extension of the linear case. We have established firm footholds low enough to remain near linear grounds.

We are surrounded and deeply involved, in the natural world, with nonlinear events which are not necessarily mathematical. For example, our emotional reaction to music or, in general, our likes and dislikes, are nonlinear. Another area of experience is our kinaesthetic reactions (for example, to heat and cold) or our complicated muscular compensation in the nonlinear problem of pedaling a bicycle up and down a hillside. On a grand scale, we can follow Volterra's model of the struggle for existence between two species, one of which preys exclusively on the other. It is interesting to note that this dynamic situation is described by nonlinear differential equations giving the time rate of change of each population (see Chapter 4).

However, the classical methods of mathematical physics have often used linearization, and we have become accustomed to idealizations of natural phenomena which are frequently made for the sake of arriving at explicit solutions. One assumes ideal homogeneous and uniform fluids, no friction, perfect insulation, isoptropic media, weightless and perfectly rigid objects, molecules that are infinitesimally small solid spheres, and collisions of zero time duration. Many a curve has been bravely approximated by a straight line in the appropriate range.

For the reader who has been thus bemused into an artificially linear conception of the universe, we recommend the interesting counterexamples given by L. D. Kovach in his article, Life Can Be So Nonlinear, *American Scientists*, vol. 48, 1960. Another article which the reader will enjoy is by C. Truesdell, Recent Advances in Rational Mechanics, *Science*, April, 1958.

It may be illuminating to examine briefly the general approach to linear and nonlinear equations. Spectral theory is used in the linear case to obtain a general resolution in terms of eigenfunctions for an arbitrary right side of an equation of the forms $(\lambda I - A)x = y$ and to give a complete understanding of the operator associated with the system. For a linear differential equation of a physical system, the eigenvalues correspond to the natural frequencies of vibration of the unforced homogeneous system $(\lambda I - A)x = 0$ (see Chapter 4). In order to solve the previous nonhomogeneous system, one must exclude these frequencies. We shall now elaborate on this.

If A is a linear operator, and if $Ax = \lambda x$ [i.e., $(\lambda I - A)x = 0$] has a nonzero solution, then λ belongs to the spectrum of the operator A. If, for example, A is a matrix, then a nontrivial solution implies that the determinant $|\lambda I - A|$ vanishes; this occurs for all the characteristic values. These are the roots of the determinantal polynomial. In the particular space in which solution of a linear operator equation is desired [e.g., $L_1(0, \infty)$ for $(dx/dt) - \lambda x = 0$, thus Re $(\lambda) < 0$] the spectrum can be divided into point, residual, and continuous parts. The resolvent set consists of those λ not in the spectrum. For these values the resolvent of A, i.e., $(\lambda I - A)^{-1}$, exists (e.g., since the determinant is not zero the matrix is invertible) so that one can, for example, solve a nonhomogeneous equation $x = (\lambda I - A)^{-1}y$. If $|\lambda| > \|A\|$, then $(\lambda I - A)^{-1} = (1/\lambda) \sum_{k=0}^{\infty} A^k/\lambda^k$ exists, i.e., the series converges and can be applied to y and evaluated.

Another useful treatment of the resolvent is found in the theory of a one-parameter semigroup of linear transformations $T(\xi)$, $0 \leq \xi < \infty$ of a Banach space X into itself. Here $T(\xi)$ is a semigroup if $T(\xi_1 + \xi_2)x = [T(\xi_1) \cdot T(\xi_2)]x$, $0 < \xi_1$, $\xi_2 < \infty$. To illustrate this treatment let it be required to find all solutions of $t^2 \, dx/dt + \lambda x = y(t)$ in the space $C[0, \infty]$ of bounded continuous functions. We shall regard this equation in the form $(\lambda - A)x = y$. We first examine the homogeneous equation with $y(t) \equiv 0$. The resulting equation has the solution $x = ae^{\lambda/t}$ which is in $C[0, \infty]$ for Re $(\lambda) < 0$ and for $\lambda = 0$. Actually λ is also in the spectrum for Re $(\lambda) = 0$ which follows from the ensuing discussion. We ask if there is a semigroup $T(\xi)$ of which the unbounded operator $A = -t^2 d/dt$ is the infinitesimal generator, i.e., satisfies

$$\frac{d}{d\xi} T(\xi)x = AT(\xi)x = T(\xi)Ax$$

for all x such that $t^2 x'(t)$ is in $C[0, \infty]$. We find that if $T(\xi)x(t) = x[t/(1 + \xi t)]$, then indeed $(d/d\xi)x[t/(1 + \xi t)] = x'[t/(1 + \xi t)](-t^2)/(1 + \xi t)^2$. This is $T(\xi)$ acting on $Ax = -t^2 x'(t)$, and hence $A = -t^2 d/dt$.

Formally from $(dT/d\xi) = AT$, we can expect that $T = \exp(\xi A)$ in some sense. We want the solution of $(\lambda - A)x = y$, which is $x = (\lambda - A)^{-1}y$. Since

$$\int_0^{\infty} e^{-p\xi} \, d\xi = 1/p$$

for Re $(p) > 0$, we have formally

$$\int_0^{\infty} \exp[-\xi(\lambda - A)] \, d\xi = \int_0^{\infty} e^{-\xi\lambda}T(\xi) \, d\xi = 1/(\lambda - A),$$

the resolvent of A. This holds for Re $(\lambda) > \lim_{\xi \to \infty} \log \|T(\xi)\|/\xi$. For our problem this gives $\int_0^{\infty} e^{-\xi\lambda}y[t/(1 + \xi t)] \, d\xi$ for Re $(\lambda) > 0$. Thus the resolvent is in the right half-plane, and since the spectrum is closed it occupies the left half-plane and the rest of the imaginary axis. Thus Re $(\lambda) \leq 0$ for λ in the spectrum. After simplification of the resolvent [put $s = t/(1 + \xi t)$ and $r = 1/s$], we have the resolvent acting on a function y is $e^{\lambda/t} \int_{1/t}^{\infty} e^{-\lambda r} y(1/r) \, dr$, Re $(\lambda) > 0$, which is the solution of the non-homogeneous equation.

If A is nonlinear, then it is often convenient to transform the equation into the equivalent form $Ax = x$; and if the contraction mapping hypotheses are satisfied, obtain the solution as the limit of iterations defined by $x_{n+1} = A(x_n)$, $n = 0, 1, 2,$ . . . In other cases one may often establish the existence of a solution (if not a sequence of approximations) by a fixed-point theorem. A recent new result concerning nonlinear operators due to G. J. Minty, which was announced in the September, 1963, issue of the *Bull. Am. Math. Soc.* (too late to include in this text) is:

If A is an operator—everywhere defined, continuous, and monotonic—and satisfies for some real M

$$\|x\| > M \text{ implies Re } [x, Ax] \geq 0$$

then the equation $Ax = 0$ has a solution which is unique if A is strictly monotonic.

An operator A which maps a subset of a Hilbert space with real or complex scalars into the space is monotone provided that if x_1 and x_2 are in the subset then

$$\text{Re } [x_1 - x_2, F(x_1) - F(x_2)] \geq 0$$

One may philosophically reflect that man has been solving many nonlinear problems without the use of formal nonlinear methods. The presence of mathematical concepts of nonlinearity has made it much easier to analyze and solve simple scientific nonlinear problems. How much promise is there in applying these methods to physical problems arising in operations? One is inclined to speculate cautiously that qualitative theories will, to some extent, surpass the present quantitative ones in mathematics. One outstanding example of the qualitative approach to nonlinear problems is Lyapunov's theory of stability, which essentially puts aside the question of constructing the solutions and investigates the stability of the solutions. Even in cases in which the solutions are explicitly given, the question of stability is the crucial one.

The subject matter of the book is such that no title would be satisfactory from every viewpoint. Hence, we settled on what is perhaps an ambitious title, since nonlinear mathematics is the mathematics for treating nonlinear problems.

We are indebted to many colleagues and associates in preparing this volume. We are above all grateful to Mrs. Laura G. Barnhill for her valuable help in preparing and typing the manuscript and to Mrs. Kay Dunn. For constructive comments and suggestions, our thanks go to

Prof. Henry Antosiewicz
Mr. E. M. L. Beale
Mr. Morton Cooper
Mr. Ralph Cooper
Prof. Rene DeVogelaere
Prof. Stephen Diliberto
Mr. Stanley W. Doroff
Dr. William S. Dorn

Prof. Nelson Dunford
Dr. Kurt Eisemann
Mr. Anthony V. Fiacco
Mr. Karl Ganow
Mr. Dhirendra Nath Gosh
Dr. Jack Hale
Dr. Robert Kalman
Mr. James E. Kelley, Jr.

Dr. Kenneth Krohn
Dr. Joseph LaSalle
Prof. George Leitmann
Prof. Lawrence Markus
Mr. Garth McCormick
Prof. J. B. Rosen
Commander D. W. Sencenbaugh

Dr. B. L. Silverstein
Prof. Edward Stiefel
Dr. Arnold Stokes
Prof. John Todd
Prof. J. Wegner
Dr. Philip Wolfe
Prof. Lotfi Zadeh

Thomas L. Saaty
Joseph Bram

CONTENTS

ONE

LINEAR AND NONLINEAR TRANSFORMATIONS

1-1 INTRODUCTION

An operator $\mathfrak{I}$ is said to be *linear* if the effect of operating on the sum of two entities, e.g., functions, is equal to the sum of the effects of operating on them separately; i.e.,

$$\mathfrak{I}(f + g) = \mathfrak{I}(f) + \mathfrak{I}(g)$$

and
$$\mathfrak{I}(af) = a\mathfrak{I}(f)$$

where a is a constant; $\mathfrak{I}$ is *nonlinear* if it is not linear.

If $\mathfrak{I}$ is a linear transformation, then $\mathfrak{I}(f) = g$ is a linear equation; if $g = 0$, it is *homogeneous;* otherwise it is *non-homogeneous.* The latter terms are of no value if $\mathfrak{I}$ is nonlinear. If $\mathfrak{I}$ is nonlinear, then $\mathfrak{I}(f) = 0$ is known as a nonlinear equation. One can define the *deficiency* of a nonlinear operator by the difference

$$\mathfrak{I}(f + g) - [\mathfrak{I}(f) + \mathfrak{I}(g)]$$

An example of a linear operator is the differential operator

$$\mathfrak{I}(f) = \frac{d^n f}{dx^n}$$

for which both linearity properties are easily verified. An example of a nonlinear operator is

$$\mathfrak{I}(f) = \left(\frac{d^n f}{dx^n}\right)^2 - f^3$$

or
$$\mathfrak{I}(f) = \int_0^1 k(f,t)\, dt$$

where $k(f,t)$ is a nonlinear function of f. The reader should determine the deficiencies of these two nonlinear operators.

Today a student interested in linear processes will gain the deepest and clearest insight by being exposed to abstract representations in linear algebra, finite-dimensional vector spaces, and abstract normed linear spaces. Equipped with this insight he is free and at ease in exploring the linear aspects of various fields of application. However, at present the subject of abstract nonlinear operators is a limited one; consequently special cases must be successfully analyzed before we can begin to formulate the abstract theory.

In the remainder of this chapter we give a brief, motivated version of essential concepts in abstract linear spaces and follow it with some of the better-known facts pertaining to nonlinear operators.

A system of simultaneous nonlinear equations, studied in Chap. 2, is encountered in the solution of problems arising in all the remaining chapters. Since the solution of such a system can be obtained by the method of steepest descent, which is an optimization technique, and since problems of inequalities and equations are frequently related, it is clear that the field in which, at present, inequalities play a significant role should be represented. Therefore we included the interesting optimization subject of nonlinear programming in Chap. 3. The topic of differential equations covered in Chap. 4 is the heart of nonlinear mathematics, and it occupies a substantial part of the book, contributing a variety of methods and ideas. From a theoretical standpoint Chap. 5 is a union of differential equations and the calculus of variations. From a practical standpoint one must introduce control systems to bring about this union. By introducing such systems one assumes additional responsibilities, such as settling stability questions.

Chapter 6 presents an aspect of probability theory concerned with useful nonlinearities. For example, in many cases, without prediction and filtering, control systems would be handicapped, if not completely ineffective.

Each chapter of the book is essentially independent and does not require familiarity with the remaining chapters. There is one exception to this, namely the dependence of an understanding of control theory on ordinary nonlinear differential equations.

The triangle of Fig. 1-1 describes the relations among three aspects of the topics to be discussed in this volume. Optimization theory, i.e., the theory of maximum, minimum, and minimax in the constrained form of ordinary calculus, in the context of variational problems, and in game theory, provides techniques for selecting the most "desirable" (in some sense) of all possible solutions of descriptive mathematical models. Such models are formulated either "deterministically" by algebraic equations, differential, difference, and integral equations, or by probability theory and stochastic processes. An application of optimization to control theory will be given in Chap. 5. Chapter 3 on nonlinear programming is concerned with a part of optimization theory which is in a state of rapid and intense development. In Chap. 2, the method of steepest descent is used to find the minimum of a function. Variational methods are found in condensed form in Chap. 5, as they are adequately treated in many recent books on the subject of the calculus of variations. The subject of differential equations itself provides an input into optimization theory in

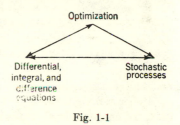

Fig. 1-1

the form of necessary conditions for a stationary solution. In this manner, for example, one obtains Euler's equation—a differential equation— in the calculus of variations. The relation of optimization theory to stochastic processes has so far been unidirectional. One attempts to minimize errors in least-square stochastic methods and looks for maximum-likelihood estimates of probability-distribution parameters. Another application of optimization theory to probability is in seeking the distribution which yields the maximum entropy in information theory.

Differential and integral equations play an important role in the theory of stochastic processes. Integral equations arise in studying joint probability distributions, differential equations in representing transition probabilities and obtaining generating functions. One often takes a probability problem and derives a differential equation from it. For example, a wide class of probability problems can be reduced to the Fokker-Planck differential equation. A recent interesting subject of research has been the study of differential equations with stochastic right sides. Thus we have an input from probability theory to differential equations.

The triangle of Fig. 1-1 can now be viewed as the base of a tetrahedron

whose apex is abstract mathematics, of which functional analysis is the
most pertinent to the triangular base.   It is for this reason that we start
with the elements of functional analysis, which provides the most satisfactory and comprehensive framework for the foregoing subjects.   The
marriage of algebra and topology that begot functional analysis has had
great success in simplifying classical results in analysis as well as providing the methods for achieving new ones.

The material of this volume provides a unity of the elements in the
triangle shown in Fig. 1-2.   However, the subjects of partial differential
equations, integral and difference equations, and numerical methods had
to be deferred to the future.   We felt that numerical methods could best
be discussed after background material had been given on subjects in
which these methods are used.

In the major part of this chapter, we shall be concerned with vector
spaces and linear transformations.   These are the most fruitful concepts

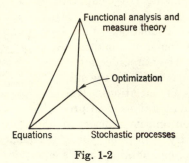

Fig. 1-2

for the study of the solvability of linear algebraic systems and their
generalizations to infinite systems, integral equations, and boundary-
value problems in differential equations, which occur so frequently in
analysis.   Needless to say, the problem of solvability must often be
preceded by a study of existence and uniqueness, and in the absence of
explicit solutions, one must examine general behavior through characterization theory, which in this case is provided by spectral theory, of
which we give a special case in Theorem 1-4.   This theorem, which
asserts that every compact self-adjoint operator can be "reduced to
diagonal form," is the basis of the functional calculus for operators, which
enables one to understand the behavior of various functions of operators
that arise frequently, for example, $\exp tA$, $(1 - \lambda A)^{-1}$, and $\sqrt{A}$.

In Secs. 1-2 to 1-5, we treat finite-dimensional vector spaces and transformations, which we soon specialize to inner-product spaces and self-
adjoint transformations.   An introduction to the infinite-dimensional
case (Hilbert space) is given in Secs. 1-6 and 1-7, and this is followed by

some applications in Sec. 1-8. Section 1-9 provides a glimpse into the more general Banach-space situation; this is used in Sec. 5-4. Section 1-10, is devoted to some of the small but useful amount of general information that we have on nonlinear mappings. The remaining section, Sec. 1-11, contains a survey of Lebesgue integration.

1-2 VECTOR SPACES; LINEAR TRANSFORMATIONS

The problem of finding the values of $x_1, x_2, \ldots, x_n$ which satisfy the n "simultaneous" linear equations

$$\sum_{j=1}^{n} a_{ij}x_j = c_i \qquad i = 1, 2, \ldots, n \qquad (1\text{-}1)$$

is an old one, and there is a complete set of answers to the questions of existence and uniqueness. If the determinant $|a_{ij}|$ is nonzero, then a unique solution exists; if the determinant is zero, then either no solution exists or infinitely many exist. When new problems arose involving infinitely many linear equations in infinitely many unknowns, and when integrals replaced sums, e.g.,

$$f(t) + \int_0^1 K(t,s)f(s)\ ds = g(t) \qquad (1\text{-}2)$$

in which $f(t)$ is the unknown function, completely new methods were naturally required, but the analogy with the finite system of equations was always a strong and fecund motivation.

In every such problem, we can imagine taking a set $\{x_i\}$ or a function $f(t)$ and doing something to the $\{x_i\}$ or $f(t)$ to get a new set of numbers using the left side of (1-1), or a new function using the left side of (1-2). If the new set of numbers is the set c_i, then (1-1) is solved; if the new function is $g(t)$, then (1-2) is solved. We call this "doing something" to the $\{x_i\}$ or $f(t)$ *transforming*, and the rule for carrying out the process is called a *transformation*, which we usually label with a capital letter; e.g.,

$$\{x_i\} \xrightarrow{A} \Big\{ \sum_{j=1}^{n} a_{ij}x_j \Big\} \qquad i = 1, \ldots, n$$

or
$$f(t) \xrightarrow{B} f(t) + \int_0^1 K(t,s)f(s)\ ds$$

which we read: $\{x_i\}$ is sent to (or carried into) $\Big\{ \sum_{j=1}^{n} a_{ij}x_j \Big\}$ by the transformation A; $f(t)$ is sent to the function $f(t) + \int_0^1 K(t,s)f(s)\ ds$ by the transformation B.

The problems we are discussing here are *linear* problems in the following sense: If $\{x_{1i}\}$ and $\{x_{2i}\}$ are sets on which A can act, then so is $\{x_{1i} + x_{2i}\}$, and $\{x_{1i} + x_{2i}\} \xrightarrow{A} A\{x_{1i}\} + A\{x_{2i}\}$, in which we have written $A\{x_{1i}\}$ for the result of transforming $\{x_{1i}\}$; that is,

$$A\{x_{1i}\} = \left\{\sum_{j=1}^{n} a_{ij}x_{1j}\right\}$$

Furthermore, $\{\lambda x_i\}$ is also a suitable subject for A if λ is any number, and $A\{\lambda x_i\} = \lambda A\{x_i\}$; that is,

$$\{\lambda x_i\} \xrightarrow{A} \left\{\lambda \sum_{j=1}^{n} a_{ij}x_j\right\}$$

Similar observations are valid for the transformation B acting on functions $f(t)$. The ideas are elementary, but the notation and phraseology are cumbersome, so we remedy the situation with some standard terminology and notation.

Guided by the familiar examples of vectors in 2 and 3 space, we define a real (complex) vector space $\mathcal{V}$ to be a collection of elements $\{x,y,z, \ldots\}$ together with a rule for "addition" (for every x, y in $\mathcal{V}$, there is a unique element $x + y$ in $\mathcal{V}$) and a rule for "scalar multiplication" by real (complex) scalars, or numbers (for every x in $\mathcal{V}$ and every scalar λ, there is a unique element λx in $\mathcal{V}$). We do not specify the rules for addition and scalar multiplication, in general, except to require that, whatever the rules are, the following axioms must be satisfied:

1. $x + (y + z) = (x + y) + z$ for all x, y, z in $\mathcal{V}$.
2. There is a unique element 0 in $\mathcal{V}$ such that $x + 0 = 0 + x = x$ for all x in $\mathcal{V}$.
3. For every x in $\mathcal{V}$, there is a unique "inverse" $-x$ such that $x + (-x) = (-x) + x = 0$. [We shall write $x - y$ for $x + (-y)$ henceforth.]
4. $x + y = y + x$ for all x, y in $\mathcal{V}$.
5. $\lambda(x + y) = \lambda x + \lambda y$ for all real (complex) λ and all x, y in $\mathcal{V}$.
6. $(\lambda + \kappa)x = \lambda x + \kappa x$ for all real (complex) λ, κ and all x in $\mathcal{V}$.
7. $0 \cdot x = 0$ for all x in $\mathcal{V}$.
8. $1 \cdot x = x$ for all x in $\mathcal{V}$.
9. $\lambda(\kappa x) = (\lambda\kappa)x$ for all λ, κ, x.

These axioms are not all independent.

Question: What is the only difference between a real vector space and a complex vector space?

Exercise 1-1 Let $\mathcal{V}$ be the collection of all real n-tuples $(x_1,x_2, \ldots .x_n)$ *Define* addition as follows: $(x_1, \ldots ,x_n) + (y_1, \ldots ,y_n) = (x_1 + y_1, \ldots , x_n + y_n)$;

define scalar multiplication as follows (λ real):

$$\lambda(x_1, \ldots, x_n) = (\lambda x_1, \ldots, \lambda x_n)$$

Verify that the axioms are all satisfied.

Exercise 1-2 Let $\mathcal{V}$ be the collection of all real-valued continuous functions $f(t)$ on the unit interval $I = [0,1]$. Define $f + g$ as that function whose value at t in I is $f(t) + g(t)$; define λf as that function whose value at t is $\lambda f(t)$, λ real. Then $\mathcal{V}$ is a vector space.

If $\mathcal{V}$ is a vector space and there is a set $e_1, e_2, \ldots, e_n$ in $\mathcal{V}$ such that (1) no linear combination $\sum\limits_{i=1}^{n} \lambda_i e_i$ of the e_i's vanishes unless the λ_i's are all zero (the e_i's are linearly independent) and (2) for every x in $\mathcal{V}$ there exist scalars $\lambda_1, \ldots, \lambda_n$ such that $x = \sum\limits_{i=1}^{n} \lambda_i e_i$ (the e_i's span $\mathcal{V}$), then e_1, $\ldots$, e_n is called a *basis* for $\mathcal{V}$, and $\mathcal{V}$ is finite-dimensional; its dimension is n. (We can define the dimension this way because any other basis in $\mathcal{V}$ will also consist of exactly n elements.) If no such set exists, $\mathcal{V}$ is called infinite-dimensional. The space $\mathcal{V}$ of Exercise 1-1 is n-dimensional; in Exercise 1-2, $\mathcal{V}$ is infinite-dimensional. The space $\mathcal{V}$ of real (complex) n-tuples is the prototype of all n-dimensional vector spaces; if one knows all about matrix theory, then he can easily prove every result in finite-dimensional vector spaces once he knows the terminology, of which there is much more than we have given so far. However, very few of the corresponding results in infinite-dimensional spaces can be proved by using matrices, and a wholly geometric (spatial) point of view becomes not only desirable but necessary.

If $\mathcal{V}$ and $\mathcal{W}$ are two (possibly the same) vector spaces, and A is a mapping from $\mathcal{V}$ to $\mathcal{W}$ (for each x in $\mathcal{V}$, there is a unique y in $\mathcal{W}$—written $y = Ax$), in symbols,

$$A : \mathcal{V} \to \mathcal{W}$$

then A is called a *linear* transformation if $A(\lambda x + \kappa y) = \lambda Ax + \kappa Ay$ for all x, y in $\mathcal{V}$ and all scalars λ, κ.

We have already observed that the transformations A and B defined by the left sides of (1-1) and (1-2) are linear transformations. The problems posed by (1-1) and (1-2) can be phrased as follows: Given the transformation A (we will henceforth drop the qualifier "linear"—all our transformations here will be linear) and the vector c, find x such that $Ax = c$. This problem leads to others related to it, e.g., uniqueness or multiplicity of solutions and possible methods of calculating solutions. To solve such problems, it is at least desirable to know all we can about the properties of (linear) transformations.

1-3   EIGENVALUES; EIGENVECTORS

Suppose we have a system of equations similar to (1-1),

$$\sum_{j=1}^{n} a_{ij}x_j = c_i \qquad i = 1, 2, \ldots, n$$

in which the matrix $A = \{a_{ij}\}$ is diagonal ($a_{ij} = 0$ if $i \neq j$). Then our problem is very simple; we have $x_i = c_i/a_{ii}$, assuming $a_{ii} \neq 0$, for each i. The corresponding linear transformation

$$\{x_i\} \xrightarrow{A} A\{x_i\}$$

is also very simple.   We have $A\{x_i\} = \{a_{ii}x_i\}$; no sums appear here. The vectors

$$e_1 = (1,0, \ldots ,0), \ e_2 = (0,1,0, \ldots ,0), \ \ldots , e_n = (0,0, \ldots ,0,1)$$

evidently form a basis for our vector space of n-tuples.   We see that $Ae_1 = a_{11}e_1, \ Ae_2 = a_{22}e_2, \ \ldots , Ae_n = a_{nn}e_n$.   So each of our basis vectors has the pleasant property that when it is acted upon by A, it is sent to a scalar multiple of itself.

Now if A is any transformation in a finite-dimensional $\mathcal{V}$, and if we can find a basis $u_1, \ldots , u_n$ of $\mathcal{V}$ for which $Au_i = \lambda_i u_i, i = 1, \ldots , n$ in which the λ_i are scalars, then we are in essentially the same situation as in the last paragraph.   If c is in $\mathcal{V}$ and we want the solution of $Ax = c$, we need only determine the scalars κ_i for which

$$c = \sum_{i=1}^{n} \kappa_i u_i$$

and then we have

$$x = \sum_{i=1}^{n} \frac{\kappa_i}{\lambda_i} u_i$$

if the λ_i are all nonzero.   The only unpleasant eventualities that may arise are that the λ_i's may be complex even though we are in a real vector space, or there may not be a basis of such vectors u_i.   We shall soon restrict ourselves to an important special type of transformation for which such eventualities are impossible.

At any rate, we are ready to define eigenvalues and eigenvectors of a transformation A.   If, for some scalar λ, there is a nonzero u in $\mathcal{V}$ such that $Au = \lambda u$, then λ is called an *eigenvalue*, and u is an *eigenvector* corresponding to λ.   Other terms for eigenvalue are *proper value, latent root*, and *characteristic root;* eigenvectors are sometimes called *proper vectors, latent vectors*, or *characteristic vectors.*

For every transformation $A : \mathcal{V} \to \mathcal{V}$, $\mathcal{V}$ finite-dimensional, and every basis $e_1, \ldots, e_n$ in $\mathcal{V}$, there is a unique matrix $M = M(A; e_1, \ldots, e_n)$ with entries a_{ij} which are defined by the equations

$$A e_j = \sum_{i=1}^{n} a_{ij} e_i \qquad j = 1, 2, \ldots, n \qquad (1\text{-}3)$$

Then, if x is any vector in $\mathcal{V}$, we can easily calculate Ax as follows: Write x in terms of $e_1, e_2, \ldots, e_n$:

$$x = \sum_{i=1}^{n} \xi_i e_i$$

Then we must have

$$Ax = \sum_{j=1}^{n} \xi_j A e_j = \sum_{j=1}^{n} \xi_j \sum_{i=1}^{n} a_{ij} e_i$$

or
$$Ax = \sum_{i=1}^{n} \sum_{j=1}^{n} a_{ij} \xi_j e_i$$

In brief, if we know $A e_i$ for $i = 1, \ldots, n$, then we also know Ax for every x. Moreover, we have an easily described algorithm for finding the components of Ax with respect to $e_1, \ldots, e_n$ (that is, the scalars in $Ax = \sum_{i=1}^{n} \eta_i e_i$); we simply write the components of $x = \sum_{i=1}^{n} \xi_i e_i$ in a column and multiply on the left by $M(A; e_1, \ldots, e_n)$:

$$\begin{pmatrix} \eta_1 \\ \cdots \\ \eta_n \end{pmatrix} = \begin{pmatrix} a_{11} & \cdots & a_{1n} \\ \cdots & \cdots & \cdots \\ a_{n1} & \cdots & a_{nn} \end{pmatrix} \begin{pmatrix} \xi_1 \\ \cdots \\ \xi_n \end{pmatrix}$$

Exercise 1-3 Let $\mathcal{V}$ be the three-dimensional vector space of all real quadratic polynomials $a + bt + ct^2$, and let A be the transformation that sends each such polynomial into its derivative $b + 2ct$. If we use the basis $e_1 = 1$, $e_2 = t$, $e_3 = t^2$, find $M(A; e_1, e_2, e_3)$. If we use $e_1 = 1 + t$, $e_2 = t - t^2$, $e_3 = t^2$, find M again.

Exercise 1-4 Let $e_1, \ldots, e_n$ and $e_1', \ldots, e_n'$ be two bases in $\mathcal{V}$. Write

$$e_j' = \sum_{i=1}^{n} p_{ij} e_i \qquad j = 1, \ldots, n$$

or, symbolically,

$$(e_1', \ldots, e_n') = (e_1, \ldots, e_n) \begin{pmatrix} p_{11} & \cdots & p_{1n} \\ \cdots & \cdots & \cdots \\ p_{n1} & \cdots & p_{nn} \end{pmatrix}$$

Then the matrix $P = \{p_{ij}\}$ is necessarily invertible (P^{-1} exists). Let

$$M_1 = M(A; e_1, \ldots, e_n) \qquad \text{and} \qquad M_2 = M(A; e_1', \ldots, e_n')$$

Express M_2 in terms of P and M_1. (*Answer:* $M_2 = P^{-1} M_1 P$.)

Let A be a transformation $A : \mathcal{U} \to \mathcal{U}$, $\mathcal{U}$ finite-dimensional. Then λ is an eigenvalue of A if there is a nonzero u such that $Au = \lambda u$. Take a basis $e_1, \ldots, e_n$ in $\mathcal{U}$, and let $M(A)$ be the matrix corresponding to A (and $e_1, \ldots, e_n$). Then λ is an eigenvalue if there is a nonzero column vector $(\xi_1, \ldots, \xi_n)$ such that

$$M(A) \begin{pmatrix} \xi_1 \\ \cdots \\ \xi_n \end{pmatrix} = \lambda \begin{pmatrix} \xi_1 \\ \cdots \\ \xi_n \end{pmatrix}$$

or

$$\begin{pmatrix} a_{11} - \lambda & a_{12} & \cdots & a_{1n} \\ a_{21} & a_{22} - \lambda & \cdots & a_{2n} \\ \cdots\cdots\cdots\cdots\cdots\cdots\cdots\cdots \\ a_{n1} & a_{n2} & \cdots & a_{nn} - \lambda \end{pmatrix} \begin{pmatrix} \xi_1 \\ \xi_2 \\ \cdots \\ \xi_n \end{pmatrix} = 0$$

But this is possible if and only if the determinant of the coefficients vanishes; in symbols,

$$|A - \lambda I| = 0 \tag{1-4}$$

in which I is the identity matrix. If we expand this determinant, in which λ is an unknown, we get a polynomial starting with $(-1)^n \lambda^n$. This polynomial [or rather $(-1)^n$ times the polynomial, which begins with λ^n] is called the *characteristic polynomial* of A; it does not depend on which basis is used. The eigenvalues of A are precisely the roots of this polynomial (of degree n). If we are in a real vector space, the coefficients of the characteristic polynomial are real, of course.

1-4 INNER - PRODUCT SPACES

Much more can be said and done for arbitrary transformations in finite-dimensional spaces, but we shall now call into play the familiar metric structure of euclidean space and specialize our transformation to a particular class of transformations that have extremely elegant properties and which arise with such frequency in applications as to merit the special attention they enjoy. The euclidean metric, or distance, between two vectors u and v in 3 space is the number

$$\|u - v\| = \Big(\sum_{i=1}^{3} |u_i - v_i|^2 \Big)^{1/2}$$

where u_i and v_i are the respective components of u and v along three pairwise-orthogonal axes. In n space, we can proceed as follows: We take a basis $e_1, \ldots, e_n$ in $\mathcal{U}$ and declare by fiat that they are pairwise-orthogonal and of unit length. We then define the inner product (or dot product) between any two vectors u and v to be the number

$$(u,v) = \sum_{i=1}^{n} u_i \bar{v}_i \qquad (1\text{-}5)$$

where u_i and v_i are the components of u and v with respect to $e_1, \ldots, e_n$, $\bar{v}_i$ is the complex conjugate of v_i, and we define the length, or norm, of a vector u to be

$$\|u\| = \sqrt{(u,u)} = \left(\sum_{i=1}^{n} |u_i|^2\right)^{\frac{1}{2}} \qquad (1\text{-}6)$$

Equations (1-5) and (1-6) are the required definitions in either real or complex space; note $(u,u) \geq 0$ always. Finally the distance between u and v is

$$\|u - v\| = (u - v, u - v)^{\frac{1}{2}}$$

[If another person took a different basis $e_1', \ldots, e_n'$ in $\mho$, he would get an essentially different structure (except in special cases—see below). His distance between vectors u and v would be different from ours.] We say that u and v are orthogonal if $(u,v) = 0$.

Exercise 1-5   Show that our basis vectors $e_1, \ldots, e_n$ satisfy

$$(e_i, e_j) = 0 \qquad \text{if} \qquad i \neq j$$
$$(e_i, e_i) = 1$$

This is not the usual way to define the euclidean norm. Usually one says something like this: An inner product in a real or complex vector space is a bilinear form satisfying the following axioms, etc.; an inner-product, euclidean, or unitary vector space is a vector space together with an inner product; the norm of a vector u is $\|u\| = \sqrt{(u,u)}$; an orthonormal basis is a basis $e_1, \ldots, e_n$ such that $(e_i, e_j) = \delta_{ij}$. The main point is that we are presented with an inner product. If we have a completely amorphous vector space, e.g., one that does not arise from a particular application, we are free to take any inner product. What we did in the preceding paragraph was to fix on one particular inner product; every inner product can be obtained in that way.

Exercise 1-6   Using our definition of the inner product, prove:

(a)
$$(\lambda_1 u_1 + \lambda_2 u_2, v) = \lambda_1(u_1,v) + \lambda_2(u_2,v)$$
and
$$(u, \lambda_1 v_1 + \lambda_2 v_2) = \bar{\lambda}_1(u,v_1) + \bar{\lambda}_2(u,v_2)$$

(The inner product is "hermitian bilinear," i.e., linear in the first argument, conjugate linear in the second.)

(b)
$$(v,u) = \overline{(u,v)}$$

(The inner product is "symmetric.")

(c) $(u,u) > 0$ unless $u = 0$, in which case $(u,u) = 0$ (the inner product is "strictly positive").

These properties can be taken as axioms by means of which the concept of an inner product is characterized.

Remark: In the interval $(0,2\pi)$, the functions e^{int},

$$n = 0, \pm 1, \pm 2, \ldots$$

have the property that

$$\frac{1}{2\pi} \int_0^{2\pi} e^{int} \overline{e^{imt}} \, dt = \delta_{nm} \tag{1-7}$$

This may suggest that a "natural" inner product for continuous functions f, g defined on $(0,2\pi)$ would be

$$(f,g) = \frac{1}{2\pi} \int_0^{2\pi} f(t)\overline{g(t)} \, dt$$

It is easy to check that all the properties listed in the previous exercise are enjoyed by our function (f,g) here, so that (f,g) is indeed an inner product.   In no application is an inner product arbitrarily or whimsically introduced.   The inner product will always be suggested by the use it is intended for.

Returning to our finite-dimensional inner-product space, let us observe that if $e_1', \ldots, e_n'$ is another basis for which $(e_i',e_j') = \delta_{ij}$ (where the inner product has been previously defined), then the same inner product is also expressible in terms of the components of u and v with respect to this second (orthonormal) basis by the same formula, Eq. (1-5).

Exercise 1-7   Prove the last statement.

1-5   SELF - ADJOINT TRANSFORMATIONS

We can now define our special class of transformations.   A transformation $A: \mathcal{V} \to \mathcal{V}$, in which $\mathcal{V}$ is a unitary space, is self-adjoint or hermitian (or symmetric in case our space is real) if $(Ax,y) = (x,Ay)$ for every x, y in $\mathcal{V}$.

Exercise 1-8   Show that A is self-adjoint if and only if $(Ae_i,e_j) = (e_i,Ae_j)$ for $i, j = 1, 2, \ldots, n$, in which $e_1, \ldots, e_n$ is any orthonormal basis.   Therefore if $M(A; e_1, \ldots, e_n) = \{a_{ij}\}$, A is self-adjoint if and only if

$$a_{ji} = \bar{a}_{ij} \qquad \text{for} \qquad i, j = 1, 2, \ldots, n$$

We can now state the well-known and important

Theorem **1-1**   Let A be a self-adjoint transformation on the unitary space $\mathcal{V}$.   Then there is an orthonormal basis $e_1, \ldots, e_n$ of eigenvectors of A; the corresponding eigenvalues are all real.

Proof: Regarding the eigenvalues, suppose

$$Ax = \lambda x \tag{1-8}$$

for some $x \neq 0$; then $(Ax,x) = \lambda(x,x)$, and since $(Ax,x) = (x,Ax) = \overline{(x,Ax)}$ is real, we have

$$\lambda = \frac{(Ax,x)}{(x,x)}$$

and λ is real.

The rest of the theorem is easily proved by induction on the dimension n of the space $\mathcal{U}$.   In fact, the theorem is true for $n = 1$.   If we suppose it is true for spaces of dimension less than n, then let us take any (orthonormal) basis in $\mathcal{U}$, of dimension n, and work with the matrix that comes from A and the basis.   There exists at least one eigenvalue λ, since the roots of the characteristic polynomial of A all are eigenvalues.   Let e_1 be a corresponding eigenvector of unit length (note that if $Ax = \lambda x$, then also $Ax/\|x\| = \lambda x/\|x\|$).   Take $e_2, e_3, \ldots, e_n$ in any way such that $(e_1, e_2, \ldots, e_n)$ is an orthonormal basis (see Exercise 1-9 below).   Then $M(A; e_1, \ldots, e_n)$ has the form

$$M(A; e_1, \ldots, e_n) = \begin{pmatrix} \lambda & a'_{12} & a'_{13} & \cdots & a'_{1n} \\ 0 & & & & \\ \cdot & & & & \\ \cdot & & A'_2 & & \\ \cdot & & & & \\ 0 & & & & \end{pmatrix} \tag{1-9}$$

in which A'_2 is an $n - 1 \times n - 1$ submatrix.   But since $M(A)$ is hermitian, we must conclude that $a'_{12} = \cdots = a'_{1n} = 0$, and

$$M(A; e_1, \ldots, e_n) = \begin{pmatrix} \lambda & 0 & \cdots & 0 \\ 0 & & & \\ \cdot & & & \\ \cdot & & A'_2 & \\ \cdot & & & \\ 0 & & & \end{pmatrix} \tag{1-10}$$

Since A'_2 can be regarded as a self-adjoint transformation on an $(n - 1)$-dimensional space, there is a basis (orthonormal) for the $(n - 1)$-tuples with respect to which A'_2 is a diagonal matrix, by our induction hypothesis, and the proof is complete.

Exercise 1-9   Let $e_1, \ldots, e_j$ be pairwise-orthogonal unit vectors in an n-dimensional space $\mathcal{V}$, with $j < n$. Show that there is a unit vector e_{j+1} that is orthogonal to $e_1, \ldots, e_j$. *Hint:* Take any orthonormal basis, and express e_i as an n-tuple $(x_{i1}, \ldots, x_{in})$, $i = 1, \ldots, j$, with respect to the chosen basis. Let the unknown e_{j+1} have the (unknown) components $(y_1, \ldots, y_n)$. Then all that is necessary is to solve the j equations

$$\sum_{k=1}^{n} y_k \bar{x}_{ik} = 0 \tag{1-11}$$

$i = 1, \ldots, j$, for the n unknowns y_k; then use λy_k [which also works in (1-11)] in which λ satisfies

$$|\lambda|^2 \sum_{k=1}^{n} |y_k|^2 = 1$$

We can now answer very easily a number of interesting questions we might ask about self-adjoint transformations. First, let us formally define "to reduce a transformation to diagonal form." This is to choose a basis for which the matrix of the transformation is a diagonal matrix. Theorem 1-1 asserts that this is always possible for a self-adjoint transformation (and the basis is orthonormal). Also, if A denotes a transformation, and $p(\lambda) = c_0 \lambda^n + \cdots + c_n$ is a polynomial in λ with real or complex c_j, then $p(A)$ is by definition $c_0 A^n + c_1 A^{n-1} + \cdots + c_n I$, in which I is the identity transformation. Here the linear combination of powers of the transformation is defined in the obvious way, or we can say that $p(A)$ is the transformation that corresponds to $p(M)$ if $M = M(A)$ is the matrix corresponding to A (and some basis). We suppose that the reader is familiar with polynomials in a matrix M (at least with scalar coefficients).

Exercise 1-10   What is the obvious way of defining $A \cdot B$ if A and B are two transformations on $\mathcal{V}$ to $\mathcal{V}$? [*Answer:* $(AB)x = A(Bx)$ for x in $\mathcal{V}$; that is, apply B first, then A.] What is the obvious way of defining $A + B$, $\alpha A + \beta B$, and $p(A)$? What about $p(A,B)$, where $p(\lambda,\mu)$ is a polynomial in λ and μ, and A and B are not necessarily commuting transformations $(AB \neq BA)$? [*Answer:* You will never see an expression like $p(A,B)$ without an explicit description of how to treat terms in AB, BA, $A^2 B$, ABA, BA^2, AB^2, etc.]

Exercise 1-11   (a) If $\lambda_1, \ldots, \lambda_n$ are the eigenvalues of the self-adjoint transformation A, what are the eigenvalues of $p(A)$, where p is a polynomial in one variable? (Either reduce A to diagonal form or else note that $\Sigma c_n A^n x = \Sigma c_n \lambda^n x$ follows directly from $Ax = \lambda x$.)

(b) If A is self-adjoint, find $\max\limits_{\|x\|=1} (Ax,x)$ and $\min\limits_{\|x\|=1} (Ax,x)$.

(c) Find all stationary values of $f(x) = (Ax,x)$ subject to $(x,x) = 1$ (A self-adjoint).

(d) A self-adjoint transformation A is defined to be positive (or positive-semidefinite) if $(Ax,x) \geq 0$ for every x in $\mathcal{V}$. If A is positive, show that the eigenvalues of A are all nonnegative.

(*e*) Show that every positive A has a positive square root (i.e., there is a positive B such that $B^2 = A$).

(*f*) The equation $Ax = \lambda x + c$, where λ and c are a given scalar and a given vector, has a solution x if and only if λ is not an eigenvalue of A, or if λ is an eigenvalue and c is orthogonal to all eigenvectors of A corresponding to λ (A self-adjoint). In the former case, the solution is unique; in the latter case, it is not. Prove this.

(*g*) Show that, if $p(\lambda)$ is the characteristic polynomial of A, then $p(A) = 0$ (A self-adjoint). This is a special case of the Cayley-Hamilton theorem, which states that every square matrix A satisfies its characteristic equation.

Exercise 1-12   (*a*) Show that if $\mathcal{U}$ is a complex n-dimensional space and A is a transformation on $\mathcal{U}$ to $\mathcal{U}$, and if the eigenvalues of A are all distinct (the characteristic polynomial of A has no repeated roots), then there is a basis $e_1, \ldots, e_n$ in $\mathcal{U}$ with respect to which $M(A; e_1, \ldots, e_n)$ is diagonal. *Hint:* Let $e_1 \ldots, e_n$ be the eigenvectors. Show that they are independent and that therefore they span $\mathcal{U}$.

(*b*) Let $\lambda_1, \ldots, \lambda_n$ denote the eigenvalues of A. Then for every function $f(\lambda)$, analytic in a region containing $\lambda_1, \ldots, \lambda_n$, we can define a transformation $f(A)$ by the equations

$$f(A)e_j = f(\lambda_j)e_j \qquad j = 1, \ldots, n$$

Show that if $f(\lambda)$ can be represented by a power series

$$f(\lambda) = \sum_{n=0}^{\infty} a_n(\lambda - b)^n$$

that converges in a disk containing $\lambda_1, \ldots, \lambda_n$, then

$$\lim_{N \to \infty} \sum_{n=0}^{N} a_n(A - bI)^n = f(A)$$

in which the limit is to be interpreted as follows (for example): B_n converges to C if $B_n x$ converges to Cx for every x in $\mathcal{U}$, $\mathcal{U}$ considered as a euclidean space, say. *Hint:* Reduce A to diagonal form.

(*c*) Let A have the distinct eigenvalues $\lambda_1, \ldots, \lambda_n$, and let $g_j(\lambda)$ be the complex polynomial of degree $n - 1$ that assumes the value 1 for $\lambda = \lambda_j$ and vanishes at all the other eigenvalues, $j = 1, \ldots, n$. Let $P_j = g_j(A)$; prove that $P_i P_j = 0$ if $i \neq j$, $P_i^2 = P_i$, and $\sum_{i=1}^{n} P_i = I$. Furthermore, prove that

$$A = \sum_{j=1}^{n} \lambda_j P_j$$

and, more generally,

$$f(A) = \sum_{j=1}^{n} f(\lambda_j)P_j$$

if $f(\lambda)$ is analytic in a region containing $\lambda_1, \ldots, \lambda_n$.

(*d*) If $n = 2$, we have

$$g_1(\lambda) = \frac{\lambda - \lambda_2}{\lambda_1 - \lambda_2} \qquad g_2(\lambda) = \frac{\lambda - \lambda_1}{\lambda_2 - \lambda_1}$$

Therefore
$$f(A) = \frac{f(\lambda_1)}{\lambda_1 - \lambda_2}(A - \lambda_2 I) + \frac{f(\lambda_2)}{\lambda_2 - \lambda_1}(A - \lambda_1 I)$$

If A has the matrix representation

$$M(A) = \begin{pmatrix} 2 & 1 \\ 1 & 2 \end{pmatrix}$$

show that
$$M(A^{100}) = \begin{pmatrix} \dfrac{3^{100} + 1}{2} & \dfrac{3^{100} - 1}{2} \\[2ex] \dfrac{3^{100} - 1}{2} & \dfrac{3^{100} + 1}{2} \end{pmatrix}$$

If
$$M(A) = \begin{pmatrix} 1 & 1 \\ 1 & -1 \end{pmatrix}$$

find $M(e^{tA})$.

(e) Let

$$M(A) = \begin{pmatrix} -1 & 1 & 1 \\ 0 & -2 & -3 \\ 0 & 0 & 1 \end{pmatrix}$$

Find $M(B)$ for which $B^4 = A^2$.

(f) All the preceding discussion is valid for any transformation which "has a diagonal form," whether or not the roots are all distinct. In particular, it is valid when A is self-adjoint. If you are familiar with the Jordan canonical form for an arbitrary complex matrix, discuss the analogous possibilities for defining $f(A)$, with $f(\lambda)$ analytic.

1-6   THE INFINITE - DIMENSIONAL CASE

In Sec. 1-2 we had an example [Eq. (1-2)] of a linear equation in a vector space that was not finite-dimensional. The equation can be written

$$f(t) + Af(t) = g(t) \tag{1-12}$$

in which the transformation A sends the function f into the function given by an integral:

$$f(t) \xrightarrow{A} \int_0^1 K(t,s)f(s)\,ds \tag{1-13}$$

Here $g(t)$ is given, and $f(t)$ is required. In problems of this sort, it is necessary to find the appropriate vector space in which to formulate the relevant theory. In the case of integral equations with a symmetric kernel [that is, $K(s,t) = \overline{K(t,s)}$], the appropriate vector space is Hilbert space, which is the infinite-dimensional analogue of our euclidean or unitary space. Here $\bar{K}$ is the complex conjugate of K.

A Hilbert space H is a real or complex vector space endowed with an inner product satisfying the axioms of Exercise 1-6, with which we are already familiar, and which is *complete;* i.e., if $\{x_n\}$ is a sequence of vectors in H such that $\lim_{m \to \infty} (x_n - x_m) = 0$, then there is an x^* in H for which $\lim_{n \to \infty} x_n = x^*$. The limits here are defined in terms of the norm, $\|x\| = \sqrt{(x,x)}$; we have, by definition, $\lim_{n \to \infty} x_n = y$ if $\lim_{n \to \infty} \|x_n - y\| = 0$ (the distance from x_n to y tends to zero as $n \to \infty$).

As an example, let H be the collection of all real or complex sequences $\{x_n\}$ for which $\sum_{n=1}^{\infty} |x_n|^2 < \infty$, that is, for which $\{x_n\}$ is *square-summable.* This is the natural generalization of n-tuples, of course. We define, for $x = \{x_n\}$, $y = \{y_n\}$, the inner product

$$(x,y) = \sum_{n=1}^{\infty} x_n \bar{y}_n \tag{1-14}$$

and it can be shown, by Schwarz's inequality (see Exercise 1-14 below) that the infinite series in (1-14) is absolutely convergent.

Exercise 1-13 If

$$A = \begin{pmatrix} a & b \\ b & c \end{pmatrix}$$

is a real symmetric 2×2 matrix, and if A is positive, that is, $x'Ax \geq 0$ for every real column vector $x = (x_1, x_2)$, prove $ac - b^2 \geq 0$. [*Hint:* Put $x = (1,v)$ and minimize

$$f(v) \equiv (1,v)A \begin{pmatrix} 1 \\ v \end{pmatrix}$$

with respect to v.]

Exercise 1-14 (*a*) For all complex $x_1, \ldots, x_N, y_1, \ldots, y_N$, show that

$$\left| \sum_{j=1}^{N} x_j \bar{y}_j \right| \leq \left\{ \sum_{j=1}^{N} |x_j|^2 \right\}^{1/2} \left\{ \sum_{j=1}^{N} |y_j|^2 \right\}^{1/2} \tag{1-15}$$

[*Hint:* $\left| \sum_{j=1}^{N} x_j \bar{y}_j \right| \leq \sum_{j=1}^{N} |x_j| \cdot |y_j|$, so it suffices to prove (1-15) for the case in which all the x_j and y_j are real and nonnegative, which we shall now suppose. Now, for each j, the matrix

$$\begin{pmatrix} x_j^2 & x_j y_j \\ x_j y_j & y_j^2 \end{pmatrix}$$

is positive. Since the sum of positive matrices is evidently positive, we now sum from 1 to N and use Exercise 1-13.]

(b) Now suppose that $\sum_1^\infty |x_n|^2 < \infty$ and $\sum_1^\infty |y_n|^2 < \infty$; show that $\sum_1^\infty x_n \bar{y}_n$ is absolutely convergent and

$$\left| \sum_1^\infty x_n y_n \right| \le \sqrt{\sum_1^\infty |x_n|^2} \sqrt{\sum_1^\infty |y_n|^2} \tag{1-16}$$

This is one form of the Schwarz inequality. (Use part a.)

Returning to our example H of square-summable sequences, if we define addition and scalar multiplication componentwise, just as in the case of n-tuples, then H becomes a vector space with an inner product. The only fact that is not obvious is that the sum of two square-summable sequences is also square-summable. But this follows from the Schwartz inequality. In fact, if $\{x_n\}$ and $\{y_n\}$ belong to H, then

$$\sum_1^\infty |x_n + y_n|^2 = \sum_1^\infty |x_n|^2 + \sum_1^\infty |y_n|^2 + \sum_1^\infty x_n \bar{y}_n + \sum_1^\infty \bar{y}_n x_n$$

which, upon application of the Schwarz inequality, is easily seen to be finite; in fact,

$$\sqrt{\sum_1^\infty |x_n + y_n|^2} \le \sqrt{\sum_1^\infty |x_n|^2} + \sqrt{\sum_1^\infty |y_n|^2} \tag{1-17}$$

which is called the *triangle inequality;* the length of the side of a triangle does not exceed the sum of the lengths of the other two sides.

To complete the discussion on this space, we must show that it is complete. Let $\{x_n{}^{(k)}\}$ be a sequence of vectors in H, $k = 1, 2, \ldots$, such that $\lim_{\substack{k \to \infty \\ j \to \infty}} \|x^{(k)} - x^{(j)}\| = 0$. Then, for each n, we must have

$$\lim_{\substack{k \to \infty \\ j \to \infty}} |x_n{}^{(k)} - x_n{}^{(j)}| = 0 \tag{1-18}$$

and consequently, there is a y_n such that

$$\lim_{k \to \infty} x_n{}^k = y_n$$

for each n. Let $\epsilon > 0$, and let p and q be any integers with $q > p$. Then there is a k_0 such that for all k, j larger than k_0 we have

$$\sum_{n=p}^q |x_n{}^{(k)} - x_n{}^{(j)}|^2 < \epsilon^2 \tag{1-19}$$

Now let $j \to \infty$ to get

$$\sum_{n=p}^{q} |x_n^{(k)} - y_n|^2 < \epsilon^2 \tag{1-20}$$

and use the triangle inequality to conclude that

$$\sqrt{\sum_p^q |y_n|^2} \leq \epsilon + \sqrt{\sum_p^q |x_n^{(k)}|^2}$$

Now we can let $q \to \infty$, and we see that $\{y_n\}$ is square-summable; $\{y_n\}$ belongs to H. Finally, we choose n_0 so large that

$$\sum_{n_0}^{\infty} |y_n|^2 < \epsilon^2$$

Then $\quad \displaystyle\sum_{n=1}^{\infty} |x_n^{(k)} - y_n|^2 = \sum_1^{n_0} |x_n^{(k)} - y_n|^2 + \sum_{n_0+1}^{\infty} |x_n^{(k)} - y_n|^2 \tag{1-21}$

But in (1-20) we can let $q \to \infty$, and we have, for $p = n_0 + 1$,

$$\sum_{n_0+1}^{\infty} |x_n^{(k)} - y_n|^2 \leq \epsilon^2 \tag{1-22}$$

for $k \geq k_0$ [the choice of k_0 for (1-19) was independent of p, q]. Now in (1-21), we let $k \to \infty$ and conclude that

$$\lim_{k \to \infty} \|x^{(k)} - y\| = 0$$

The preceding discussion can be summarized as follows:

Theorem 1-2 Let H be the collection of square-summable sequences $\{x_n\}$, real or complex. If addition and scalar multiplication are defined componentwise, and if the inner product is defined by (1-14), then H is a complete inner-product space, i.e., a Hilbert space.

Let us turn now to functions $f(t)$ defined on a bounded t interval, say $0 \leq t \leq 2\pi$. Let H_0 be the collection of complex functions $f(t)$, Riemann-integrable or "improperly" integrable, for which

$$\int_0^{2\pi} |f(t)|^2 \, dt < \infty$$

We define the sum of two such functions, or vectors, pointwise,

$$(f + g)(t) = f(t) + g(t)$$

and similarly for scalar multiplication. We define the inner product by

$$(f,g) = \int_0^{2\pi} f(t)\overline{g(t)} \, dt \tag{1-23}$$

Then we can show, by using the Riemann approximating sums, that the integral in (1-23) is absolutely integrable; i.e.,

$$\int_0^{2\pi} |f(t)\overline{g(t)}| \, dt < \infty$$

and

$$\int_0^{2\pi} f(t)\overline{g(t)} \, dt \,| \leq \sqrt{\int_0^{2\pi} |f(t)|^2 \, dt} \sqrt{\int_0^{2\pi} |g(t)|^2 \, dt} \qquad (1\text{-}24)$$

(see Exercise 1-14).   This is another instance of the Schwartz inequality, and, just as in the case of sequences, it enables us to conclude that $f + g$ is square-integrable if f and g are.   The only thing lacking is completeness; the inner-product space H_0 is not complete.   It is possible to have $\|f_n - f_m\| \to 0$ as $m, n \to \infty$, and yet there is no function in H_0 to which $\{f_n\}$ converges in norm.   (Recall that in a metric space, i.e., a space in which there is a "distance" between pairs of points, $x_n \to x$ means the distance from x_n to x converges to 0.)   In the theory of the Lebesgue integral, which subsumes the Riemann integral, we do have completeness, but we shall not assume that the reader is familiar with that theory.

Now there is an intimate connection between the space H_0 just defined (or rather its completion, that is, H_0 together with its Lebesgue square-integrable limits) and the complex sequence space H defined earlier. Observe that the functions $u_n(t) = (1/\sqrt{2\pi})e^{int}$, $n = 0, \pm 1, \pm 2, \ldots$, satisfy

$$(u_n, u_m) = \delta_{mn} \qquad (1\text{-}24a)$$

for every pair m, n, and, moreover, for every f in H_0 and every $\epsilon > 0$ there is a linear combination $s_N(t)$ of the u_n such that

$$\|s_N - f\| < \epsilon$$

In fact, if

$$c_n = \int_0^{2\pi} f(t) \, \frac{e^{-int}}{\sqrt{2\pi}} \, dt$$

$n = 0, \pm 1, \pm 2, \ldots$, denotes the nth Fourier coefficient of f, then the partial sums

$$s_N(t) = \sum_{-N}^{N} c_n \, \frac{e^{int}}{\sqrt{2\pi}}$$

satisfy

$$\lim_{N \to \infty} \|s_N - f\| = 0 \qquad (1\text{-}25)$$

(see Ref. 3).   A family $\{u_n\}$ of vectors in an inner-product space is called a *complete orthonormal set* (CON set) if the vectors satisfy (1-24a) and if for every f in the space, we have

$$\|f\|^2 = \sum_{n=-\infty}^{\infty} |(f, u_n)|^2 \qquad (1\text{-}26)$$

(Parseval's equation).   Equations (1-25) and (1-26) are equivalent; i.e., each implies the other.   The family $u_n(t) = e^{int}/\sqrt{2\pi}$ is a CON set on the interval $[0, 2\pi]$ (or any interval of length 2π).

Exercise 1-15   Prove the equivalence of (1-25) and (1-26).   [*Hint:* $\|f - s_n\|^2 = \|f\|^2 - (f, s_N) - (s_N, f) + \|s_N\|^2$; also if $s_N = \sum_{-N}^{N} (f, u_n) u_n$, then $\|s_N\|^2 = \sum_{-N}^{N} |(f, u_n)|^2$.]

It follows from (1-26) that if we associate with each f in H_0 the sequence $\{c_n\}$, $n = 0, \pm 1, \pm 2, \ldots$, in which $c_n = (f, u_n)$, then the correspondence (or mapping)

$$f \rightarrow \{c_n\}$$

is norm-preserving, i.e.,

$$\|f\| = \|\{c_n\}\| \tag{1-26a}$$

and, of course, sums and scalar multiples go into the corresponding sums and scalar multiples.   Finally, it follows from (1-26a) that if, in addition,

$$g \rightarrow \{d_n\}$$

then

$$(f, g) = (\{c_n\}, \{d_n\}) = \sum_{-\infty}^{\infty} c_n \bar{d}_n$$

All this means that there is essentially no difference (with one exception) between the function space and the sequence space.   The only exception is that the latter space is complete.   We remark that our sequence space here really consists of "bisequences" $\{x_n\}$, $n = 0, \pm 1, \pm 2, \ldots$, instead of sequences, but this makes no real difference.   Finally, we observe that the same situation obtains for functions defined on any bounded interval; the appropriate collection of exponentials (or sines, cosines) will form a CON set.

For the interval $0 \leq t < \infty$, the functions

$$u_n(t) = L_n(t) = \frac{1}{n!} e^{t/2} \frac{d^n}{dt^n} (t^n e^{-t}) \tag{1-27}$$

$n = 0, 1, 2, \ldots$, form a CON set[3] for the set H_0 of square-integrable functions $f(t)$ on $(0, \infty)$.   These are essentially the Laguerre functions. Again, the functions

$$u_n(t) = \frac{e^{t^2/2}}{\sqrt{2^n n!} \sqrt{\pi}} \frac{d^n}{dt^n} e^{-t^2} \tag{1-28}$$

$n = 0, 1, 2, \ldots$ constitute a CON set for the functions $f(t)$ square-integrable on $-\infty < t < \infty$; these are essentially the Hermite functions.[3]

From the Hilbert-space viewpoint, all these function spaces are equivalent to a dense subset of the sequence space (or the whole sequence space if we allow Lebesgue-integrable functions).

Every Hilbert space contains a maximal orthonormal set $\{e_\lambda\}$, that is, a collection of pairwise-orthogonal unit vectors which cannot be enlarged by the adjunction of another unit vector and still be an orthonormal set. We shall restrict ourselves to those spaces for which this maximal set is denumerable.

Exercise 1-16   Show that, if $\{e_n\}$, $n = 1, 2, 3, \ldots$, is a maximal orthonormal set and x is in H, then $\|x\|^2 = \sum_{n=1}^{\infty} |(x,e_n)|^2$ (Parseval's equation), so that $\{e_n\}$ is a CON set.   Use (a) and (b).

(a) Let $y_N = \sum_{j=1}^{N} (x,e_n)e_n$.   Then

$$(x - y_N, e_1) = (x - y_N, e_2) = \cdots = (x - y_N, e_N) = 0$$

Therefore

$$(x - y_N, y_N) = 0 \quad \text{and} \quad \|x\|^2 = \|(x - y_N) + y_N\|^2$$
$$= \|x - y_N\|^2 + \|y_N\|^2 \geq \|y_N\|^2 = \sum_{n=1}^{N} |(x,e_n)|^2$$

Therefore $\sum_{n=1}^{\infty} |(x,e_n)|^2 \leq \|x\|^2$ (Bessel's inequality).

(b) Now let $y = \sum_{n=1}^{\infty} (x,e_n)e_n$. $\left[\textit{Note:} \|y_N - y_M\|^2 = \sum_{N+1}^{M} |(x,e_n)|^2 \rightarrow 0 \text{ as } M, \right.$ $N \rightarrow \infty$, and since H is complete, there is a y such that $y_N \rightarrow y$, $y = \sum_{1}^{\infty} (x,e_n)e_n.$ $\Big]$ Then $(y,e_n) = (x,e_n)$ for every n, or $(y - x, e_n) = 0$ for every n. Since $\{e_n\}$ is maximal, $y - x = 0$, or $y = x$. This means that $y_N \rightarrow x$ because $y_N \rightarrow y$; hence $\|x - y_N\|^2 \rightarrow 0$, and, by (a), $\|y_N\|^2 \rightarrow \|x\|^2$.

Exercise 1-17   Prove Schwarz's inequality,

$$|(x,y)| \leq \|x\| \cdot \|y\|$$

Use the fact that

$$0 \leq \|\alpha x + \beta y\|^2$$

for all complex α, β, getting a positive-definite 2×2 hermitian matrix (see Exercise 1-13).

A subspace $\mathfrak{M}$ of a Hilbert space H is a subset of H which is also a vector space and which is closed; that is, $\mathfrak{M}$ contains all its limit points. So $\mathfrak{M}$ is also a Hilbert space.   If $\mathcal{E}$ is any collection of vectors of H, then $\mho(\mathcal{E})$ denotes the subspace generated by $\mathcal{E}$; this is the smallest subspace $\mathfrak{M}$

of H that contains ε.   It consists of all linear combinations of vectors in ε and all limits of such vectors.

In every "least-squares" problem, a vector x and a set ε of vectors are given, and the problem is to express x as a linear combination $\hat{x}$ of vectors in ε, or a limit $\hat{x}$ of such combinations, in such a way that $\|x - \hat{x}\|$ is as small as possible.   This is evidently equivalent to finding an $\hat{x}$ in $\mathcal{V}(\varepsilon) = \mathfrak{M}$ such that

$$\|x - \hat{x}\| = \min_{y \in \mathfrak{M}} \|x - y\|$$

where ϵ means "belongs to."   It is an important and useful fact that this problem always has a solution; the minimum always is attained.

To prove this, let

$$\alpha = \inf_{y \in \mathfrak{M}} \|x - y\|$$

and let $y_n \epsilon \mathfrak{M}$ with $\|x - y_n\|$ converging to α as $n \to \infty$.   Since we always have

$$\|u + v\|^2 + \|u - v\|^2 = 2\|u\|^2 + 2\|v\|^2$$

we set $u = x - y_n$, $v = x - y_m$ and get

$$\|2x - y_n - y_m\|^2 + \|y_n - y_m\|^2 = 2\|x - y_n\|^2 + 2\|x - y_m\|^2$$

or $\qquad 2\|x - y_n\|^2 + 2\|x - y_m\|^2 \geq 4\alpha^2 + \|y_n - y_m\|^2$

Since the left-hand side converges to $4\alpha^2$ as $m, n \to \infty$, it follows that $\|y_n - y_m\| \to 0$, and since $\mathfrak{M}$ is closed, there is an $\hat{x}$ in $\mathfrak{M}$ such that $y_n \to \hat{x}$ as $n \to \infty$.   Then also $\|x - y_n\| \to \|x - \hat{x}\| = \alpha$:

$$\|x - \hat{x}\| = \min_{y \in \mathfrak{M}} \|x - y\|$$

Furthermore, we have

$$(x - \hat{x}, y) = 0$$

for every y in $\mathfrak{M}$.   In fact, we have

$$\|x - \hat{x}\|^2 \leq \|x - \hat{x} + \lambda y\|^2$$

for every y in $\mathfrak{M}$ and every complex scalar λ, so that

$$0 \leq \lambda(y, x - \hat{x}) + \bar{\lambda}(x - \hat{x}, y) + |\lambda|^2\|y\|^2$$

Since this is true for all λ, in particular for arbitrarily small λ, we must have

$$(x - \hat{x}, y) = 0$$

for every y in $\mathfrak{M}$.   Therefore, we also have

$$(x - \hat{x}, \hat{x}) = 0$$

and $\qquad\qquad \|x\|^2 = \|x - \hat{x}\|^2 + \|\hat{x}\|^2$

and $\qquad\qquad \min_{y \in \mathfrak{M}} \|x - y\|^2 = \|x\|^2 - \|\hat{x}\|^2$

If we denote by $\mathfrak{M}^\perp$ the subspace of all vectors z that are orthogonal to every vector in $\mathfrak{M}$, we see that $x - \hat{x} \in \mathfrak{M}^\perp$, $\hat{x} \in \mathfrak{M}$. We summarize these facts as follows:

Let $\mathfrak{M}$ be a subspace of H, and let $x \in H$. Then x can be expressed uniquely as a sum

$$x = x_1 + x_2$$

with x_1 in $\mathfrak{M}$ *and* x_2 in $\mathfrak{M}^\perp$.

The uniqueness follows from the fact that if $z_1 + z_2 = 0$ with z_1 in $\mathfrak{M}$ and z_2 in $\mathfrak{M}^\perp$, then

$$z_1 = z_2 = 0$$

Hence, in any least-squares problem in which x and $\mathcal{E}$ are given, the vector $\hat{x}$ in $\mathcal{V}(\mathcal{E}) = \mathfrak{M}$ which is the solution is characterized by the equation

$$(x,y) = (\hat{x},y)$$

for every y in $\mathcal{E}$.

1-7 OPERATORS

Corresponding to the situation in finite-dimensional spaces, we might expect that if we knew what a transformation A of H into H does to elements of a CON set, then we would know A completely, and moreover that we could assign values Au_n arbitrarily to each element of such a CON set $\{u_n\}$. That this is not so can be seen if we try to define A by the equations

$$Au_n = nu_n \qquad n = 1, 2, 3, \ldots$$

Let $x = \sum_1^\infty (1/n)u_n$. Since $\sum_1^\infty 1/n^2 < \infty$, x is in H. What is Ax? The natural guess would be

$$\lim_{N \to \infty} \sum_1^N \frac{1}{n} Au_n = \lim_{N \to \infty} \sum_1^N u_n$$

but this limit does not exist.

This brings us to the notion of continuity, which was not a problem in the finite-dimensional case. We shall call a linear transformation A of H into H *continuous* if whenever x_n converges to x in H, then also Ax_n converges to Ax. A linear transformation is called *bounded* if there is a constant M such that

$$\|Ax\| \leq M\|x\| \tag{1-29}$$

for every x in H. It can be shown that A is bounded if and only if A is continuous. The smallest M for which (1-29) holds is called the norm of A, written $\|A\|$;

$$\|Ax\| \leq \|A\| \cdot \|x\|$$

for all x. An operator is a bounded linear transformation.

Exercise 1-18 Prove that a linear transformation A is continuous if and only if A is bounded. (*Hint:* If A is continuous and not bounded, then there is a sequence $\{x_n\}$ in H for which $\|Ax_n\| > n\|x_n\|$. Set $y_n = x_n/n\|x_n\|$; then $\|Ay_n\| > 1$ and $Ay_n \to 0$. The reverse implication is easy.)

An operator A is called *self-adjoint* (or hermitian) if $(Ax,y) = (x,Ay)$ for every x, y in H. For self-adjoint operators, a generalized sort of "diagonalization" is possible, but the statement and proof of this *spectral* theorem is beyond the scope of this book. We shall therefore limit ourselves to a special case which is still general enough to include some classical problems. This is the case of the self-adjoint *compact* operator, to be defined below.

Let us note an important difference between infinite- and finite-dimensional Hilbert spaces. If $\{x_n\}$ is a bounded sequence in a finite-dimensional space, i.e.,

$$\|x_n\| \leq M$$

for all n, then there is a limit point x^* of $\{x_n\}$. However, if we take $x_n = u_n$, where $\{u_n\}$ is a CON set, in an infinite-dimensional H, then no limit point exists, for if there were such a point, a subsequence of $\{x_n\}$ would converge to it, and the distance between successive elements of the subsequence would converge to zero, which cannot happen in our example.

But there is a more general kind of convergence that will serve us for some purposes: We say that a sequence x_n converges to x^* *weakly* if, for every y in H, we have

$$\lim_{n \to \infty} (x_n,y) = (x^*,y) \tag{1-30}$$

Exercise 1-19 Show that, if $x_n \to x^*$, then also $x_n \to x^*$ weakly.

For weak convergence, we have the following result:

Theorem 1-3 Let $\{x_n\}$ be a bounded sequence in H (H assumed complete). Then a subsequence of $\{x_n\}$ converges weakly.

Proof: Let $e_1, e_2, \ldots, e_n, \ldots$ be a CON set in H. Since the sequence of complex numbers (x_n,e_1) is bounded (Schwartz inequality), there is a subsequence $\{y_n\}$ of $\{x_n\}$ such that (y_n,e_1) converges. Since (y_n,e_2) is bounded, there is a subsequence $\{z_n\}$ of $\{y_n\}$ such that (z_n,e_2) converges. In every case we make sure that the first element y_1 is farther along in the sequence than x_1, z_1 is farther along than y_1, etc. We obtain in this way a family of sequences, each of which is a subsequence of the preceding and each of which "converges on the next e_j," that is, (y_n,e_1) converges, (z_n,e_2) converges, . . . , (q_n,e_j) converges, ... Then the sequence $x_1, y_1, z_1, \ldots, q_1, \ldots,$ in which we take the first element of each subsequence, converges on every e_j. Call this new subsequence $\{u_n\}$. Then (u_n,e_j) converges as $n \to \infty$ for each fixed j. It follows that (u_n,y) converges for each finite linear combination $y = \sum\limits_{j=1}^{N} \alpha_j e_j$. Finally, for arbitrary y, $y = \sum\limits_{j=1}^{N} \alpha_j e_j + \sum\limits_{N+1}^{\infty} \alpha_j e_j$, in which N is chosen so large that

$$\left\| \sum_{N+1}^{\infty} \alpha_j e_j \right\| < \epsilon$$

We use the fact that $\|u_n\|$ is bounded and the Schwartz inequality, and conclude that (u_n,y) converges for every y in H.

Let $\lim\limits_{n \to \infty} (u_n,e_j) = \xi_j$, $j = 1, 2, \ldots$. Let M be a bound for $\|u_n\|^2$; then

$$M \geq \|u_n\|^2 = \sum_{j=1}^{\infty} |(u_n,e_j)|^2$$

for all n (see Exercise 1-16). Therefore

$$M \geq \sum_{j=1}^{N} |(u_n,e_j)|^2$$

and we may let $n \to \infty$ and conclude that

$$M \geq \sum_{j=1}^{N} |\xi_j|^2$$

so that $$M \geq \sum_{j=1}^{\infty} |\xi_j|^2$$

Now let $u^* = \sum\limits_{j=1}^{\infty} \xi_j e_j$. Then $u_n \to u^*$ weakly, and the proof is complete.

We are now ready for the definition: An operator A is called *compact* or *completely continuous* if, for every weakly convergent sequence $\{x_n\}$, the sequence $\{Ax_n\}$ is *strongly convergent*, that is, Ax_n converges in norm.

Exercise 1-20 Prove that an operator A is compact if and only if, for every bounded sequence $\{x_n\}$, the sequence $\{Ax_n\}$ contains a convergent subsequence. Thus the latter condition can also be taken as the defining property of a compact operator. (*Hint:* If $\{x_n\}$ is bounded, it has a weakly convergent subsequence by Theorem 1-3.)

The analogue of the finite-dimensional "diagonalization" theorem is as follows (H is assumed complete, of course):

Theorem 1-4 Let A be self-adjoint and compact. Let N be the subspace of all vectors x for which $Ax = 0$ (N may consist only of the zero vector). Let M be the orthogonal complement of N (that is, the subspace of vectors x that are orthogonal to every vector in N). Then A sends M into M; there is a CON set $\{e_n\}$ of eigenvectors in M, $Ae_n = \lambda_n e_n$, $n = 1, 2, \ldots$; the λ_n are real and nonzero; the sequence $\{\lambda_n\}$ is finite or else $\lambda_n \to 0$ as $n \to \infty$.

Proof: To show that A sends M into M, let x be in M, and let y be any vector in N. Then $(Ax,y) = (x,Ay) = 0$, which shows that Ax is orthogonal to every vector in N; Ax is in M. That every eigenvalue λ is real is proved exactly as in the finite-dimensional case. To show that $\lambda_n \neq 0$ if e_n is in M, we observe that $Ae_n = \lambda_n e_n$ with $\lambda_n = 0$ implies e_n is in N, that is, e_n is in N and M, which is impossible since $e_n \neq 0$.

Now suppose that we have found eigenvectors $e_1, \ldots, e_{n-1}$, $Ae_j = \lambda_j e_j$, $j = 1, 2, \ldots, n-1$, or else that we have not yet found any. Let

$$K_n = \sup \{|(Ax,x)| : \|x\| \leq 1; (x,e_j) = 0, j = 1, 2, \ldots n-1\} \quad (1\text{-}31)$$

[This is the supremum of all the values of $|(Ax,x)|$ for which $\|x\| \leq 1$ and x is orthogonal to $e_1, \ldots, e_{n-1}$.] Then the real numbers (Ax,x) for which x satisfies the same conditions either have the supremum K_n or else the infimum $-K_n$, or both. Suppose for definiteness that the supremum is K_n, the reasoning in the other case being similar. Then there is a sequence $\{x_k\}$, $k = 1, 2, \ldots$, for which $\|x_k\| \leq 1$ and x_k is orthogonal to $e_1, \ldots, e_{n-1}$ and such that

$$\lim_{k \to \infty} (Ax_k, x_k) = K_n$$

By Theorem 1-3, a subsequence $\{y_k\}$ of $\{x_k\}$ converges weakly, say $y_k \to u$ weakly. Then $\|u\| \leq 1$. Now

$$|(Ay_k, y_k) - (Au,u)| \leq |(Ay_k, y_k) - (Au, y_k)| + |(Au, y_k) - (Au,u)|$$

The second absolute value tends to zero because $y_j \to u$ weakly. The first absolute value does likewise because it is bounded by

$$\|Ay_k - Au\| \cdot \|y_k\|$$

and $Ay_k \to Au$ strongly and $\|y_k\| \leq 1$. The conclusion is that

$$(Au,u) = K_n \quad (1\text{-}32)$$

Note that we must have $\|u\| = 1$, because if $\|u\| < 1$, then $\|cu\| = 1$ for a certain $c > 1$ and K_n would no longer be the supremum defined in (1-31). Write

$$Au - K_n u = z \tag{1-33}$$

We shall show that $z = 0$. Now $(z,u) = 0$, so z is orthogonal to u. Also, z is orthogonal to $e_1, \ldots, e_{n-1}$ since u is. [Note that whenever x is orthogonal to $e_1, \ldots, e_{n-1}$, then Ax is also, because

$$(Ax, e_j) = (x, Ae_j) = (x, \lambda_j e_j) = 0]$$

By (1-31), for any real α we have

$$(A(u + \alpha z), (u + \alpha z)) \leq K_n \|u + \alpha z\|^2$$

and since $(Au, z) = \|z\|^2$ by (1-33), we have

$$K_n + 2\alpha \|z\|^2 + \alpha^2(Az, z) \leq K_n(1 + \alpha^2 \|z\|^2)$$

But from (1-31)

$$-K_n \|z\|^2 \leq (Az, z)$$

and, with the previous inequality, we obtain

$$2\alpha \|z\|^2 - \alpha^2 K_n \|z\|^2 \leq K_n \alpha^2 \|z\|^2$$

For $\alpha > 0$, we see that

$$2\|z\|^2 \leq 2K_n \alpha \|z\|^2$$

For α small enough, this is impossible unless $\|z\| = 0$. The conclusion is that $z = 0$, and (1-33) becomes

$$Au = K_n u$$

We now set $e_n = u$, $\lambda_n = K_n$. If we were in the other case, in which we had $-K_n$ equal to the infimum of (Ax, x), we would have $Au = -K_n u$, and we would take $e_n = u$, $\lambda_n = -K_n$.

This procedure, applied for $n = 1, 2, \ldots$, will result in a collection of pairwise-orthogonal unit vectors $e_1, \ldots, e_n, \ldots$ with eigenvalues $\lambda_1, \lambda_2, \ldots$ such that $|\lambda_1| \geq |\lambda_2| \geq \ldots$, because the supremum K_n defined in (1-31) is taken over an ever-decreasing set as n increases. Now, if $K_n = 0$ for some n, then $e_1, \ldots, e_{n-1}$ span M. In fact, if x is orthogonal to $e_1, \ldots, e_{n-1}$, then so is Ax, as remarked above and by (1-31),

$$(A(x + Ax), (x + Ax)) = 0$$

which, when expanded, coupled with the fact that $(Ax, x) = 0$,

$$(AAx, Ax) = 0$$

results in

$$(Ax, Ax) = 0$$

or x belongs to N as well as M, so $x = 0$.

In the opposite case, $K_n > 0$ for all n. Then $\lambda_n \to 0$ as $n \to \infty$ because the set $\{e_n\}$ converges weakly to zero (use Bessel's inequality—see Exercise 1-16a—which does not require that $\{e_n\}$ be maximal), so that $Ae_n \to 0$ strongly; that is, $\lambda_n e_n \to 0$ (and $\|e_n\| = 1$). In this case also, the set $\{e_n\}$ spans M. The same argument as was used in the last paragraph is also valid here, because if x is orthogonal to all the e_n, then

$$|(A(x + Ax), (x + Ax))| \leq K_n \|x + Ax\|^2$$

for every n, and $K_n \to 0$. The proof is now complete.

Exercise 1-21 Prove, for self-adjoint compact operators, the results of Exercise 1-11, except for part g.

1-8 APPLICATIONS

Let H be the Hilbert space of complex square-summable sequences $x = \{x_m\}$;

$$\|x\| = \Big(\sum_{m=1}^{\infty} |x_m|^2 \Big)^{1/2}$$

If $A = \{a_{m,m}\}$, m, $n = 1, 2, \ldots$, is an infinite complex matrix such that

$$\sum_{m=1}^{\infty} \sum_{n=1}^{\infty} |a_{mn}|^2 < \infty \tag{1-34}$$

then the transformation

$$x \xrightarrow{A} y = \{y_m\}$$

with

$$y_m = \sum_{n=1}^{\infty} a_{mn} y_n \tag{1-35}$$

is a bounded operator belonging to the *Hilbert-Schmidt* class. We shall see that such an A is a compact operator (but not self-adjoint unless $\bar{a}_{mn} = a_{nm}$).

Theorem 1-5 Every Hilbert-Schmidt operator is compact.

Proof: Let $\{x_m^{(k)}\}$ converge to zero weakly as $k \to \infty$. We must show that the sequence $\{y_m^{(k)}\}$ converges to zero strongly, where

$$y_m^{(k)} = \sum_{n=1}^{\infty} a_{mn} x_n^{(k)} \tag{1-36}$$

Now each row of A is a square-integrable sequence, and consequently

$y_m{}^{(k)} \to 0$ for each m as $k \to \infty$. Furthermore, since $\{x_m{}^{(k)}\} \to 0$ weakly as $k \to \infty$, it follows that there is a constant M such that

$$\|\{x_m{}^{(k)}\}\|^2 = \sum_{m=1}^{\infty} |x_m{}^{(k)}|^2 \leq M^2$$

for every k (see Exercise 1-22 below).

Let $\epsilon > 0$ and choose N_0 so large that

$$\sum_{m=N_0}^{\infty} \sum_{n=1}^{\infty} |a_{mn}|^2 < \epsilon^2 \tag{1-37}$$

Then $\displaystyle\sum_{m=N_0}^{\infty} |y_m{}^{(k)}|^2 \leq \sum_{m=N_0}^{\infty} \sum_{n=1}^{\infty} |a_{mn}|^2 \sum_{n=1}^{\infty} |x_n{}^{(k)}|^2 \leq \epsilon^2 M^2$

for every k, by the Schwarz inequality. Since $y_m{}^{(k)} \to 0$ as $k \to \infty$, we have $\displaystyle\sum_{m=1}^{N_0} |y_m{}^{(k)}|^2 \to 0$ as $k \to \infty$, and combining the last two results yields

$$\sum_{m=1}^{\infty} |y_m{}^{(k)}|^2 \to 0$$

as $k \to \infty$. The proof is now complete.

Exercise 1-22 Show that, if $x_k \to 0$ weakly as $k \to \infty$, then $\{x_k\}$ is bounded; i.e., there is an M such that $\|x_k\| \leq M$ for every k. Use (a) and (b). ·

(a) It is enough to show that $\{x_k\}$ is bounded on some sphere; i.e., for some center z_0 and some radius r, if S is the sphere consisting of all vectors u for which $\|u - z_0\| \leq r$, then there is a constant K such that

$$|(x_k, u)| \leq K$$

for every u in S. *Hint:* If $\{x_k\}$ is indeed bounded on S, then let $K' = \max_k |(x_k, z_0)|$ $[(x_k, z_0) \to 0$, of course$]$. Then if $\|y\| \leq r$, the vector $y + z_0$ belongs to S, so that

$$|(x_k, y + z_0)| \leq K \qquad \text{for all } k$$

and consequently

$$|(x_k, y)| \leq K + K' \qquad \text{for all } k$$

whenever $\|y\| \leq r$. But $rx_k/\|x_k\|$ has norm less than or equal to r; therefore $\|x_k\| \leq (K + K')/r$ for every k.

(b) Suppose then that there is no sphere on which $\{x_k\}$ is bounded. Let S_1 be any sphere. Then there are a k_1 and a u_1 in S_1 such that $|(x_{k_1}, u_1)| > 2$, and therefore there is a sphere S_2 about u_1, lying inside S_1, with radius less than 1, such that $|(x_{k_1}, u)| > 1$ for all u in S_2. Similarly, there is a $k_2 > k_1$ and a sphere S_3 lying within S_2 with radius less than $\frac{1}{2}$, such that $|(x_{k_2}, u)| > 2$ for all u in S_3, and, in general, there is a $k_n > k_{n-1}$ and a sphere S_{n+1} contained in S_n with radius less than $1/n$ such that $|(x_{k_n}, u)| > n$ for all u in S_n. But H is complete, so that there is a u^* that belongs to all the spheres S_n. We have $|(x_{k_n}, u^*)| > n$ for all n, while $(x_{k_n}, u^*) \to 0$ by hypothesis.

Now let H_0 be the space of complex square-integrable functions defined on $a < t < b$, with a and b finite. As we remarked earlier, H_0 is not complete if we restrict ourselves to Riemann-integrable functions. Let us consider the integral operator A with the (hermitian) symmetric kernel $K(s,t) = \overline{K(t,s)}$, with $K(s,t)$ continuous on the closed square $a \leq s$, $t \leq b$. If $\{\varphi_n(t)\}$ is a CON set in H_0, then there is an infinite matrix $A = \{a_{mn}\}$ corresponding to $K(s,t)$ such that if

$$f(t) = \sum_{n=1}^{\infty} x_n \varphi_n(t)$$

and f is sent into g by $K(s,t)$, that is,

$$g(s) = \int_a^b K(s,t) f(t) \, dt$$

then

$$g(t) = \sum_{n=1}^{\infty} y_n \varphi_n(t)$$

with

$$y_m = \sum_{n=1}^{\infty} a_{mn} x_n$$

Evidently, we must have

$$a_{mn} = \int\!\!\int_a^b K(s,t) \varphi_n(t) \, \overline{\varphi_m(s)} \, dt \, ds$$

Since $\{\varphi_m(t)\}$ is a CON set on $[a,b]$, the collection $\{\overline{\varphi_m(t)}\varphi_n(s)\}$ is a CON set of the square $a \leq s$, $t \leq b$, and since $K(s,t)$ is evidently square-integrable on this square, we can use Parseval's equality (Exercise 1-16) to show that

$$\int\!\!\int_a^b |K(s,t)|^2 \, ds \, dt = \sum_{m=1}^{\infty} \sum_{n=1}^{\infty} |a_{mn}|^2 < \infty$$

It follows that our matrix A is a Hilbert-Schmidt matrix, and the integral operator is a hermitian compact operator. Therefore, by Theorem 1-4, there is an orthonormal set $\{e_n\}$ of eigenvectors, $Ae_n = \lambda_n e_n$. However we do not know yet that these functions $e_n(t)$ are Riemann-integrable. But since our present kernel $K(s,t)$ is assumed to be continuous on the closed square, the functions $e_n(t)$ are in fact continuous. It is enough to observe that in the proof of Theorem 1-4, the supremum defined in (1-31) would be the same if we imposed the additional condition on the x's that they be continuous when we are in a function space. We would then have a sequence $\{y_k(t)\}$ that converges weakly, such that

$$z_k(t) = \int_a^b K(t,s) y_k(s) \, ds$$

converges strongly. Now the sequence $\{z_k(t)\}$ here is equicontinuous [in the ϵ, δ definition of continuity of each $z_k(t)$, δ does not depend on k; see page 47 for a formal definition] because, for all k,

$$|z_k(t) - z_k(t')| \leq \int_a^b |K(t,s) - K(t',s)| \cdot |y_k(s)| \, ds$$
$$\leq M \left[\int_a^b |K(t,s) - K(t',s)|^2 \, ds \right]^{1/2}$$

by the Schwartz inequality and the fact that $\{y_k\}$ is bounded in norm (Exercise 1-22). The result is that a subsequence of $\{z_k(t)\}$ converges uniformly. The limit $u(t)$, which is a continuous function, is also the limit in the Hilbert-space norm, so it must coincide with the u in the proof of Theorem 1-4; i.e.,

$$\int_a^b K(s,t)u(t) \, dt = \lambda u(s)$$

All our eigenvectors are continuous.

A real Sturm-Liouville self-adjoint system is a second-order linear differential equation with a (boundary) condition at each end point of a bounded interval:

$$\frac{d}{dt}\left[p(t)\frac{dy}{dt} \right] + r(t)y(t) = f(t)$$
$$c_1 y(a) + c_2 y'(a) = 0 \qquad (1\text{-}38)$$
$$c_3 y(b) + c_4 y'(b) = 0$$

in which $p(t) > 0$ in $[a,b]$, $r(t)$ is real and continuous, and $c_1, \ldots, c_4$ are real constants, $|c_1| + |c_2| \neq 0$, $|c_3| + |c_4| \neq 0$. (See Ref. 9 for more general boundary conditions.) The differential expression in (1-38) may be regarded as a linear transformation on smooth functions $y(t)$, but it is not a bounded operator. However there is a Green's function $G(s,t,\lambda_0)$ for every $\lambda_0 > \max_{a \leq t \leq b} |r(t)|$ such that if we put

$$g(t) = \int_a^b G(t,s,\lambda_0)f(s) \, ds \qquad (1\text{-}39)$$

then $g(t)$ satisfies the boundary conditions of (1-38), $g(t)$ is twice continuously differentiable if $p(t)$, $r(t)$, and $f(t)$ are continuous, and

$$\frac{d}{dt}(pg') + [r(t) - \lambda_0]g(t) = f(t) \qquad (1\text{-}40)$$

$G(s,t,\lambda_0)$ is continuous and symmetric in (s,t). Equation (1-39) is an integral equation equivalent to (1-40) with the boundary conditions in (1-38), and our previous theory is applicable. There is a collection of eigenvalues K_n and eigenfunctions $e_n(t)$ such that

$$\int_a^b G(t,s,\lambda_0)e_n(s) \, ds = K_n e_n(t)$$

$e_n(t)$ satisfies the boundary conditions, and

$$(pe_n')' + [r(t) - \lambda_0]e_n = K_n^{-1}e_n$$

from (1-40). The equation in (1-38) has a unique solution satisfying the boundary conditions if and only if $\lambda_0 \neq -K_n^{-1}$ for every n. Much more can be said on this problem, but we shall stop here.

1-9 BANACH SPACES AND LINEAR FUNCTIONALS

We have discussed some of the elementary features of Hilbert space and operators in such a space. Hilbert space is an important example of a more general concept: the complete normed vector space, or Banach space. A real or complex vector space E is called a normed space if there is given a length or norm $\|x\|$ for every x in E such that

1. $\|x\| \geq 0$
2. $\|x + y\| \leq \|x\| + \|y\|$
3. $\|\lambda x\| = |\lambda| \cdot \|x\|$ for every scalar λ
4. $\|x\| = 0$ only if $x = 0$

[Hilbert space is such a space, in which $\|x\| = \sqrt{(x,x)}$, but in a general normed space we do not have an inner product.] We already know what a complete normed space is: If $\|x_n - x_m\| \to 0$ as $n, m \to \infty$, then there is an x^* in E such that $x_n \to x^*$ as $n \to \infty$ (that is, $\|x_n - x^*\| \to 0$).

An important example of a Banach space is the space $C[a,b]$, the space of all real (or all complex) continuous functions on the closed interval $[a,b]$. The norm in $C[a,b]$ is always defined by

$$\|x(t)\| = \max_{a \leq t \leq b} |x(t)|$$

Exercise 1-23 Show that $C[a,b]$ is a Banach space.

Linear transformations are defined in the same way as before. There is no nice "diagonalizing" theorem here.

Let E be a real (or complex) Banach space. If $\varphi(x)$ is a real (or complex) valued function on E, and φ is linear, i.e.,

$$\varphi(\alpha x + \beta y) = \alpha\varphi(x) + \beta\varphi(y)$$

then φ is called a linear functional. Just as in the case of linear transformations (of which φ is a special case), φ will be continuous in E if and only if φ is bounded, i.e., there is an M for which

$$|\varphi(x)| \leq M\|x\| \tag{1-41}$$

for every x in E. In what follows, all linear functionals will be bounded.

The smallest M for which (1-41) holds is called the norm of φ and is denoted $\|\varphi\|$.

If φ and ψ are two linear functionals on E, then the map

$$x \rightarrow \varphi(x) + \psi(x)$$

is a linear functional. It is called the *sum* of φ and ψ. Evidently

$$\|\varphi + \psi\| \leq \|\varphi\| + \|\psi\|$$

Similarly, if λ is a scalar, then the map

$$x \rightarrow \lambda\varphi(x)$$

is also a linear functional. We have

$$\|\lambda\varphi\| = |\lambda| \cdot \|\varphi\|$$

It follows that the set of all linear functionals on E forms a normed vector space; this space is called the *dual* space or the *conjugate* space of E and is denoted E^*. It is easy to see that E^* is complete (the same is true even if we start with an incomplete normed space E). So E^* is a Banach space.

There are a number of standard Banach spaces for which the dual spaces can be explicitly described. For example, let $E = C[a,b]$, the space of real-valued continuous functions on the bounded interval $a \leq t \leq b$, with

$$\|x\| = \max_{a \leq t \leq b} |x(t)|$$

If φ is a linear functional on E, then there is a function $F(t)$ of bounded variation on $[a,b]$, continuous from the right, such that[6]

$$\varphi(x) = \int_a^b x(t)\, dF(t)$$

for every x in E. In this case,

$$\|\varphi\| = \sup_{t_1 < \cdots < t_N} \sum_{j=1}^{N} |F(t_j) - F(t_{j-1})| \tag{1-42}$$

the total variation of F. If $F(t)$ corresponds to φ and $G(t)$ corresponds to ψ, then clearly $\alpha F + \beta G$ corresponds to $\alpha\varphi + \beta\psi$. It is therefore natural to identify the linear functionals here with the collection of functions $F(t)$ of bounded variation which are right-continuous, with $\|F\|$ given by (1-42). We then say a linear functional φ on $E = C[a,b]$ *is* a function of bounded variation on $[a,b]$, and we write

$$\varphi(x) = \int_a^b x(t)\, d\varphi(t)$$

Additional examples are afforded by the L_p spaces. Let us take a fixed interval $a \leq t \leq b$ in which a and b may be finite or infinite. For each fixed p such that $1 < p < \infty$, let E be the real (or complex) vector space of functions $x(t)$ for which

$$\int_a^b |x(t)|^p \, dt \; < \; \infty \tag{1-43}$$

It can be shown that if q satisfies $p^{-1} + q^{-1} = 1$, and if

$$\int_a^b |y(t)|^q \, dt \; < \; \infty$$

then the integral $\int_a^b x(t)y(t) \, dt$ is absolutely convergent, and

$$\left| \int_a^b x(t)y(t) \, dt \right| \leq \left[\int_a^b |x(t)|^p \, dt \right]^{1/p} \left[\int_a^b |y(t)|^q \, dt \right]^{1/q} \tag{1-44}$$

This important inequality is the *Hölder inequality*. Note that if $p = 2$, then also $q = 2$, and we get the Schwartz inequality as a special case of Hölder's inequality. Another classical result here is the *Minkowski inequality*: If $x(t)$ and $y(t)$ both satisfy (1-43), then so does $x(t) + y(t)$, and

$$\left[\int_a^b |x(t) + y(t)|^p \, dt \right]^{1/p} \leq \left[\int_a^b |x(t)|^p \, dt \right]^{1/p} + \left[\int_a^b |y(t)|^p \, dt \right]^{1/p} \tag{1-45}$$

This is also a generalization of the case in which $p = q = 2$.

It follows that our space E is a normed vector space if we set

$$\|x\| = \|x\|_p = \left[\int_a^b |x(t)|^p \, dt \right]^{1/p}$$

for every x in E. This space is called $L_p(a,b)$; it is complete if we include Lebesgue-integrable functions, and incomplete if we restrict ourselves to Riemann-integrable functions.

It is not hard to show, for finite a and b, that if $x(t)$ is bounded on (a,b), then the limit

$$\lim_{p \to \infty} \left[\int_a^b |x(t)|^p \, dt \right]^{1/p} = M$$

exists, is finite, and

$$|x(t)| \leq M \tag{1-46}$$

for all t in (a,b) except possibly for a "small" set of t's—more precisely, except possibly for a set of t's of Lebesgue measure zero. Furthermore, M is the smallest number for which (1-46) holds. [If $x(t)$ is continuous except for a finite number of points at which x has left- and right-hand limits, then (1-46) is valid for all t.] The number M is called the *essential supremum* of x and is denoted by $\|x\|_\infty$.

The set E of all measurable functions $x(t)$ bounded on (a,b) is a Banach space if we use the norm just defined:

$$\|x\|_\infty = \lim_{p \to \infty} \left[\int_a^b |x(t)|^p \, dt \right]^{1/p} \tag{1-47}$$

in which a and b are finite (see Sec. 1-11).

The results given above for $1 < p < \infty$ can now be extended to the two additional cases $p = 1$ and $p = \infty$, in which the corresponding q's are $q = \infty$ and $q = 1$.

Let $E = L_p(a,b)$, $1 \leq p < \infty$. Then if φ is a linear functional on E, there is a function $g(t)$ in $L_q(a,b)$ such that

$$\varphi(x) = \int_a^b x(t)g(t) \, dt \tag{1-48}$$

for every x in E. Just as in the case in which E was $C[a,b]$, we identify the linear functional φ on L_p with the corresponding L_q function, and we say that E^* *is* the set of all L_q functions. In brief, if $E = L_p(a,b)$, then $E^* = L_q(a,b)$, $1 \leq p < \infty$.

The same is not true for $p = \infty$. The dual space E^* of $E = L_\infty$ contains the L_1 functions, but there are also many other linear functionals in E^* which are not L_1 functions.

Let us now consider the set of m-tuples $(x_1, \ldots, x_m)$ whose elements belong to a Banach space E. (This will be used in Sec. 5-4.) The set E_m of all m-tuples $x = (x_1, \ldots, x_m)$ in which each x_j belongs to the Banach space E is itself a Banach space. The norm in E_m can be taken to be

$$\|x\| = \left[\sum_{j=1}^m \|x_j\|^p \right]^{1/p} \tag{1-49}$$

for any (fixed) p satisfying $1 \leq p \leq \infty$. Then $(E_m)^*$, the dual space of E_m, consists of all m-tuples $\varphi = (\varphi_1, \ldots, \varphi_m)$, where φ_j is in E^* for each j. The result is that $(E_m)^* = E_m^*$. We have

$$\varphi(x) = \sum_{j=1}^m \varphi_j(x_j) \tag{1-50}$$

for each x in E_m and each φ in E_m^*. Also

$$\|\varphi\| = \left[\sum_{j=1}^m \|\varphi_j\|^q \right]^{1/q} \tag{1-51}$$

in which $p^{-1} + q^{-1} = 1$.

Finally let us remark that notationally it is convenient to write

$$(x,\varphi) = \varphi(x) \tag{1-52}$$

for x in E and φ in E^*. The application of φ to x is linear in x and φ

separately, and the notation in (1-52) is suggested by the inner-product notation in Hilbert space. It follows that (1-50) can also be written

$$(x,\varphi) = \sum_{j=1}^{m} (x_j,\varphi_j)$$

We state without proof the important Hahn-Banach theorem: Let E be a real or complex Banach space. Let F be a subspace of E, and suppose φ_0 is a linear functional defined on F with norm $\|\varphi_0\|$. Then φ_0 can be extended to a linear functional φ on E such that $\|\varphi\| = \|\varphi_0\|$ (that is, there is a functional φ on E such that $\|\varphi\| = \|\varphi_0\|$, and φ coincides with φ_0 on F).

1-10   FIXED - POINT THEOREMS AND APPLICATIONS

Our main purpose in this section is to define an important kind of (not necessarily linear) transformation, the contraction operator. Let E be a Banach space and let A send an open subset U of E into U; suppose also that there is a constant K, $0 < K < 1$, such that for every x_1, x_2 in U, we have

$$\|A(x_1) - A(x_2)\| \le K\|x_1 - x_2\|$$

(that is, A satisfies a Lipchitz condition with $K < 1$). Then A is called a contraction (operator) on U. The following theorem has many applications.

Theorem 1-6   Let A be a contraction on U in a Banach space E with Lipschitz constant K. Suppose there is a sphere $S:\{\|x - x_0\| \le a\}$ that is contained in U, and that

$$\|A(x_0) - x_0\| < (1 - K)a$$

Then there is a unique fixed point x^* in S, that is, a point x^* for which $A(x^*) = x^*$.

Proof:[6]   Define the sequence $\{x_n\}$ which is given recursively by

$$x_n = A(x_{n-1}) \qquad n = 1, 2, \ldots \qquad (1\text{-}53)$$

(and in which x_0 is the vector of our hypothesis). By hypothesis, x_0 and x_1 lie in the sphere S. If $\|x_j - x_0\| < (1 - K^j)a$ for $j = 1, 2, \ldots , n$ (so that $x_1, x_2, \ldots , x_n$ lie in S), then

$$\|x_{n+1} - x_0\| \le \|x_{n+1} - x_n\| + \|x_n - x_0\|$$
$$\le \|A(x_n) - A(x_{n-1})\| + (1 - K^n)a$$

But

$$\|A(x_n) - A(x_{n-1})\| \le K\|x_n - x_{n-1}\| = K\|A(x_{n-1}) - A(x_{n-2})\|$$
$$\le K^2\|x_{n-1} - x_{n-2}\| \le \cdots \le K^n\|x_1 - x_0\| \le K^n(1 - K)a$$

It follows that

$$\|x_{n+1} - x_0\| \le (1 - K^{n+1})a$$

so that x_n lies in S for every n.   We see also that because $\|x_{n+1} - x_n\| \le K^n \|x_1 - x_0\|$, we have

$$\|x_{n+r} - x_n\| \le \sum_{j=n}^{n+r-1} \|x_{j+1} - x_j\| \le \sum_{j=n}^{\infty} \|x_{j+1} - x_j\| \le \frac{K^n}{1 - K} \|x_1 - x_0\|$$

for $r \ge 1$—the sequence $\{x_n\}$ is a Cauchy sequence, $x_n \to x^*$ as $n \to \infty$, and x^* belongs to S.   Since $A(x)$ is continuous on U (by the Lipschitz condition), we now let $n \to \infty$ in (1-53) and get

$$x^* = A(x^*)$$

There cannot be a second point y in U for which $y = A(y)$, for if there were we would have

$$\|x^* - y\| = \|A(x^*) - A(y)\| \le K\|x^* - y\| < \|x^* - y\|$$

which is impossible.   The proof is now complete.

We shall apply this theorem first in a finite-dimensional context. Let R_n denote the vector space of real n-tuples $(y_1, \ldots, y_n)$.   We shall take the norm in R_n to be

$$\|y\| = \max_{1 \le i \le n} |y_i| \tag{1-54}$$

the absolute value of the largest component.   [The reader should verify that (1-54) really defines a norm.]

If A is an n-rowed square matrix with real entries, then A sends n-tuples, as column vectors, to n-tuples and represents a linear transformation of R_n into R_n.   Let A have elements a_{ij}.   Then if $z = Ay$, we have

$$z_i = \sum_{j=1}^{n} a_{ij} y_j$$

$$|z_i| \le \sum_{j=1}^{n} |a_{ij}| \cdot |y_j| \le \left(\sum_{j=1}^{n} |a_{ij}| \right) \|y\|$$

and as a result

$$\|z\| \le \left(\max_{1 \le i \le n} \sum_{j=1}^{n} |a_{ij}| \right) \|y\| \tag{1-55}$$

Suppose now that the elements a_{ij} of A are continuous functions of y in a region of R_n, and of x in a region of R_m,

$$a_{ij} = a_{ij}(x,y)$$

and that for some x_0, y_0 we have $a_{ij}(x_0,y_0) = 0$, i, $j = 1, 2, \ldots, n$. Then for every $\epsilon > 0$, there is a neighborhood $N_1(x_0)$ in R_m and a neighborhood $N_2(y_0)$ in R_n such that

$$\max_{1 \leq i \leq n} \sum_{j=1}^{n} |a_{ij}(x,y)| < \epsilon$$

for all x in $N_1(x_0)$ and all y in $N_2(y_0)$. We shall need this elementary observation in the following *implicit-function* theorem.

Theorem 1-7 Let $g(x,y) = \{g_1, \ldots, g_n\}$ be continuous for x in a neighborhood of x_0 in R_m and for y in a neighborhood of y_0 in R_n, with $g(x_0,y_0) = 0$. Suppose g is continuously differentiable in y and that the determinant

$$\left| \left\{ \frac{\partial g_i}{\partial y_j}(x_0,y_0) \right\} \right| \neq 0$$

Then there is a neighborhood $N_1(x_0)$ in R_m and a neighborhood $N_2(y_0)$ in R_n such that, for every x in N_1, there is a unique $y = \varphi(x)$ in N_2 for which $g(x,\varphi(x)) = 0$. If $g(x,y)$ is k times continuously differentiable in x and y, then $\varphi(x)$ is k times continuously differentiable ($k \geq 1$).

Proof: We shall write $J(x,y)$ for the jacobian matrix

$$J(x,y) = \left\{ \frac{\partial g_i}{\partial y_j}(x,y) \right\}$$

Then, by hypothesis, $J(x_0,y_0)^{-1}$ exists. Define the function f on $R_m \times R_n$ by

$$f(x,y) = y - J(x_0,y_0)^{-1}g(x,y)$$

Then $f(x,y) = y$ if and only if $g(x,y) = 0$; in particular, $f(x_0,y_0) = y_0$. Note that f is continuously differentiable in y; its jacobian matrix is

$$\left\{ \frac{\partial f_i}{\partial y_j} \right\} = \{\delta_{ij}\} - J(x_0,y_0)^{-1}\left\{ \frac{\partial g_i(x,y)}{\partial y_j} \right\}$$

Consequently, $\left\{ \frac{\partial f_i}{\partial y_j}(x_0,y_0) \right\} = 0$. By our previous remarks, there is a neighborhood of x_0 and a neighborhood of y_0 such that

$$\max_{1 \leq i \leq n} \sum_{j=1}^{n} \left| \frac{\partial f_i}{\partial y_j}(x,y) \right| < K \tag{1-56}$$

for all x and y in the respective neighborhoods, where K is a constant, $0 < K < 1$, chosen beforehand.

Now let y' and y'' be any two vectors in our y neighborhood, and define, for fixed x,

$$h(\lambda) = f(x, y' + \lambda(y'' - y')) - f(x,y') \qquad 0 \leq \lambda \leq 1$$

The vector $h(\lambda)$ is a differentiable function of λ with $h(0) = 0$ and

$$h(1) = f(x,y'') - f(x,y') \tag{1-57}$$

By the mean-value theorem we have, for each component h_i,

$$h_i(1) = \frac{dh_i}{d\lambda}(\bar{\lambda}_i) \tag{1-58}$$

for some $\bar{\lambda}_i$ in $0 < \bar{\lambda}_i < 1$. But

$$\frac{dh}{d\lambda} = \left\{\frac{\partial f_i}{\partial y_j}\right\}(y'' - y')$$

(a matrix times a vector), and, by (1-55) through (1-58), we have

$$\|f(x,y'') - f(x,y')\| < K\|y'' - y'\|$$

valid for all x, y', and y'' in the respective neighborhoods of x_0 and y_0 for which (1-56) holds. Let $\{\|x - x_0\| < a\}$ be contained in our x neighborhood. Now take a smaller x neighborhood, if necessary, so that in this smaller neighborhood we have

$$\|f(x,y_0) - y_0\| < (1 - K)a$$

[recall that $f(x_0,y_0) = y_0$]. We can now apply Theorem 1-6 and conclude that there is a unique $y = \varphi(x)$ such that $f(x,y) = y$, or

$$g(x,\varphi(x)) = 0 \tag{1-59}$$

The final statement of the theorem can be obtained in the standard way: Let x_r be replaced by $x_r + \Delta x_r$, r fixed, $r = 1, 2, \ldots, m$, and use the mean-value theorem on (1-59). We obtain

$$\left\{\frac{\partial \varphi_i}{\partial x_r}\right\} = -\left\{\frac{\partial g_i}{\partial y_j}\right\}^{-1}\left\{\frac{\partial g_i}{\partial x_r}\right\} \qquad i, j = 1, \ldots, n \qquad r = 1, \ldots, m$$

and each of the matrices on the right is $k - 1$ times continuously differentiable. The proof is now complete.

For another application of Theorem 1-6 let us consider the system of ordinary differential equations

$$y_1'(t) = f_1(t,y_1, \ldots, y_n)$$
$$\cdots\cdots\cdots\cdots\cdots\cdots \tag{1-60}$$
$$y_n'(t) = f_n(t,y_1, \ldots, y_n)$$

with initial conditions $y_1(0) = c_1, \ldots, y_n(0) = c_n$. This system is evidently equivalent to the set of integral equations

$$y_1(t) = c_1 + \int_0^t f_1(s,y_1(s), \ldots, y_n(s))\, ds$$
$$\cdots\cdots\cdots\cdots\cdots\cdots\cdots\cdots\cdots \tag{1-61}$$
$$y_n(t) = c_n + \int_0^t f_n(s,y_1(s), \ldots, y_n(s))\, ds$$

Let us use the following vector notation: We write $y(t)$ for the column vector with components $y_1(t), \ldots, y_n(t)$ and c for the vector with components $c_1, \ldots, c_n$; similarly $f(s,y(s))$ has components

$$f_i(s,y_1(s), \ldots, y_n(s))$$

When we write $\int_0^t z(s) \, ds$, where $z(s)$ is a vector with components $z_i(s)$, we mean the vector with components $\int_0^t z_i(s) \, ds$. Then (1-61) can be written

$$y(t) = c + \int_0^t f(s,y(s)) \, ds \qquad (1\text{-}62)$$

If $\delta > 0$, then we can take the space of all n-tuples $(y_1(t), \ldots, y_n(t))$ in which $y_i(t)$ is continuous for $0 \leq t \leq \delta$ and define a norm in the space by the formula

$$\|y\| = \max_{1 \leq i \leq n} \; \max_{0 \leq t \leq \delta} \; |y_i(t)|$$

We can now present the following existence theorem for systems of first-order ordinary differential equations.

Theorem 1-8   Let the functions $f_i(t,y_1, \ldots, y_n)$ of $n+1$ real variables be continuous in a neighborhood of $t = 0$, $y_i = c_i$, $i = 1, \ldots, n$, and suppose they satisfy a Lipschitz condition

$$|f_i(t,y_1, \ldots, y_n) - f_i(t,\bar{y}_1, \ldots, \bar{y}_n)| \leq M \max_{1 \leq j \leq n} |y_j - \bar{y}_j|$$

for $0 \leq t \leq \delta$, $|y_i - c_i| < a$, $|\bar{y}_i - c_i| < a$, $i = 1, 2, \ldots, n$. Then there is a $\delta' > 0$ such that on the interval $0 \leq t \leq \delta'$, the system

$$\frac{dy_i}{dt} = f_i(t,y_1(t), \ldots, y_n(t)) \qquad i = 1, \ldots, n$$

has a unique solution $y_i(t)$ for which $y_i(0) = c_i$.

Proof: Choose K such that $0 < K < 1$. Choose L such that, for $0 \leq t \leq \delta$, we have

$$|f_i(t,c_1, \ldots, c_n)| \leq L \qquad i = 1, \ldots, n$$

Choose δ' such that $0 < \delta' \leq \delta$,

$$\delta' \leq \frac{1}{L}(1 - K)a$$

and

$$\delta' \leq \frac{1}{M} K$$

where M and a are given in the hypothesis. Our Banach space will consist of n-tuples $(y_1(t), \ldots, y_n(t))$ of functions continuous on $0 \leq t \leq \delta'$, with the norm described above.

Define the vector function $A(y)$ by

$$A(y)(t) = c + \int_0^t f(s, y(s)) \, ds$$

for $0 \leq t \leq \delta'$. Then if $\|y - c\| < a$ and $\|\bar{y} - c\| < a$, we have

$$A(y)(t) - A(\bar{y})(t) = \int_0^t [f(s, y(s)) - f(s, \bar{y}(s))] \, ds$$
$$\|A(y) - A(\bar{y})\| \leq \int_0^{\delta'} \|f(s, y(s)) - f(s, \bar{y}(s))\| \, ds$$
$$\leq M\|y - \bar{y}\|\delta' \leq K\|y - \bar{y}\|$$

Furthermore, the ith component of $A(c) - c$ is

$$\int_0^t f_i(s, c_1, \ldots, c_n) \, ds \qquad i = 1, 2, \ldots, n$$

which for $t \leq \delta' \leq \delta$ does not exceed $L\delta'$ in absolute value. Therefore

$$\|A(c) - c\| \leq L\delta' \leq (1 - K)a$$

By Theorem 1-6, we conclude that there is a unique y such that $\|y - c\| \leq a$, and

$$A(y) = y$$

This completes the proof.

It should be clear that the choice of $t = 0$ was of no importance. Any initial value of t can play the role of our $t = 0$ if the hypotheses are appropriately modified.

The requirement in Theorem 1-6 that $A(x)$ be a contraction is a strong one, and there are more general situations in which we can assert the existence of a fixed point. The following, the *Brouwer fixed-point theorem*, can be found in Graves[6] and Dunford and Schwartz:[4]

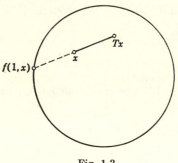

Fig. 1-3

Theorem 1-9 Let T be a continuous mapping of the closed unit sphere S of R_n into itself. Then there is a point x in S such that $Tx = x$.

Proof: We shall suppose at first that T is infinitely differentiable. This restriction will be removed at the end of the proof. The proof is indirect; we suppose that $Tx \neq x$ for every x in S and derive a contradiction.

Since $Tx \neq x$ for every x in S, we can determine a unique straight line from Tx to x and continue it to the boundary Γ of S (see Fig. 1-3). The resulting point in the boundary is of the form

$$y = x + a(x)(x - Tx)$$

in which $a(x)$ is a nonnegative scalar determined by the requirement that

$$\|y\|^2 = \sum_{j=1}^{n} y_j{}^2 = 1$$

This yields

$$\|x\|^2 + 2a(x, x - Tx) + a^2\|x - Tx\|^2 = 1$$

from which we may solve for $a(x)$:

$$a(x) = \frac{-(x, x - Tx) + [(x, x - Tx)^2 + \|x - Tx\|^2(1 - \|x\|)^2]^{\frac{1}{2}}}{\|x - Tx\|^2}$$

We see that the discriminant is (strictly) positive for every x in S; indeed, if it is zero, then $\|x\| = 1$ and $(Tx,x) = (x,x) = 1$, but the latter is impossible since $\|Tx\| \leq 1$ and $Tx \neq x$. It follows that the indicated square root is an infinitely differentiable function of x, and $a(x)$ is infinitely differentiable. Note also that $a(x) = 0$ if $\|x\| = 1$, that is, if x is in Γ. (This is also clear from the geometry of the construction.)

For every t in $0 \leq t \leq 1$, let

$$f(t,x) = x + ta(x)(x - Tx)$$

Then $f(0,x) = 1$ for all x in S, and $f(1,x) = x + a(x)(x - Tx)$, so that $f(1,x)$ is in Γ. It follows, first, that $f_x(0,x) = I$ (the n-rowed identity matrix) for all x in S, and, secondly, that for $t = 1$, the n components $f^1(1,x), \ldots, f^n(1,x)$ of f are functionally dependent, since

$$\sum_{i=1}^{n} [f^i(1,x)]^2 = 1$$

for all x. Therefore, the jacobian $|\partial f^i / \partial x_j|$ vanishes identically when $t = 1$.

Let $D_0, D_1, \ldots, D_n$ denote the cofactors of $\epsilon_0, \epsilon_1, \ldots, \epsilon_n$ in the $(n + 1)$-rowed square matrix

$$M = \begin{pmatrix} \epsilon_0 & \epsilon_1 & \epsilon_2 & \cdots & \epsilon_n \\[2mm] \dfrac{\partial f^1}{\partial t} & \dfrac{\partial f^1}{\partial x_1} & \dfrac{\partial f^1}{\partial x_2} & \cdots & \dfrac{\partial f^1}{\partial x_n} \\[2mm] \cdots & \cdots & \cdots & \cdots & \cdots \\[2mm] \dfrac{\partial f^n}{\partial t} & \dfrac{\partial f^n}{\partial x_1} & \dfrac{\partial f^n}{\partial x_2} & \cdots & \dfrac{\partial f^n}{\partial x_n} \end{pmatrix}$$

Then

$$\frac{\partial D_0}{\partial t} + \frac{\partial D_1}{\partial x_1} + \cdots + \frac{\partial D_n}{\partial x_n} = 0$$

for all x in S, $0 \leq t \leq 1$. Indeed, the indicated sum can be obtained formally by replacing the first row of M with the differential operators $\partial/\partial t, \ldots, \partial/\partial x_n$ and "expanding the determinant"; if we write x_0 for t, a typical one of the $(n + 1)!$ terms has the form

$$\pm \frac{\partial}{\partial x_k}\left(\frac{\partial f^1}{\partial x_i} \cdots \frac{\partial f^j}{\partial x_l} \cdots \frac{\partial f^n}{\partial x_q}\right)$$

and we see that for every term in the expansion of the form

$$\pm \frac{\partial f^1}{\partial x_i} \cdots \frac{\partial^2 f^j}{\partial x_k \partial x_l} \cdots \frac{\partial f^n}{\partial x_q} \qquad i, j, \ldots, q = 0, 1, \ldots, n$$

there is another term of exactly the same form with k and l interchanged, and hence with the opposite sign. The expansion therefore vanishes.

We can now derive our desired contradiction. Let

$$I(t) = \int_S D_0(t,x) \, dx_1 \cdots dx_n$$

Then $I(0) = \int_S dx_1 \cdots dx_n > 0$, since the jacobian matrix is the identity, $f_x(0,x) = I$, as we noted earlier. Also, $I(1) = 0$, since the jacobian vanishes identically in x for $t = 1$. The contradiction will be achieved by showing that $I'(t) = 0$ for all t.

We have

$$I'(t) = \int_S \frac{\partial D_0}{\partial t} \, dx_1 \cdots dx_n$$

or

$$I'(t) = -\int_S \sum_{j=1}^{n} \frac{\partial D_j}{\partial x_j} \, dx_1 \cdots dx_n$$

But now we can use the Divergence Theorem, which holds in R_n as well as R_2 and R_3 (with the same proof). We obtain

$$I'(t) = -\int_\Gamma \sum_{j=1}^{n} D_j n_j \, d\sigma$$

in which $\{n_j\}$ denotes the outward normal to S. Since D_j, for $j = 1, \ldots, n$ contains the column $\{\partial f^i/\partial t\}$, which vanishes on Γ because

$$f_t(t,x) = a(x)(x - Tx)$$

and $a(x) = 0$ on Γ, we have $D_j = 0$ on Γ, and $I'(t) = 0$ for all t. This completes the proof under the assumption that T is infinitely differentiable.

Now every continuous mapping T can be approached uniformly by polynomials in $x_1, \ldots, x_n$ by the Weierstrass theorem. So let $T_n \to T$, T_n infinitely differentiable. Let x_n be a fixed point of T_n, $T_n x_n = x_n$. The sequence $\{x_n\}$ has a limit point x in S, and a subsequence of x_n

converges to x. It follows easily that x is a fixed point of T, $Tx = x$, and the proof is complete.

A set K is said to have the *fixed-point property* if, for every continuous mapping T of K into K, there is an x in K such that $Tx = x$. Theorem 1-9 states that every closed sphere in R_n has the fixed-point property. So does every closed bounded convex set, as the following corollary shows.

Corollary: If K is a closed bounded convex subset of R_n, then K has the fixed-point property.

Proof: Let T be a continuous map of K into itself. Let S be a closed sphere containing K. For every x in S, let $N(x)$ denote the unique point in K nearest to x [if y_1 and y_2 in K have the same distance to x, then $(y_1 + y_2)/2$, also in K, is closer to x]. The mapping N of S onto K is continuous, for if $x_n \to x$ and $N(x_n) \nrightarrow N(x)$, then a subsequence of $N(x_n)$ converges to a point $y \neq N(x)$. But $\|x_n - N(x_n)\| \leq \|x_n - N(x)\| \to \|x - N(x)\|$ as $n \to \infty$, and a subsequence of $x_n - N(x_n)$ converges to $x - y$, so that $\|x - y\| \leq \|x - N(x)\|$, a contradiction unless $y = N(x)$.

Now the function $T(N(x))$ maps S into K, a part of S, and by Theorem 1-9 there is an x such that $T(N(x)) = x$. Since x is in K, $N(x) = x$. It follows that $T(x) = x$, and the proof is complete.

This result has been generalized to infinite-dimensional vector spaces by various writers. We give next a proof that a compact convex subset K of a Banach space has the fixed-point property; the theorem was first proved by Schauder[15a] and later generalized by Leray. Let us recall that a subset K of a metric space is compact if every infinite sequence in K has a limit point in K.

Theorem 1-10   Let X be a Banach space, and let K be a compact convex subset of X. Then K has the fixed-point property.

Proof: Let T be a continuous map of K into K. For every $\epsilon > 0$, there exist $x_1, \ldots, x_N$ in K such that every point x in K is within ϵ of some x_j, for otherwise we could obtain recursively a sequence $x_1, \ldots, x_N, \ldots$ such that every x_N and x_M are farther apart than ϵ. But then there would be a limit point x^* of the sequence, x^* in K, and within a sphere about x^* of radius $\epsilon/3$ there would be infinitely many points x_j of our sequence, and every pair of them would be closer than $2\epsilon/3$, a contradiction.

With $x_1, \ldots, x_N$ in K as defined above, define for $j = 1, 2, \ldots, N$,

$$g_j(x) = \begin{cases} \epsilon - \|x - x_j\| & \text{if} \quad \|x - x_j\| < \epsilon \\ 0 & \text{if} \quad \|x - x_j\| \geq \epsilon \end{cases}$$

Then each $g_j(x)$ is continuous. Now set

$$h_j(x) = \frac{g_j(x)}{\sum\limits_{i=1}^{N} g_i(x)}$$

Since every x is within ϵ of some x_i, the denominator is always positive. Therefore $h_j(x)$ is continuous and nonnegative, and $\sum_{j=1}^{N} h_j(x) = 1$ for all x in K. Furthermore, $h_j(x) = 0$ if $\|x - x_j\| \geq \epsilon$.

For every x in K, let

$$V(x) = \sum_{j=1}^{N} h_j(x)x_j$$

Then V is a continuous map from K into the convex hull of $x_1, \ldots, x_N$ (the set of all convex linear combinations of $x_1, \ldots, x_N$). For every x in K we have

$$x - V(x) = \sum_{j=1}^{N} h_j(x)(x - x_j)$$

and the sum need only be taken over those j for which $\|x - x_j\| < \epsilon$, as $h_j(x) = 0$ otherwise. Consequently,

$$\|x - V(x)\| \leq \Sigma h_j(x)\|x - x_j\| < \epsilon$$

for every x in K.

Let K_ϵ denote the convex hull of $x_1, \ldots, x_N$. Then K_ϵ lies in an $(N - 1)$-dimensional space. The mapping

$$x \to V(T(x))$$

in terms of our original map T, sends K_ϵ into itself continuously. (We are tacitly using the fact that the metric in any finite-dimensional Banach space is equivalent to a euclidean metric.) By the last corollary, there is a fixed point x_ϵ in K_ϵ:

$$V(T(x_\epsilon)) = x_\epsilon$$

Finally, for $n = 1, 2, \ldots$, let $\epsilon = 1/n$. By the previous construction, there is an x_n in K such that

$$V_n(T(x_n)) = x_n$$

for each n, and $\|V_n y - y\| < 1/n$ for every n. The sequence $\{x_n\}$ contains a convergent subsequence; call it $\{x_n\}$ again. Then $x_n \to x^*$, say, as $n \to \infty$. We have

$$T(x^*) - x^* = [T(x^*) - T(x_n)] + [T(x_n) - V_n(T(x_n))] + [x_n - x^*]$$

The first and third bracketed quantities tend to zero, and the middle one is less in norm than $1/n$; the proof is complete.

The reader may consult Ref. 4 for a further generalization to locally convex linear spaces, due to Tychonoff.

To apply Theorem 1-10, we must verify that our set K is convex and compact. Convexity is easily treated, but to verify compactness for the space of continuous functions, we need a criterion by which we may recognize a compact subset, which in metric spaces means that every infinite set has a convergent subsequence. Ascoli's theorem[6] gives us such a criterion. In order to state the theorem, we must define two elementary concepts. Let E be a closed bounded subset of R_n, and let $\mathfrak{F}$ be a class of real-valued functions defined on E. Then $\mathfrak{F}$ is called *uniformly bounded* if there is an $M > 0$ such that

$$|f(x)| \leq M$$

for every f in $\mathfrak{F}$ and every x in E. Also, $\mathfrak{F}$ is called *equicontinuous* if, for every $\epsilon > 0$, there is a $\delta > 0$ such that, whenever $\|x - x'\| < \delta$, we have

$$|f(x) - f(x')| < \epsilon$$

for every f in $\mathfrak{F}$.

The theorem of Ascoli states:

If $\mathfrak{F}$ is an infinite set of functions defined on the closed bounded set E in R_n, and if $\mathfrak{F}$ is uniformly bounded and equicontinuous, then $\mathfrak{F}$ contains a uniformly convergent sequence $\{f_n\}$.

We shall merely sketch the proof. Choose a countable dense set $\{x_n\}$ in E. Choose an infinite sequence $\{g_n\}$ from $\mathfrak{F}$. Since $|g_n(x_m)| \leq M$ for all m, n, a subsequence of $\{g_n\}$ converges at x_1, a subsequence of this subsequence converges at x_2, etc. (See also the beginning of the proof of Theorem 1-3.) By taking the first element of each subsequence, we get a new subsequence that converges at each x_m. But then the equicontinuity condition enables us to conclude that the convergence is uniform on E.

As an example of the application of Theorem 1-10, let $F(s,t,u)$ be a continuous, bounded, real-valued function defined for $0 \leq s$, $t \leq 1$, $-\infty < u < \infty$, and consider the integral equation

$$g(s) = f(s) + \int_0^1 F(s,t,f(t))\, dt$$

in which $g(s)$ is real and continuous for $0 \leq s \leq 1$, and f is unknown. Theorem 1-10 enables us to assert the existence of a continuous f satisfying our equation. We proceed as follows:

Let us take as our Banach space $X = C[0,1]$ (see Sec. 1-8). Let M be a bound for $F(s,t,u)$. Our equation is equivalent to

$$f(s) = Tf(s)$$

in which
$$Tf(s) = g(s) - \int_0^1 F(s,t,f(t))\, dt$$

It follows that

$$|Tf(s)| \leq |g(s)| + M$$
$$\|Tf\| \leq \|g\| + M = M'$$

Let S denote the closed sphere in $C[0,1]$ of radius M'. Then T sends S into S. It is easy to see that T is a continuous mapping.

Let K_0 denote the image of S under T, $K_0 = T(S)$. Then one can show that K_0 is uniformly bounded and equicontinuous. Consequently, the same is true for the convex hull K_0^* of K_0. It follows that the closure K of K_0^* (the set consisting of K_0^* and its limit points) is compact. But K_0^* is convex, so that K is also; K is compact and convex. Since K_0 lies in S, which is convex, K_0^* lies in S; since S is closed, K, the closure of K_0^*, lies in S. Therefore $T(K)$ is part of $T(S) = K_0$, which is part of K; that is, T sends K into K. By Theorem 1-10, there is an f in K such that $Tf = f$.

Of course, such an existence proof does not tell us how to construct a solution or a sequence of approximations converging to a solution. For this, different methods are required.

Exercise 1-24   Verify that the assumptions of Ascoli's theorem are true for the set K_0^* in the last example.

Note that if we attempt to imitate the construction used in Theorem 1-6 as exemplified in Chap. 4 in the existence proof for ordinary differential equations (Picard's theorem), we would write

$$f_{n+1}(s) = g(s) - \int_0^1 F(s,t,f_n(t)) \, dt$$

But here we cannot apply Theorem 1-6 and conclude that $f_n(s)$ converges. Even the application of Ascoli's theorem here, which asserts that a subsequence of $\{f_n\}$ converges uniformly, is not enough in itself to produce a limit that satisfies our integral equation. The fixed-point theorem does enable us to conclude the existence of a solution, i.e., a fixed point. Ascoli's theorem was needed to apply the fixed-point theorem.

1-11   LEBESGUE INTEGRATION, A SURVEY

We present here, without proofs, some of the main facts in the theory of Lebesgue integration to amplify some earlier allusions. Lebesgue measure is a generalization of the length of an interval. Let E be any subset of the real line, and consider any countable (finite or enumerable) collection $\{I_n\}$ of intervals that cover E (every point in E belongs to some I_n) and the corresponding sum of the lengths of the I_n's, which could be infinite. The greatest lower bound of all such sums is called the *outer measure of E*, and denoted $m^*(E)$. If E is an interval, $m^*(E)$ is clearly its length.

We define the inner measure $m_*(E)$ first for bounded sets. If E is contained in a bounded interval I, we define

$$m_*(E) = m^*(I) - m^*(I - E)$$

where $I - E$ is the set of points in I and not in E. For arbitrary E, let E_n be the set of points in E that do not exceed n in absolute value. Then, by definition,

$$m_*(E) = \lim_{n \to \infty} m_*(E_n)$$

A set E is called *measurable* if $m^*(E) = m_*(E)$; its *measure* is this common value and is denoted $m(E)$. Not every set is measurable. Most important are the facts that if E and F are measurable, then so is $E - F$; if $\{E_n\}$ is a sequence of measurable sets, so is their union E (the set of points belonging to some E_n); and if no two of the E_n's have points in common, then

$$m(E) = \sum_{n=1}^{\infty} m(E_n)$$

A real-valued function $f(x)$ defined on a measurable set E is called *measurable* if, for every real λ, the set of points for which $f(x) \leq \lambda$ is measurable. If f and g are two such function, then $\alpha f + \beta g$ is also measurable, α, β being real constants. If $\{f_n(x)\}$ is a sequence of measurable functions and $\lim_{n \to \infty} f_n(x) = f(x)$ exists for every x in E, then $f(x)$ is measurable. Every Riemann-integrable function is measurable.

Suppose now that $f(x)$ is bounded and measurable on a set E of finite measure. If $a < f(x) \leq b$ for every x in E, take

$$a = \lambda_0 < \lambda_1 < \cdots < \lambda_n = b \tag{1-63}$$

and let E_j be the set of points in E for which $\lambda_{j-1} < f(x) \leq \lambda_j$, $j = 1, 2, \ldots, n$. It follows from the preceding that each E_j is measurable. If we put

$$S = \sum_{j=1}^{n} \lambda_j m(E_j)$$

then S is an approximating sum for the Lebesgue integral. If, in Eq. (1-63), we let $n \to \infty$ and $\max_{1 \leq j \leq n} (\lambda_j - \lambda_{j-1}) \to 0$ as $n \to \infty$, then the corresponding sums S converge to a finite limit, called the *integral of f over E*, and written $\int_E f(x)\, dx$. Every bounded measurable function possesses a well-defined integral over a set of finite measure. The Lebesgue and the Riemann integrals of every Riemann-integrable (hence bounded) function are equal.

If E is any measurable set, and $f(x)$ is finite but not necessarily bounded, we set

$$E_n = \{x \text{ in } E : |f(x)| \leq n \quad \text{and} \quad |x| \leq n\}$$

Then f is called *integrable* over E if

$$\lim_{n \to \infty} \int_{E_n} |f(x)|\, dx$$

is finite.   In this case,

$$\lim_{n \to \infty} \int_{E_n} f(x)\, dx$$

exists and is finite, and we define $\int_E f(x)\, dx$ to be this limit.   Thus f is integrable over E if and only if $|f|$ is.

If f and g are measurable functions that are equal except for a set of points x which has measure zero, we say that $f(x) = g(x)$ almost everywhere.   It is not hard to see that for any measurable set E, f is integrable over E if and only if g is; and if so, then $\int_E f\, dx = \int_E g\, dx$.   We now allow a function f to be integrable over E if f is $\pm \infty$ on a subset of E of measure zero, provided that the slightly altered function $g(x)$, equal to $f(x)$ if $f(x)$ is finite and equal to zero where $f(x)$ is infinite, is integrable over E.

The following *Lebesgue bounded-convergence* theorem is important and useful.

Theorem   If $\{f_n(x)\}$ is a sequence of functions integrable over E, if $f_n(x) \to f(x)$ as $n \to \infty$ almost everywhere (i.e., except for a set of measure zero), and if there is an integrable function $g(x)$ on E such that $|f_n(x)| \le g(x)$ almost everywhere, then $f(x)$ is integrable over E and

$$\lim_{n \to \infty} \int_E f_n(x)\, dx = \int_E f(x)\, dx$$

This theorem enables us to interchange the integration and the limit operation when the hypotheses are satisfied.

Let E be a fixed measurable set.   Let $L_p(E)$, we mean the set of all functions $f(x)$ defined on E such that $|f|^p$ is integrable, $p \ge 1$.   We can define a norm in L_p by

$$\|f\| = \|f\|_p = \left\{ \int_E |f(x)|^p\, dx \right\}^{1/p}$$

as in Sec. 1-9.   There is one minor difficulty that is easily treated.   If $f = g$ almost everywhere, then $\|f - g\| = 0$.   But we demanded of a norm that $\|f - g\| = 0$ be true if and only if $f = g$.   We get around this by identifying f and g if they are equal almost everywhere.   This means that $L_p(E)$ is really a set of function *classes;* i.e., the class $[f]$ is the class of all functions g that are equal to f almost everywhere.   However, we always write f, and we can think of functions instead of function classes if we keep in mind the identification that has been made.

We have already alluded to the important result that $L_p(E)$ is complete if $p \ge 1$.   Let us state it formally.

Theorem If $\{f_n\}$ is a sequence of functions in $L_p(E)$ such that $\lim\limits_{\substack{m\to\infty \\ n\to\infty}} \|f_n - f_m\| = 0$, then there is a function f in $L_p(E)$ such that

$$\lim_{n\to\infty} \|f_n - f\| = 0$$

If $1 \le p < \infty$, then for every f in $L_p(E)$ there is a sequence $\{f_n(x)\}$ with f_n continuous and in $L_p(E)$ such that $\|f_n - f\| \to 0$ as $n \to \infty$. In particular, the f_n's are Riemann-integrable; the Riemann-integrable L_p functions are dense in $L_p(E)$.

For bounded measurable functions another norm is possible, the essential supremum. If $f(x)$ is defined on E, and there is an M such that

$$\{x \text{ in } E: \quad |f(x)| > M\}$$

has measure zero, then we say that f is essentially bounded on E, and we define the essential supremum of $|f|$ to be the smallest such M. This is denoted by $\|f\|_\infty$. Again, we must identify two such functions if they are equal almost everywhere. The measure of E may be finite or infinite. This generalizes the definition of L_∞ of Sec. 1-9.

For complex valued functions f, we write $f = f_1 + if_2$, with f_1 and f_2 real. Then f is measurable (or integrable) if and only if f_1 and f_2 are, by definition, For real n-space R_n, the same theory is valid. The definitions and theorems are unchanged.

One of the best complete and modern treatments of the theory is Halmos's "Measure Theory." For a brief and quite general account, Cramer's "Mathematical Methods of Statistics" is recommended.

REFERENCES

1. Bartle, R. G.: Newton's Method in Banach Spaces, *Proc. Am. Math. Soc.*, vol. 6, pp. 827–831, 1955.

2. ——— and L. M. Graves: Mappings between Function Spaces, *Trans. Am. Math. Soc.*, vol. 72, pp. 400–413, 1952.

3. Courant, R., and D. Hilbert: "Methods of Mathematical Physics," vol. 1, Interscience Publishers, Inc., New York, 1953.

4. Dunford, N., and J. T. Schwartz: "Linear Operators," pt. I, General Theory, Interscience Publishers, Inc., New York, 1958.

5. Frazer, R. A., W. J. Duncan, and A. R. Collar: "Elementary Matrices," Cambridge University Press, London, 1946.

6. Graves, L. M.: "The Theory of Functions of Real Variables," 2d ed., McGraw-Hill Book Company, Inc., New York, 1956.

7. Halmos, P. R.: "Finite Dimensional Vector Spaces," D. Van Nostrand Company, Inc., Princeton, N.J., 1958.

8. Halmos, P. R.: "Introduction to Hilbert Space and the Theory of Spectral Multiplicity," Chelsea Publishing Company, New York, 1951.

9. Ince, E. L.: "Ordinary Differential Equations," Longmans, Green & Co., Ltd., London, 1926 (reprinted by Dover Publications, Inc., New York).

10.   Kolmogorov, A. N., and S. V. Fomin: "Elements of the Theory of Functions and Functional Analysis," vol. 2, Graylock Press, 1961.

11.   Krasnosel'skii, M. A.: "Topical Methods in the Theory of Nonlinear Integral Equations," Gosudarstv. Izdat. Tehn. Teor. Lit., Moscow, 1956.

12.   Loomis, L. H.: "An Introduction to Abstract Harmonic Analysis," D. Van Nostrand Company, Inc., Princeton, N.J., 1953.

13.   Rickart, C. E.: "General Theory of Banach Algebras," D. Van Nostrand Company, Inc., Princeton, N.J., 1960.

14.   Riesz, F., and B. Sz.-Nagy: "Lecons d'Analyse Fonctionnelle," Akademiai Kiado, Budapest, 1952.

15.   Saaty, T. L.: "Mathematical Methods of Operations Research," McGraw-Hill Book Company, Inc., New York, 1959.

15a.   Schauder, J.: Der Fixpunctsatz in Funktionalräumen, *Studia Math.*, vol. 2 pp. 171–180, 1930.

16.   Stone, M. H.: Linear Transformations in Hilbert Space and Their Applications to Analysis, *Am. Math. Soc.*, 1932.

17.   Sz.-Nagy, B. von: "Spektraldarstellung Linear Transformationen des Hilbertschen Raumes," Ergebnisse der Mathematik, vol. 5, J. Springer-Verlag OHG, Berlin, 1942; reprinted by J. W. Edwards, Publisher, Incorporated, Ann Arbor, Mich., 1947.

18.   Taylor, A. E.: "Introduction to Functional Analysis," John Wiley & Sons, Inc., New York, 1958.

TWO

NONLINEAR ALGEBRAIC AND TRANSCENDENTAL EQUATIONS

2-1 INTRODUCTION

In this chapter we shall discuss some basic methods used in solving nonlinear equations of the form $f_i(x_1, \ldots, x_n) = 0$, $i = 1, \ldots, m$. The f_i are assumed to be functions of real variables $-\infty < x_j < \infty$. We shall see later that such systems of equations arise in many ways. An expression of the form $y = f(x_1, \ldots, x_n)$ is an *algebraic* function of x_j, $j = 1, \ldots, n$, if y is a zero of a polynomial in y whose coefficients are rational functions (ratio of two polynomials) in x_j. A function that is not algebraic is called *transcendental*. Let us remark that a system of complex equations in several complex variables can be replaced by an equivalent system containing twice as many real equations and real variables.

Most analytic methods used in solving systems of nonlinear equations are iterative. Thus, based on an initial point (which usually must fall near the desired solution), a new point is computed through the algorithm used.   The process is continued by successively calculating points that yield improved approximations to the solution. A convergence proof of such a method, together with an estimate of the error of approximation for the nth iteration, are usually supplied after ensuring the existence, and sometimes the uniqueness, of the solution.

Frequently, for computation purposes, methods are adapted into forms that shorten the time required to obtain an answer.   For example, it will be seen later that the iterations of the Newton-Raphson method can

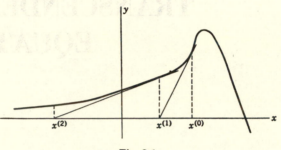

Fig. 2-1

be modified in more than one way.   In modifying a method, we look for a form that is more convenient for computation and still produces rapid convergence.

The problem of choosing the initial point is nontrivial, as an error in the choice may lead away from the desired solution, as illustrated by Fig. 2-1, where the tangent to the graph is used to go from one approximation to the next.   For a system of equations representing a physical situation already in progress, the existing solution may be adopted as the initial approximation.   In general, exploration for an initial value may be required.

It is important to be familiar with the geometric ideas related to the problem of solving a system of equations, and hence we first briefly mention graphic methods for solving systems of equations.[1]

Such methods are only useful for the case of two variables and occasionally for a case of three or more variables where by means of successive projections it is possible to study the problem in the plane.

For example, graphic methods can be used to obtain the solution of a single equation $f(x) = 0$ by plotting the graph of $y = f(x)$ and noting the points of intersection with the real axis or by splitting $f(x)$ into a sum (or difference) $f(x) = f_1(x) - f_2(x) = 0$ and selecting the point of intersection of the two curves $y = f_1(x)$ and $y = f_2(x)$.   Thus solutions of the equation $e^{2x} \sin x = 1$ may be obtained from the intersections of $y = \sin x$ with $y = e^{-2x}$.

Exercise 2-1 Obtain graphically the solutions of the foregoing problem between zero and 4π.

Exercise 2-2 (a) Solve graphically the equation $\tan x = ax$. Show that (1) $x = 0$ is the only real root for $0 \leq x \leq \pi/2$, $0 < a \leq 1$, (2) there are no roots for $0 \leq x \leq \pi/2$ for $a < 0$, and (3) for large x and $|a| > 0$ the roots tend asymptotically to $(n + \frac{1}{2})\pi$.

(b) Discuss the solution of $\tan z = az/(z^2 + b)$ required in a classical Sturm-Liouville problem. (Cauchy, "Exercise d'Analyse," vol. 1, p. 306.)

(c) Solve graphically the equation $ab^x - cx^2 = 0$, where $a, b, c > 0$ are constants.

We note that $f_i = 0$ define surfaces in n space.

1. If $m < n$ the system of m equations has no unique point satisfying it. Thus, for example, the intersection of the plane $x + y + z = 1$ with the sphere $x^2 + y^2 + z^2 = 4$ yields a circle.

If $m < n$, and if there are m variables, say $x_1, \ldots, x_m$, such that the matrix $(\partial f_i/\partial x_j)$, $i, j = 1, 2, \ldots, m$, is nonsingular for some values of the variables, then the system can be solved for $x_1, \ldots, x_m$ in terms of the remaining $n - m$ variables, $x_{m+1}, \ldots, x_n$,

$$x_i = \varphi_i(x_{m+1}, \ldots, x_n) \qquad i = 1, \ldots, m$$

Then every point $(x_1, \ldots, x_n)$ for which

$$x_i = \varphi_i(x_{m+1}, \ldots, x_n) \qquad i = 1, \ldots, m$$

satisfies the system. The process of solving for the m variables in a neighborhood of a point which satisfies the equations is studied through implicit-function theory (see Theorem 1-7).

In the case of a system of inequalities

$$f_i(x_1, \ldots, x_n) \leq 0 \qquad i = 1, \ldots, m$$

one need not insist that $m \leq n$. The number of inequalities may be arbitrary since a set of inequalities usually defines a region or regions over which all the inequalities are satisfied, and the introduction of additional inequalities simply reduces the size of the region. Inequalities can be reduced to equations by adding the square of a new variable to each one, and hence the problem reduces to one in which the number of equations is less than the number of variables. This remark does not include all we have to say about inequalities. The latter play a very important role in the nonlinear maximization or minimization studied in the next chapter.

2. If $m = n$, then with proper restrictions a solution, real (e.g., of

$x + y = 4$, $xy = 1$) or complex (e.g., of $x^2 + y^2 = 1$, $y = 2$), exists. Usually one proves that there is a unique solution having certain properties. However, as in the examples given even when $m = n$ the solution need not be unique. Most of this chapter will be devoted to studying iterative methods of solution for the case $m = n$.

3. If $m > n$, then in general the system has no solution, e.g., $x + y = 2$, $xy = 1$, $x + y = 5$. However, the exceptional situation may arise in which all the surfaces pass through a common point. The task of demonstrating the existence of a solution in this case is more difficult and must be studied for each problem; i.e., there is no sufficiently useful theorem to cover a variety of cases.

It is important to point out that all the techniques described below require that the functions involved have derivatives of various orders, i.e., be sufficiently smooth. There are cases of practical importance in which these functions need not have derivatives everywhere. They may even be discontinuous. This remark serves to point out the severe restrictions that calculus methods impose on advancing a general theory for the solution of nonlinear equations. Obviously, no such restrictions are required for linear systems (which are inherently smooth) whose solutions are obtained algebraically by the use of Cramer's rule or by elimination methods.

In addition to the Newton-Raphson method and the gradient method which we study in this chapter, Picard's method and perturbation methods will appear in later chapters. The book contains a large number of ideas related to nonlinear problems. The fact that our approach begins with algebraic equations and iterative techniques should not be construed as comprising a cornerstone for the rest of the book. It constitutes one aspect of the whole by exposing the student to existence and uniqueness proofs and to convergence proofs. Also it points to possible areas for inventiveness in devising iterative procedures which might possibly converge more rapidly than known classical methods.

In the remainder of this chapter we study two basic methods used in solving the system $f_i = 0$ when $m = n$, and in a set of exercises we develop ideas used in successive-approximation procedures.

2-2 THE NEWTON - RAPHSON METHOD

1 *Motivation*

The Newton-Raphson method for computing the root of an equation is a successive-approximation procedure. In the case of an equation in a single independent variable $y = f(x)$ one starts with an estimate of x near the root of $f(x) = 0$ which one wishes to determine. One then computes the intersection of the tangent line to the graph at this estimate with the x axis and uses the intersection as the abscissa of the new estimate. The process is then repeated. This procedure is equivalent to using the linear part of the series expansion of the function for successive iterations, i.e.,

$$f(x) = f(x^{(0)}) + f'(x^{(0)})(x - x^{(0)}) + \cdots$$

where $x^{(0)}$ is the initial estimate. The number of primes used indicates the order of differentiation. To obtain $x^{(1)}$ we equate the linear part of the right side to zero and solve for x. This gives the new estimate denoted by $x^{(1)}$, that is,

$$x^{(1)} = x^{(0)} - \frac{f(x^{(0)})}{f'(x^{(0)})} \tag{2-1}$$

For the second iteration the expansion is effected about $x^{(1)}$ to yield $x^{(2)}$, and in general for the nth iteration we have

$$x^{(n)} = x^{(n-1)} - \frac{f(x^{(n-1)})}{f'(x^{(n-1)})} \tag{2-2}$$

Exercise 2-3 Derive Eq. (2-2) by computing the intersection of the tangent line to $f(x)$ at $x^{(n-1)}$ and the x axis.

Exercise 2-4 The function $f(x) = x^3 - 2x - 5$ has a zero between $x = 2$ and $x = 3$. Let $x^{(0)} = 2$, and show to three decimal places that $x^{(3)} = 2.095$.

The Newton-Raphson method can also be used to solve a system of algebraic equations

$$f_i(x_1, \ldots, x_n) = 0 \qquad i = 1, \ldots, n$$

In this case, starting with the estimate $x^{(0)} = (x_1^{(0)}, \ldots, x_n^{(0)})$, we solve for $x_j^{(1)}$ (using Cramer's rule) in the expression obtained by equating to zero the linear parts of the series expansions of the f_i about $x^{(0)}$. We have

$$f_i(x_1^{(k-1)}, \ldots, x_n^{(k-1)}) +$$
$$\sum_{j=1}^{n} \frac{\partial f_i}{\partial x_j}\bigg|_{x^{(n-1)}} (x_j^{(k)} + x_j^{(k-1)}) = 0 \qquad i = 1, \ldots, n \tag{2-3}$$

Exercise 2-5 Give the geometric interpretation which leads to the use of this iterative procedure.

Exercise 2-6 Using Cramer's rule, write, explicitly, the approximations by Newton's method for $x_1^{(k)}$ and $x_2^{(k)}$ in terms of $x_1^{(k-1)}$ and $x_2^{(k-1)}$ for $f_i(x_1, x_2) = 0$, $i = 1, 2$.

As is the case with any successive-approximation procedure, a convergence proof and an estimate of the error incurred by stopping at the nth iteration are needed. One must show that the solution exists and, under appropriate restrictions, is unique, and that the sequence of successive steps converges to a limit which is the desired solution.

The following theorem is due to Ostrowski. It has been generalized to equations in abstract spaces by Kantorovich. Here we use the finite-dimensional version of Kantorovich's proof. Generally, the original version can be regained by using norms instead of absolute values and using inverse transformations. The results of the first theorem are used in a subsequent theorem for application to a system of equations.

2   *Existence and Convergence Proof*

To simplify the notation, we use subscripts on x to indicate the iterations of the process.

Theorem 2-1   In order that the equation $f(x) = 0$ have a solution $\bar{x}$, and that the successive iterations of Newton's method converge to this solution, it is sufficient that, for an initial value x_0, $1/f'(x_0)$ exists and

(1) $$\left| \frac{1}{f'(x_0)} \right| \leq a_0 \qquad \text{for some constant } a_0 > 0$$

(2) $$\left| \frac{f(x_0)}{f'(x_0)} \right| \leq b_0 \qquad \text{for some constant } b_0 > 0$$

and for all x satisfying

(3) $$|x - x_0| \leq \frac{1 - \sqrt{1 - 2c_0}}{c_0} b_0 \equiv g(c_0)b_0$$

where c_0 is defined by and satisfies

(4) $$c_0 \equiv a_0 b_0 A \leq \tfrac{1}{2}$$

we have

(5) $$|f''(x)| \leq A$$

In addition, the solution $\bar{x}$ satisfies (3) and the rapidity of convergence is estimated from

$$|x_n - \bar{x}| \leq \frac{(2c_0)^{2^n - 1}}{2^{n-1}} b_0$$

Proof (Kantorovich):   (Note that the theorem ensures the existence of a solution by making an initial choice which satisfies the conditions.)

Define
$$F_0(x) \equiv x - d_0 f(x) \qquad (2\text{-}4)$$

and
$$d_0 \equiv \frac{1}{f'(x_0)} \qquad (2\text{-}5)$$

Then
$$x_1 = F_0(x_0) \qquad (2\text{-}6)$$

We shall now show that conditions analogous to those given in the theorem are satisfied if x_0 is replaced by x_1.

Note that
$$|x_1 - x_0| = |d_0 f(x_0)| \leq b_0 \qquad (2\text{-}7)$$
$$|d_0[f'(x_0) - f'(x_1)]| \leq |d_0| \, |f'(x_0) - f'(x_1)|$$
$$\leq a_0 (\sup_{\bar{x} = x_0 + \theta(x_1 - x_0)} |f''(\bar{x})|) |x_1 - x_0|$$
$$\leq a_0 A b_0 = c_0 < 1 \qquad 0 \leq \theta \leq 1 \qquad (2\text{-}8)$$

because
$$|\bar{x} - x_0| = \theta |x_1 - x_0| \leq \theta b_0 \leq g(c_0) b_0$$

since $g(c_0) \geq 1$ when $0 \leq c_0 \leq \frac{1}{2}$, and thus
$$|f''(\bar{x})| \leq A$$

To show that condition 1 is satisfied we note that
$$d_1 \equiv \frac{1}{f'(x_1)} = \frac{1/f'(x_0)}{1 - [1/f'(x_0)][f'(x_0) - f'(x_1)]} \qquad (2\text{-}9)$$

from which we have, using (2-8) and the series expansion of the denominator,
$$|d_1| \leq \frac{a_0}{1 - c_0} \equiv a_1 \qquad (2\text{-}10)$$

Thus a bound for $|d_1|$ is known, and condition 1 is satisfied. To satisfy condition 2 we have, using (2-4) and (2-5),
$$F_0'(x_0) = 1 - d_0 f'(x_0) = 0$$

and from (2-6), coupled with the fact that $F_0'(x_0) = 0$, we have
$$d_0 f(x_1) = x_1 - F_0(x_1) = F_0(x_0) - F_0(x_1) + F_0'(x_0)(x_1 - x_0)$$

Now Taylor's formula with remainder gives, for a function $\varphi(x)$,
$$\varphi(x_0 + h) = \varphi(x_0) + \varphi'(x_0)h + \cdots$$
$$+ \frac{\varphi^{n-1}(x_0)}{(n-1)!} h^{n-1} + \frac{\varphi^n(x_0 + \theta h)}{n!} h^n \qquad 0 < \theta < 1$$

Using the substitutions $\varphi = F_0$, $h = x_1 - x_0$, and $n = 2$, we are led to the estimate
$$|d_0 f(x_1)| \leq \frac{1}{2} \sup_{\bar{x} = x_0 + \theta(x_1 - x_0)} |F_0''(\bar{x})| \, |x_1 - x_0|^2$$
$$= \frac{1}{2} \sup_{\bar{x} = x_0 + \theta(x_1 - x_0)} |d_0 f''(\bar{x})| \, |x_1 - x_0|^2 \leq \frac{1}{2} a_0 A b_0^2 = \frac{1}{2} c_0 b_0 \qquad (2\text{-}11)$$

Thus $\qquad |d_1 f(x_1)| = \left| \dfrac{d_1}{d_0} d_0 f(x_1) \right| \le \dfrac{c_0 b_0}{2(1 - c_0)} \equiv b_1 < b_0 \qquad$ (2-12)

The proof of condition 3 for x_1, that is, that the new interval does not lie beyond the previous interval, will be a special case of a proof given later for x_n.

Condition 5 follows from

$$c_1 = a_1 b_1 A = \frac{a_0}{1 - c_0} \frac{c_0 b_0}{2(1 - c_0)} A = \frac{1}{2} \frac{c_0{}^2}{(1 - c_0)^2} \le 2c_0{}^2 \le \frac{1}{2} \quad (2\text{-}13)$$

Thus the main four conditions (with the postponement of a proof for condition 3) are satisfied for $x = x_1$ with a_1, b_1, c_1 replacing a_0, b_0, c_0. The process can now be continued for x_n, resulting in numbers which satisfy

$$a_n = \frac{a_{n-1}}{1 - c_{n-1}} \tag{2-14}$$

$$b_n = \frac{c_{n-1} b_{n-1}}{2(1 - c_{n-1})} \tag{2-15}$$

$$c_n = \frac{c_{n-1}{}^2}{2(1 - c_{n-1})^2} \tag{2-16}$$

We also have

$$|x_n - x_{n-1}| \le b_{n-1} \tag{2-17}$$

$$c_2 \le 2c_1{}^2 \le 8c_0{}^4, \ldots, c_n \le \tfrac{1}{2}(2c_0)^{2^n} \tag{2-18}$$

$$b_n \le \frac{c_{n-1} b_{n-1}}{2(1 - c_{n-1})} \le c_{n-1} b_{n-1} \le \cdots \le c_{n-1} c_{n-2} \cdots c_0 b_0$$

$$\le \frac{1}{2^n} (2c_0)^{2^{n-1}} (2c_0)^{2^{n-2}} \cdots (2c_0) b_0 = \frac{1}{2^n} (2c_0)^{2^n - 1} b_0 \tag{2-19}$$

To prove that x_n satisfies condition 3, we show that

$$|x - x_n| \le g(c_n) b_n \tag{2-20}$$

does not extend beyond the interval given in condition 3. Let x be in the interval described in (2-20); then

$$|x - x_0| \le |x_n - x_0| + |x - x_n| \le [b_0 g(c_0) - b_n g(c_n)] + b_n g(c_n) = b_0 g(c_0)$$

and hence x is in the interval of condition 3.

The following identity will be used in the proof of the existence of a solution:

$$b_n g(c_n) - b_{n+1} g(c_{n+1}) = b_n \tag{2-21}$$

where $g(\ )$ is obtained from $g(c_0)$ by replacing c_0 with the indicated argument. It may readily be verified by direct substitution.

Exercise 2-7   Verify the foregoing identity.

To prove the existence of a limit to our sequence of iterations, we have

$$|x_{n+k} - x_n| \leq b_n + b_{n+1} + \cdots + b_{n+k-1} = b_n g(c_n) - b_{n+k} g(c_{n+k})$$
$$\leq b_n g(c_n) \leq 2b_n \leq \frac{(2c_0)^{2^n - 1} b_0}{2^{n-1}} \tag{2-22}$$

Thus $\lim\limits_{n \to \infty} x_n = \bar{x}$ exists.

To show that this limit satisfies condition 3, let $n = 0$ and $k \to \infty$ in the fourth expression of (2-22). Then

$$|\bar{x} - x_0| \leq b_0 g(c_0)$$

To show that $\bar{x}$ satisfies $f(x) = 0$, note from

$$f'(x_n)(x_{n+1} - x_n) + f(x_n) = 0 \tag{2-23}$$

that since
$$|x_{n+1} - x_n| \to 0$$

and since

$$|f'(x_n)| \leq |f'(x_0)| + |f'(x_n) - f'(x_0)| \leq |f'(x_0)| + A|x_n - x_0|$$
$$\leq |f'(x_0)| + Ag(c_0)b_0$$

the first term of (2-23) tends to zero, yielding

$$f(\bar{x}) = \lim_{n \to \infty} f(x_n) = 0$$

***Theorem* 2-2** With the conditions of the previous theorem satisfied and with the modification that

$$|f''(x)| \leq A$$
for $$|x - x_0| < \frac{1 + \sqrt{1 - 2c_0}}{c_0} b_0 \equiv h(c_0)b_0 \tag{2-24}$$

$\bar{x}$ is unique (if $c_0 = \frac{1}{2}$, then $<$ must be replaced by $\leq$).

Proof:

1. Let $c_0 < \frac{1}{2}$, and let $\bar{x}$ be a solution satisfying

$$|\bar{x} - x_0| = \theta h(c_0)b_0 \qquad 0 \leq \theta < 1$$

Thus $F_0(\bar{x}) = \bar{x}$ since $f(\bar{x}) = 0$. Also, using $F_0'(x_0) = 0$, we have

$$|\bar{x} - x_1| = |F_0(\bar{x}) - F_0(x_0)| = |F_0(\bar{x}) - F_0(x_0) - F_0'(x_0)(\bar{x} - x_0)|$$
$$\leq \frac{1}{2}a_0 A |\bar{x} - x_0|^2 = \frac{1}{2}a_0 A \theta^2 h^2(c_0)b_0^2 = \theta^2 h(c_1)b_1$$

Exercise 2-8 Verify the last equality.

In general,
$$|\bar{x} - x_n| \leq \theta^{2^n} h(c_n)b_n$$
But $$h(c_n)b_n = \frac{1 + \sqrt{1 - 2c_n}}{c_n} b_n \leq \frac{2b_n}{c_n} = \frac{2}{a_n A} \tag{2-25}$$

and, since $a_n > a_0$, we have

$$|\hat{x} - x_n| \leq \theta^{2^n} \frac{2}{a_0 A} \tag{2-26}$$

Consequently

$$\lim_{n \to \infty} x_n = \hat{x}$$

from which we have $\hat{x} = \bar{x}$, and the solution is unique.

2. If $c_0 = \frac{1}{2}$, then it is possible to have $\theta = 1$, in which case, from (2-14), $a_1 = a_0/(1 - c_0) = 2a_0$, $a_2 = 2a_1 = 4a_0$, . . . , $a_n = 2^n a_0$ and, as in (2-25),

$$|\hat{x} - x_n| \leq \frac{2}{a_n A} = \frac{1}{2^{n-1} a_0 A} \tag{2-27}$$

and again $\lim\limits_{n \to \infty} x_n = \hat{x}$.

Exercise 2-9 Let $f(x) = \frac{1}{2}x^2 - x + \alpha$, $\alpha > 0$, $x_0 = 0$. Show that $a_0 = A = 1$, $b_0 = \alpha$ and that its roots are real only for $c_0 = \alpha \leq \frac{1}{2}$. Note from this that the bounds in conditions 3 and 5 and in formula (2-24) cannot be improved even for a real equation. The smaller root is on the boundary of condition 3, and the larger one is on the boundary of (2-24), and thus the first cannot be narrowed nor the second enlarged without loss of existence and uniqueness.

Exercise 2-10 Show that if instead of x_0 one begins with x_0' and generates a sequence x_n', then one has convergence to $\hat{x}$ if

$$|x_0' - x_0| \leq \frac{1 - 2c_0}{4c_0} b_0$$

Remark: If the iterations are described by the modified Newton method

$$x_{n+1} = x_n - \frac{f(x_n)}{f'(x_0)} \tag{2-28}$$

in order to avoid computing $f'(x_n)$ at each step, the process can be shown to converge to a solution for $c_0 < \frac{1}{2}$ with a rapidity given by

$$|x_n - \bar{x}| \leq (1 - \sqrt{1 - 2c_0})^{n-1}|x_1 - \bar{x}| \tag{2-29}$$

The proof of this fact depends on showing that if x satisfies

$$|x - \bar{x}| \leq |x_1 - \bar{x}|$$
$$|x - x_0| \leq g(c_0)b_0$$

then for $x' = F_0(x)$ one has

$$|x' - \bar{x}| \leq (1 - \sqrt{1 - 2c_0})^{n-1}|x - \bar{x}| \tag{2-30}$$
$$|x' - x_0| \leq g(c_0)b_0$$

In the following theorem we shall denote the point $(x_1, \ldots, x_n)$ by x, and if the components have a superscript then so does x.

Theorem 2-3   If the left members of the system

$$f_i(x_1, \ldots, x_n) = 0 \qquad i = 1, \ldots, n$$

satisfy

(1) $$|f_i(x_1{}^0, \ldots, x_n{}^0)| \leq A \qquad i = 1, \ldots, n$$

(2) The matrix $(\partial f_i / \partial x_j)_{x=x^0}$ has a nonvanishing determinant D (with absolute value $|D|$) and we have, for the absolute value $|A_{ij}|$ of its cofactors A_{ij},

$$\max_i \frac{1}{|D|} \sum_{j=1}^{n} |A_{ij}| \leq B$$

(3) $$\left| \frac{\partial^2 f_i}{\partial x_j \partial x_k} \right| \leq C \qquad i, j, k = 1, \ldots, n$$

in the region

(4) $$\max |x_i - x_i{}^0| \leq \frac{1 - \sqrt{1 - 2c_0}}{c_0} AB$$

where c_0 is defined by and satisfies

(5) $$c_0 \equiv B^2 A C n^2 \leq \tfrac{1}{2}$$

then the system $f_i(x) = 0$ has a solution which can be obtained by Newton's method.

The proof of this theorem would be an easy consequence of Theorem 2-1 had that theorem been formulated in the abstract version we alluded to before its proof. Now that the reader has seen the more accessible proof we gave, we shall attempt to introduce the concepts needed to subsume Theorem 2-3 under Theorem 2-1.

Note first that the iterations of Newton's method in the single-variable case give

$$x_{n+1} = x_n - \frac{f(x_n)}{f'(x_n)} = x_n - [f'(x_n)]^{-1}f(x_n)$$

and in the general case, solving (2-3) by Cramer's rule,

$$x^{(n+1)} = x^{(n)} - J^{-1}(x^{(n)})f(x^{(n)})$$

where J^{-1} is the inverse of the jacobian matrix, $x^{(n)}$ and $x^{(n+1)}$ are vectors, and

$$f(x) = \begin{pmatrix} f_1(x) \\ \cdot\ \cdot\ \cdot \\ f_n(x) \end{pmatrix}$$

is a vector-valued function.

To prove that the iterations converge in either case to a value $\bar{x}$, one must keep the factors $|[f'(x_n)]^{-1}f(x_n)|$ or $|J^{-1}(x^{(n)})f(x^{(n)})|$ (the second being a generalization of the first) under control by carrying convenient bounds imposed on them at the initial point over successive iterations and showing that as $n \to \infty$ the sequence of bounds goes to zero. To impose bounds on the absolute values is essentially to keep a running account of the distance between x_{n+1} and x_n since

$$|x_{n+1} - x_n| = |[f'(x_n)]^{-1}f(x_n)|$$

Note that we also impose a condition on the absolute value of $f''(x)$, a bound which we need in the proof where we apply the mean-value theorem to the difference between the values of the derivative at two points. We wish to keep this difference under control.

However, in the general case there are several convenient ways to measure the contribution of the right side. The idea of absolute value is not meaningful when applied to $J^{-1}(x^n)f(x^{(n)})$, the product of a matrix and a vector, which of course is a vector and may be regarded as a transformation of one vector f to a new vector $J^{-1}f$. Thus we need a metric which yields information on the contribution of all the components of this vector taken together. We regard the single-variable case and the several-variable case as two formulations which differ only in dimension; i.e., one is a vector space of a single dimension whereas the other is a vector space of n dimensions. Thus we introduce a metric or a norm for vectors which enables us to obtain the appropriate generalization of the single-dimensional case. Whatever the metric may be, it should enable us to write the conditions of Theorem 2-3 along lines similar to those of Theorem 2-1. The definition of absolute value must be a special case of the metric.

The norm of a vector x, which we denote by $\|x\|$, must satisfy the conditions of any metric; these are, from Chap. 1,

1. $\|\alpha f\| = |\alpha|\,\|f\|$ for complex α
2. $\|f + g\| \leq \|f\| + \|g\|$
3. $\|f\| = 0$ if and only if $f = 0$

There are many norms which one can define. We shall give below and in the next theorem useful norms which do not differ significantly from familiar concepts of distance. Whatever norm we choose for a vector, we still must determine for our problem the norm of $J^{-1}f$, that is, $\|J^{-1}f\|$, and bounds on the transformation J^{-1}, that is, $\|J^{-1}\|$, which is a generalization of $|[f'(x)]^{-1}|$. Since J^{-1} is applied at each iteration, it is desirable to have from step to step an estimate of the size of the new vector $J^{-1}f$ compared to the old vector f. Thus we should be able to find a bound $M(x)$ (since J^{-1} depends on x) such that

$$\max_{f \neq 0} \frac{\|J^{-1}f\|}{\|f\|} \leq M(x)$$

We define

$$\|f\| = \max_i |f_i|$$

and attempt to determine M such that at $x^{(0)}$

$$\|J^{-1}f\| \leq M\|f\|$$

for all f. The smallest M for which this is valid for every f is by definition $\|J^{-1}\|$. Note that by taking the maximum of $|f_i|$ we account for the worst possibility; i.e., the bound on this component will automatically apply to the remaining ones. Note also that J^{-1} is a matrix whose elements are A_{ij}/D, where the A_{ij} are the cofactors of J, and hence from our definition of norm,

$$\|J^{-1}f\| = \max_i \left| \sum_j \frac{A_{ij}f_j}{|D|} \right|$$

$$\leq \max_i \sum_{j=1}^n \frac{|A_{ij}| \, |f_j|}{|D|}$$

$$\leq \max_i \frac{1}{|D|} \sum_{j=1}^n |A_{ij}| \max_j |f_j|$$

as the reader may readily verify.

On the right we have $\|f\|$. If we divide by it we have a bound on $\|J^{-1}\|$:

$$\frac{\|J^{-1}f\|}{\|f\|} \leq \frac{\|J^{-1}\| \, \|f\|}{\|f\|} = \|J^{-1}\| \leq \max_i \frac{1}{|D|} \sum_{j=1}^n |A_{ij}| \leq B$$

This yields a bound on the inverse of the jacobian matrix. Since this bound is a function of x, we require that it in turn be bounded by a constant B, analogous to a_0 of Theorem 2-1.

The first condition of Theorem 2-3 yields a bound on f similar to that obtained in Theorem 2-1 if the second condition is divided by the first in Theorem 2-1. Obviously if $\|f\| \leq A$ at $x^{(0)}$, then all the components are less than or equal to A as the theorem requires; condition 3 of Theorem 2-3 does not use the norm since, if the mean-value theorem is applied in the general case, the second-order partials appear in a sum, and hence it is sufficient and adequate to impose a bound simply on their absolute values. Conditions 4 and 5 are immediate generalizations of their counterparts in Theorem 2-1, and the analogy is complete. Now if we used norms as explained here in the proof of Theorem 2-1, we would have the finite-dimensional version of that theorem, which is Theorem 2-3.

The following is another theorem utilizing an approach similar to that indicated above.

Theorem If

(1) The jacobian matrix $(\partial f_i/\partial x_j)_{x=x^0}$ has an inverse (A_{ij}/D), where

$$\frac{1}{|D|} \left(\sum_{i,j=1}^{n} A_{ij}^2 \right)^{1/2} \leq B$$

(2)

$$\sum_{i=1}^{n} (x_i^1 - x_i^0)^2 \leq A^2$$

(3)

$$\sum_{i,j,k=1}^{n} \left(\frac{\partial^2 f_i}{\partial x_j \partial x_k} \right)^2 \leq K^2$$

(one can choose $K = n\sqrt{n}\, C$, where C is the same as in Theorem 2-3)

(4)

$$c_0 = BKA \leq \tfrac{1}{2}$$

then Newton's process converges, and the solution $\bar{x} = (\bar{x}_1, \ldots, \bar{x}_n)$ satisfies

$$|\bar{x} - x_0| = \left[\sum_{i=1}^{n} (\bar{x}_i - x_i^0)^2 \right]^{1/2} \leq \frac{1 - \sqrt{1 - 2c_0}}{c_0} A$$

The justification in terms of the main theorem and its proof lies in the use of norms different from those used in Theorem 2-3.

Exercise 2-11 (Ostrowski) Specialize and work out in full the case of two equations in two unknowns, and, hence, starting with $x^0 = (1.2, 1.7)$, show that $x^1 = (1.2349, 1.661)$ for the system

$$2x^3 - y^2 - 1 = 0 \qquad xy^3 - y - 4 = 0$$

Using the last theorem, show that $B = .11$, $K \leq 2.514$, $A = 0.0524$, and $c_0 < 0.5$, and, therefore, deduce convergence.

Exercise 2-12 The following formula has been developed as a modification of the Newton-Raphson method by H. W. Milnes[19] for approximating a root of multiplicity m of a polynomial:

$$x_{n+1} = x_n - \left[\frac{(m-k)!f^{(k)}(x_n)}{f^{(m)}(x_n)} \right]^{1/(m-k)} \qquad k = 0, 1, \ldots, m-1$$

The case $k = m - 1$ is particularly recommended for calculations. Apply this method to

$$f(x) = x^3 - 5x^2 - 8x - 4$$

Exercise 2-13 H. S. Wall[25] has studied a modification of Newton's method. The equation of the parabola through $(x_n, f(x_n))$ having the same first and second derivative at $x = x_n$ as $y = f(x)$ is

$$y = f(x_n) + (x - x_n)f'(x_n) + \tfrac{1}{2}(x - x_n)^2 f''(x_n)$$

Let x_{n+1} be a solution of the equation obtained by putting $y = 0$. Then

$$x_{n+1} = x_n - \frac{f(x_n)}{f'(x_n) + \tfrac{1}{2}(x_{n+1} - x_n)f''(x_n)}$$

If we substitute in the denominator $x_{n+1} - x_n = -f(x_n)/f'(x_n)$, then

$$x_{n+1} = x_n - \frac{f(x_n)}{f'(x_n) - f(x_n)f''(x_n)/2f'(x_n)}$$

which may give more rapid convergence.

Let $f(x) = x^k - a$. Newton's formula gives

$$x_{n+1} = \frac{a/x_n^{k-1} + (k-1)x_n}{k}$$

and the above formula gives

$$x_{n+1} = \frac{(k-1)x_n^k + (k+1)a}{(k+1)x_n^k + (k-1)a}$$

Starting with $x_0 = 1$, compute values up to x_3 using both formulas.

For $x^3 + bx - c = 0$, b, c real, $b \neq 0$, $c > 0$, Newton's formula gives

$$x_{n+1} = \frac{2x_n^3 + c}{3x_n^2 + b}$$

and the new formula gives

$$x_{n+1} = \frac{3x_n^5 - bx_n^3 + 6cx_n^2 + bc}{6x_n^4 + 3bx_n^2 + 3cx_n + b^2}$$

Let $b = 2$ and $c = 20$, and starting with $x_0 = 2$, show that Newton's formula gives $x_1 = 2.46$, $x_2 = 2.46954551$, whereas by the modified approximation method $x_1 = 2.6$, $x_2 = 2.47$, $x_3 = 2.469546$. Wall gives, for the actual root value to nine decimal places, 2.469545649.

L. R. Ford[9] examines the solution of a system $f_i(x_1, \ldots, x_n) = 0$, $i = 1, \ldots, n$ by first studying the solution of the equation $x = f(x)$ by successive approximation as we shall see in the following. (All this is contained in Theorem 1-6, but the present discussion requires no knowledge of Banach spaces.)

Exercise 2-14 Prove the following theorem:
Theorem Let $\bar{x}$ be a solution of $x = f(x)$, and let

$$|f'(x)| < M < 1 \qquad \text{in} \qquad I : \bar{x} - h \leq x \leq \bar{x} + h$$

If x_0 is in this interval, and if

$$x_1 = f(x_0), \; x_2 = f(x_1), \; \ldots, \; x_n = f(x_{n-1})$$

then $\lim\limits_{n \to \infty} x_n = \bar{x}$.

To prove this theorem we have, using the law of the mean,

$$x_n - \bar{x} = f(x_{n-1}) - f(\bar{x}) = f'(\xi_n)(x_{n-1} - \bar{x})$$

where ξ_n is between $\bar{x}$ and x_{n-1}, and hence ξ_n is in I.   Conclude that if x_{n-1} is in I, then so is x_n, and hence, since x_0 is in I, by induction all iterations are in I.

To complete the proof show that $\lim_{n \to \infty} |x_n - \bar{x}| = 0$.

Note from $f'(\xi_n)(x_{n-1} - \bar{x})$ that if f' is negative in the interval, then $x_{n-1} - \bar{x}$ and $x_n - \bar{x}$ differ in sign, and hence the approximations are alternately greater than and less than the root.   If, on the other hand, f' is positive in I, the approximations are either all greater than or all less than $\bar{x}$, depending on the choice of x_0.

Show that errors on the nth approximation may be estimated from

$$|x_n - \bar{x}| \leq M^n h$$

or, if f' is positive, from $x_n - \bar{x} = f'(\xi_n)[x_n - \bar{x} - (x_n - x_{n-1})]$,

$$|x_n - \bar{x}| < \frac{M}{1 - M} |x_n - x_{n-1}|.$$

and from

$$|x_n - \bar{x}| < M|x_n - x_{n-1}|$$

if f' is negative in I.

Exercise 2-15   (Existence and uniqueness)   Show by using the law of the mean on $\bar{x} - \bar{\bar{x}} = f(\bar{x}) - f(\bar{\bar{x}})$ that two solutions $\bar{x}$ and $\bar{\bar{x}}$ must be identical in I. Show that if $|f'(x)| < M < 1$ in $x' - h \leq x \leq x' + h$ and if $|f'(x) - x'| < h(1 - M)$ then $x = f(x)$ has a unique solution, and the method of successive approximations— with x_0 in the interval— gives it.   Do this by showing that there must be at least one root.   If not, then $x - f(x)$ has the same sign in the interval; suppose it is positive. At x' and $x' - h$, we have

$$0 < x' - f(x') < h(1 - M)$$
$$f'(x - h) - (x' - h) < 0$$

Add and apply the law of the mean, obtaining the contradiction that the derivative is greater than M for some value in the interval, etc.   Then show that x_n obtained by successive approximations is in the interval whenever x_{n-1} is in it (apply the mean-value theorem to $x_n - x'$).

Exercise 2-16   Let $F(x) = 0$ have the root $\bar{x}$, let $F'(x)$ be positive for I:$\bar{x} - h \leq x \leq \bar{x} + h$, and let $0 < a < F'(x) < b$.   Since $F'(x) \neq 0$, x is a simple root in the interval.   If $0 < c < 2/b$, then $x = x - cF(x)$ satisfies the conditions of the theorem in Exercise 2-14.   Let $f(x) = x - cF(x)$; then $f'(x) = 1 - cF'(x)$.   Show that if M is the greater of $|1 - cb|$ and $|1 - ca|$, then $|f'(x)| < M < 1$, and this holds on a subinterval even if c is replaced by a function $\varphi(x)$ which satisfies $0 < \varphi(x) < 2b$ in I and where $|\varphi'(x)| < Q$.   Verify that the theorem is satisfied in this case.   Let $\varphi(x) = 1/F'(x)$, suppose that $0 < 1/F'(x) < 2/b$, and let $F''(x)$ be bounded.   Show that on a suitable subinterval

$$f'(x) = \frac{F''(x)F(x)}{[F'(x)]^2}$$

and that the theorem in Exercise 2-14 is satisfied.

Exercise 2-17   Prove the following theorem:[9]

Theorem   Let $\bar{x}$ be a solution of $x = f(x)$, and let $|f'(x)| < M < 1$ in the interval $I: \bar{x} - h \leq x \leq \bar{x} + h$.   Let $\{f_n(x)\}$, $n = 1, 2, \ldots$ be a sequence of functions such that

$$|f_n(x) - f(x)| < (1 - M)h$$

in the interval and

$$\lim_{n \to \infty} f_n(x) = f(x)$$

uniformly in I.

If x_0 is in I, and if $x_1, x_2, \ldots$ are given by

$$x_1 = f_1(x_0),\ x_2 = f_2(x_1),\ \ldots,\ x_n = f_n(x_{n-1})$$

then

$$\lim_{n \to \infty} x_n = \bar{x}$$

Hint: Show that if x_{n-1} is in I, so is x_n.   Then prove convergence by using ϵ_n for the least upper bound of $|f_n(x) - f(x)|$ in I; that is, repeat the relation

$$|x_n - \bar{x}| < \epsilon_n + M|x_{n-1} - \bar{x}|$$

Exercise 2-18   Prove a theorem similar to that in Exercise 2-17 with the conditions in I,

$$|f_n'(x)| < M < 1$$
$$|f_n(x) - f(x)| < h(1 - M)$$
$$\lim_{n \to \infty} f_n(x) = f(x)$$

Exercise 2-19   By analogy with the theorem in Exercise 2-17, show that the following holds:

Theorem   If $\bar{x} = (\bar{x}_1, \ldots, \bar{x}_n)$ is a solution of

$$x_i = f_i(x_1, \ldots, x_n)$$

in $R: \bar{x}_i - h \leq x_i \leq \bar{x}_i + h$, $i = 1, \ldots, n$, and if

$$\left| \frac{\partial f_i}{\partial x_j} \right| < M_{ij}$$

where

$$\sum_{j=1}^{n} M_{ij} < r < 1 \qquad i = 1, \ldots, n$$

and if $(x_1{}^0, \ldots, x_n{}^0)$ is in R and

$$x_i{}^k = f_i(x_1{}^{k-1}, \ldots, x_n{}^{k-1})$$

then

$$\lim_{k \to \infty} x_i{}^k = \bar{x}_i$$

is the desired solution.

Use the law of the mean for a function of n variables to write

$$x_i{}^k - \bar{x}_k = \sum_{j=1}^{n} \frac{\partial f_i}{\partial x_j}(x_j{}^{k-1} - \bar{x}_j)$$

in which $\partial f_i / \partial x_j$ is evaluated at $x_1 = \xi_1{}^k, \ldots, x_n = \xi_n{}^k$. Let N_{k-1} be the largest of $|x_i{}^{k-1} - \bar{x}_i|$; show that

$$|x_i{}^k - \bar{x}_0| < N_k < r N_{k-1} < \cdots < r^k N_0$$

and hence that

$$\lim_{k \to \infty} x_i{}^k = \bar{x}_i$$

Exercise 2-20 Show that the system $F_i(x_1, \ldots, x_n) = 0$, $i = 1, \ldots, n$, can be written (in many ways) in the form $x_i = f_i(x_1, \ldots, x_n)$ and that the theorem in Exercise 2-19 is satisfied if the jacobian of the F_i is positive in the neighborhood of the solution.

Remark: Hans Maehly[18c] has studied the problem of approximating to the complex roots of a polynomial by a method superior to the Newton-Raphson method. Let

$$f(x) = \sum_{j=0}^{n} c_j x^j$$

and let $x_1, \ldots, x_n$ be its zeros, which may be complex, and write

$$s_m = s_m(x) \equiv \sum_{k=1}^{n} \frac{1}{(x_k - x)^m} = \frac{-1}{(m-1)!} \frac{d^m}{dx^m} \log f \qquad m = 1, 2, \ldots$$

where x is an initial approximation. Define $u_k = (x_k - x)^{-1}$; then if, for example, $|u_1| \gg |u_k|$ for $k > 1$, and if all other u_k are equal to a common value v, we have $s_m = u^m + (n-1)v^m$. This is an equation in two unknowns, u and v. Therefore, we put $m = 1, 2$ and solve for u. This gives

$$u = \frac{1}{n} \{s_1 \pm [(n-1)(n s_2 - s_1{}^2)]^{1/2}\} \tag{2-31}$$

which is known as *Laguerre's formula*.

From this value of u, one can obtain the new estimate for x and repeat the step.

Exercise 2-20a Let $u_1 = u_2 = \cdots = u_p$ and $u_{p+1} = u_{p+2} = \cdots = u_{p+q}$ with $|u_1| \gg |u_k|$, $k > p + q$ and $|u_{p+1}| \gg |u_k|$, $k > p + q$; show from $s_m = p u^m + q v^m$ that $u = (p s_1 \pm \{p q[(p + q)s_2 - s_1{}^2]\}^{1/2})/p(p + q)$.

If p and q are not individually known but $p + q = s$ is known, take $m = 1, 2, 3$ and obtain a quadratic in u in terms of s_0, s_1, s_2, and s_3. If $p + q$ is not known, take $m = 1, 2, 3, 4$ and obtain a quadratic.

2-3 THE METHOD OF STEEPEST DESCENT

A system of real equations $f_i(x_1, \ldots, x_n) = 0$, $i = 1, \ldots, n$, may be reduced to a single equation $\sum_{i=1}^{n} f_i{}^2 = 0$.

Exercise 2-21 Show that a solution of the system is also a solution of the equation, and conversely.

The method of steepest descent which we study below is concerned with finding the maximum or minimum of a function. It can, therefore, be applied to find the value of x for which $\sum_{i=1}^{n} f_i^2 = 0$. Clearly zero is the minimum value of $\sum_{i=1}^{n} f_i^2$. Thus the vector x for which the minimum is attained is also a solution for the system $f_i = 0$, if one exists.

1 The Gradient

The gradient of the function $z = f(x_1, \ldots, x_n)$ is the vector

$$\nabla f = \left(\frac{\partial f}{\partial x_1}, \cdots, \frac{\partial f}{\partial x_n} \right) \tag{2-32}$$

Exercise 2-22 Write the equation of the plane tangent to $z = f(x)$ in $n + 1$ space at an arbitrary point, and indicate the directional numbers of the normal line. (Note that the normal line is in $n + 1$ space, whereas ∇f is strictly in n space; see Fig. 2-2.)

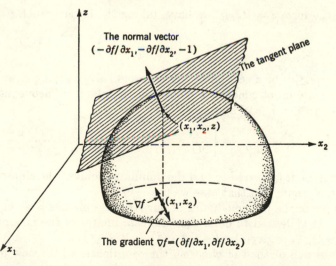

Fig. 2-2

As we shall see in the theorem below, the gradient ∇f points locally in the direction of maximum increase or steepest ascent, and hence $-\nabla f$ points in the direction of maximum decrease or steepest descent. Thus, at least in theory, one can pursue a maximum or a minimum by taking infinitesimal steps from one point to the next along the gradient. To appreciate this fact, it is useful to go into the geometry of the problem.

To study this geometrically, we examine the function

$$z = f(x_1, \ldots, x_n)$$

which represents a surface in $n + 1$ space. Assigning a constant value c to z yields a section (a contour surface) by a hyperplane which is perpendicular to the z axis and cuts it at $z = c$. By varying c, we obtain a family of contour surfaces.

Exercise 2-22 shows that the gradient at a point is perpendicular to any contour surface passing through the point. We seek the maximum by moving along the gradient from one contour surface to the next. The size of the step, i.e., how far a distance one must move along the gradient from one iteration to the next, will be discussed later. It is natural to take large steps in order to obtain rapid convergence, but the question is: How large?

Consider the ellipsoid with center at the origin,

$$\frac{x^2}{9} + \frac{y^2}{4} + z^2 = 1$$

or, in the form $z = f(x,y)$, we have, taking the lower branch

$$z = -\left(1 - \frac{x^2}{9} - \frac{y^2}{4}\right)^{1/2}$$

Let $z = c \leq 0$ be a section of this ellipsoid. By varying c, we obtain a set of contours of concentric ellipses in the plane. Their equations are $x^2/9 + y^2/4 = 1 - c^2$ and

$$-\nabla\left(1 - \frac{x^2}{9} - \frac{y^2}{4}\right)^{1/2} = \left[\frac{x}{9(1 - x^2/9 - y^2/4)^{1/2}}, \frac{y}{4(1 - x^2/9 - y^2/4)^{1/2}}\right]$$

Suppose it is desired to find the minimum point of the ellipsoid, which is at the origin. In that case we use $-\nabla f$.

Suppose the previous point of our stepwise process is given by $(-2, -1)$. It is desired to determine the coordinates of the new point. At $(-2, -1)$, ∇f is given by $(-4/3 \sqrt{11}, -3/2 \sqrt{11})$.

Thus one method of choosing the coordinates of the new point is to subtract a unit of gradient from the coordinates of the previous point. This gives, for the new point, $(-2 + 4/3 \sqrt{11}, -1 + 3/2 \sqrt{11})$. The process is then repeated. Note that to obtain the best convergence, i.e. to attain rapidly a cutoff point which would give an approximate value to the minimum, we could have used scalar multiples of the gradient

The problem remains: What is the best feasible multiple to use? The analytical motivation given below shows how to determine the scalar multiple of the gradient.

Theorem 2-4 The direction of $\nabla f(x)$ is the direction of maximum increase (i.e., rate of change) of $f(x)$.

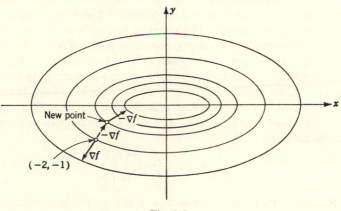

Fig. 2-3

Proof: It is instructive to examine first a proof of this theorem in the two-dimensional case. The proof is then readily generalized.

Let s be the parameter giving the distance along any path through any point (x,y) in the xy plane. We have $ds^2 = dx^2 + dy^2$. Also

$$\frac{df}{ds} = \frac{\partial f}{\partial x}\frac{dx}{ds} + \frac{\partial f}{\partial y}\frac{dy}{ds}$$

But this is the scalar product of ∇f and the vector tangent to the path at this point. (The *scalar* or *dot* product of two vectors is the sum of the products of corresponding components of the two vectors. It is also equal to the product of the magnitudes of the two vectors multiplied by the cosine of the angle between them.) Here $dx/ds = \cos\alpha$, where α is the angle which the tangent line makes with the x axis.

Now the *directional derivative* df/ds, when evaluated at our point (x_0, y_0), gives the rate of change of $f(x,y)$ with respect to the distance s traversed by a point moving on that path. It has a value for each of the directions passing through the given point. At this given point, the directional derivative is a function of the angle α. That α which yields the maximum value of the directional derivative may be determined by equating to zero the derivative of the above equation with respect to α, solving the resulting equation in α, and using that value of α which satisfies the second-derivative test for a maximum. There is a unique value α_0 between 0 and 2π which yields the maximum. This gives

$$\tan \alpha_0 = \frac{f_y(x_0,y_0)}{f_x(x_0,y_0)}$$

Actually, the point (x_0,y_0) may be replaced by an arbitrary point (x,y). By considering a right triangle with $\tan \alpha_0$ as given above, one may determine $\cos \alpha_0$ and $\sin \alpha_0$ and substitute these values in the expression for df/ds. The value of the latter in the direction α_0 is $\sqrt{f_x{}^2 + f_y{}^2}$, which is the magnitude of the gradient vector. Thus the directional derivative has the largest value when it lies along the gradient vector.

The directional derivative in the direction α_0 is normal to the contour through the given point; i.e., $f(x,y) = c$. This follows from the fact that when the directional derivative at (x,y) is equated to zero, a value of α, say α_1, is obtained which gives the direction of the tangent to the contour, and hence its slope. This leads to

$$-\tan \alpha_1 = \frac{f_x}{f_y} = \frac{1}{\tan \alpha_0}$$

Therefore, on combining the extremes of this relation, one has

$$\alpha_0 = \frac{\pi}{2} + \alpha_1$$

Hence, the gradient is normal to the contour. The method presented below utilizes the foregoing results by moving along the gradient towards a solution.

We now briefly give a proof of the general case.

The total derivative which indicates the change in $f(x_1, \ldots , x_n) = c$ is given by

$$df = \frac{\partial f}{\partial x_1} dx_1 + \cdots + \frac{\partial f}{\partial x_n} dx_n = 0 \tag{2-33}$$

From this scalar product it is seen that ∇f is orthogonal to the vector $dx \equiv (dx_1, \ldots , dx_n)$ which is taken along the surface $f = c$.

If, on the other hand, the vector dx is taken in an arbitrary direction (that is, f need not be constant), then

$$|df| = |dx| \, |\nabla f(x)| \cos \theta \tag{2-34}$$

It is obvious that $|df|$ is largest when dx points in the direction of ∇f giving $\theta = 0$. Thus df has its largest value along the gradient, and the proof is complete.

Note that at a maximum or a minimum the gradient is zero because the derivatives vanish there.

Theorem 2-4 shows that the greatest increase in f (locally) is obtained by moving along the gradient a small distance.

2 *Use of the Gradient*

In the following analytical motivation we obtain a method for choosing the scalar multiple of the gradient.

Let us consider the function $f(x + \lambda u)$, where λ is a parameter, $x = (x_1, \ldots, x_n)$, and u is a vector $u = (u_1, \ldots, u_n)$. Assume we are at a point x and we wish to move along (parallel to) some vector u a certain distance λ. Our objective is to successively approach the minimum. However, to achieve this, i.e., to obtain convergence (as rapidly as possible), we must decide in which direction u must point and how to choose λ. Let our initial estimate be x^0.

We have, by expanding about x in Taylor's series to second-order terms and neglecting the higher-order terms (valid in the neighborhood of x),

$$f(x + \lambda u) = f(x) + \lambda \sum_i \frac{\partial f}{\partial x_i} u_i + \tfrac{1}{2}\lambda^2 \sum_{i,j} \frac{\partial^2 f}{\partial x_i \partial x_j} u_i u_j \qquad (2\text{-}35)$$

If x is the previous iteration point, and we wish to compute the new iteration point $x + \lambda u$ which is nearer to the minimum, then we must have

$$f(x + \lambda u) - f(x) \leq 0$$

Now to make the two terms in λ as negative as possible, we note that the coefficient of λ is the scalar product $\nabla f \cdot u$; since ∇f points in the direction of maximum increase, to achieve our objective we take $u = -\nabla f$ in order to obtain the largest (in absolute value) decrease in this coefficient. Note that by the Schwarz inequality

$$\left| \sum_i \frac{\partial f}{\partial x_i} u_i \right| \leq \left[\sum_i \left(\frac{\partial f}{\partial x_i} \right)^2 \right]^{1/2} \left(\sum u_i^2 \right)^{1/2} \qquad (2\text{-}36)$$

and equality is obtained by putting $u_i = \partial f / \partial x_i$, at which the coefficient attains its largest value.

It is assumed that the coefficient of λ^2 is always positive in a small neighborhood of x, that is, that the surface in that neighborhood is strictly convex. [$f(x)$ is convex if $f[\alpha x + (1 - \alpha)y] \leq \alpha f(x) + (1 - \alpha)f(y)$ holds for all x and y and for $0 \leq \alpha \leq 1$. It is strictly convex if strict inequality holds. More will be said about convex functions in the next chapter.] Without such an assumption our argument would not be valid.

Now the series expansion up to the second-order terms is a quadratic function of λ.   In order to choose a value of λ which minimizes $f(x + \lambda u)$, the derivative of the right side with respect to λ is equated to zero.   This gives

$$\lambda = \frac{\sum_i (\partial f/\partial x_i)^2}{\sum_{i,j} [(\partial^2 f/\partial x_i \partial x_j)(\partial f/\partial x_i)(\partial f/\partial x_j)]} \qquad (2\text{-}37)$$

Note that all derivatives are calculated at the present iteration point, which we denote by $x^{(k)}$, and which in turn defines the kth iteration value of λ, denoted by λ_k.

Thus the new point is given by

$$x^{(k+1)} = x^{(k)} - \lambda_k u^k \qquad (2\text{-}38)$$

where
$$u^k \equiv \nabla f \Big|_{x = x^{(k)}} \qquad (2\text{-}39)$$

As in the case of the Newton-Raphson method, we shall be concerned with the question of convergence of the method.

Exercise 2-23   Let $f(x) = x'cx + b'x + a$, where c is a positive-definite matrix, x' is the transpose of x (a column vector), b is a column vector, and a is a constant. Show that $u^{(k)} = 2cx^{(k)} + b$, and write the expression for λ_k.

Exercise 2-24   Apply the gradient method to the simple case

$$z = x^2 + y^2 - 2xy$$

By starting with $x^{(0)} = (1,0)$, obtain the solution at $x^{(1)} = (\tfrac{1}{2}, \tfrac{1}{2})$.

3   *Existence and Convergence Proofs*

Here we shall describe two versions of the steepest-descent method and indicate how it may be applied to the solution of systems of linear or nonlinear equations.

The first procedure examines discrete step convergence of the gradient method for a quadratic function with positive-definite matrix and then for a more general function.   It provides a sequence $x^{(k)}$ of vectors such that $x^{(k)} \to \bar{x}$ and $f(\bar{x})$ is the minimum.   A polygonal path is obtained. For each k, the increment $x^{(k+1)} - x^{(k)}$ is a vector in the direction of the most rapid decrease in f.   The second procedure results in a smooth parametrized path $x(t) = [x_1(t), \ldots, x_n(t)]$ such that $\lim_{t \to \infty} x(t) = \bar{x}$ exists and provides the minimum $f(\bar{x})$.   At each point of the path, $f(x)$ decreases most rapidly in the direction of the tangent to the path.   (The vector space will be taken throughout to be real and finite-dimensional, although the results carry over to the infinite-dimensional Hilbert space as well.)

The function $f(x)$ will be assumed to have continuous derivatives up to the second order in the region of interest. In the statement and proof of the next theorem, we shall use the following vector and matrix notations: The vector space will be R_n, the space of real n-tuples. A vector x with components $x_1, \ldots, x_n$ will be thought of as a column vector. A prime will denote transposition, so that x' is a row vector. Then $x'x = \sum\limits_{i=1}^{n} x_i{}^2$, and the euclidean length of the vector x is $||x|| = \sqrt{x'x}$. The gradient of a function $f(x)$ is the vector $\nabla f(x)$ with components $f_{x_1}(x), \ldots, f_{x_n}(x)$. The hessian matrix of $f(x)$ is the symmetric matrix $H(x) = (f_{x_ix_j}(x))$.

A *convex set* is a set which contains, with any two points, the line segment joining them. An *open set* is a set containing, with every point of the set, a sphere about that point. A *convex domain* is an open convex set.

Theorem 2-5 Let $f(x_1, \ldots, x_n)$ be a quadratic function of $x_1, \ldots, x_n$, that is,

$$f(x) = a + 2 \sum_{j=1}^{n} b_j x_j + \sum_{j=1}^{n} \sum_{k=1}^{n} c_{jk} x_j x_k \qquad (2\text{-}40)$$

and suppose the (symmetric) matrix $C = (c_{jk})$ is positive-definite. Then:
(1) $f(x)$ has a unique minimum $\bar{x}$.
(2) The iterations given by

$$x^{(k+1)} = x^{(k)} - \lambda_k u^{(k)} \qquad (2\text{-}41)$$

where $x^{(k)}$ and $u^{(k)}$ are vectors, $x^{(0)}$ is chosen arbitrarily, $u^{(k)} = Cx^{(k)} + b$ (where b is the column vector with components $b_1, \ldots, b_n$), and λ_k is the scalar given by

$$\lambda_k = \frac{u^{(k)\prime} u^{(k)}}{u^{(k)\prime} C u^{(k)}} \qquad (2\text{-}42)$$

(where primes denote transposes), converge to $\bar{x}$; that is, $\lim\limits_{k \to \infty} x^{(k)} = \bar{x}$. (In the unlikely event that $u^{(k)\prime} C u^{(k)} = 0$, we must also have $u^{(k)} = 0$, and the process is finished; see below.)

(3) The rate of convergence is that of a geometric progression.

Proof: To prove (1) we use matrix notation and write

$$f(x) = a + 2x'b + x'Cx \qquad (2\text{-}43)$$

Since C is positive-definite, there is a nonsingular P such that $C = P'P$. If we now use the new vector $y = Px$ (so $x = P^{-1}y$), we have

$$\begin{aligned}
f(x) &= x'P'Px + 2x'b + a \\
&= y'y + 2y'P^{-1}b + a \\
&= (y' + b'P^{-1})(y + P^{-1}b) + (a - b'P^{-1}P^{-1}b) \\
&= (y' + b'P^{-1})(y + P^{-1}b) + (a - b'C^{-1}b) \qquad (2\text{-}44)
\end{aligned}$$

so that $f(x) \geq a - b'C^{-1}b$ for all x and $f(x) = a - b'C^{-1}b$ if and only if $y = -P^{-1}b$, that is, $x = -P^{-1}P^{-1}b = -C^{-1}b$; thus the minimum exists and is unique.

Remark: If for some k we have $u^{(k)'}Cu^{(k)} = 0$, then $u^{(k)'}P'Pu^{(k)} = 0$ or $Pu^{(k)} = 0$, so that $u^{(k)} = 0$ and $x^{(k)} = -C^{-1}b$; the minimum is achieved in k steps. So we suppose from now on that $u^{(k)} \neq 0$ for all k.

To prove (2) we have

$$
\begin{aligned}
f(x^{(k+1)}) &= f(x^{(k)} - \lambda_k u^{(k)}) \\
&= a + 2(x^{(k)'} - \lambda_k u^{(k)'})b + (x^{(k)'} - \lambda_k u^{(k)'})C(x^{(k)} - \lambda_k u^{(k)}) \\
&= f(x^{(k)}) - 2\lambda_k u^{(k)'}(b + Cx^{(k)}) + \lambda_k^2 u^{(k)'}Cu^{(k)} \\
&= f(x^{(k)}) - 2\lambda_k u^{(k)'}u^{(k)} + \lambda_k^2 u^{(k)'}Cu^{(k)}
\end{aligned}
\tag{2-45}
$$

Recalling the expression for λ_k, we find that

$$
f(x^{(k+1)}) - f(x^{(k)}) = -\frac{(u^{(k)'}u^{(k)})^2}{u^{(k)'}Cu^{(k)}}
$$

Since C is positive-definite, we see that the sequence $\{f(x^{(k)})\}$ is strictly decreasing.

Let $\bar{x} = -C^{-1}b$ be the value of x that gives the minimum. We shall show that

$$
f(x^{(k)}) - f(\bar{x}) \leq \beta[f(x^{(k-1)}) - f(\bar{x})]
$$

for some $\beta < 1$ and all k. Let

$$
s^{(k)} = x^{(k)} - \bar{x}
$$

Then $f(x^{(k)}) - f(\bar{x}) = f(\bar{x} + s^{(k)}) - f(\bar{x}) = s^{(k)'}Cs^{(k)}$

since $C\bar{x} = -b$. Therefore, using the fact that

$$
Cs^{(k)} = u^{(k)} \qquad s^{(k)} = C^{-1}u^{(k)}
$$

we have $\dfrac{f(x^{(k)}) - f(x^{(k+1)})}{f(x^{(k)}) - f(\bar{x})} = \dfrac{(u^{(k)'}u^{(k)})^2}{(u^{(k)'}Cu^{(k)})(u^{(k)'}C^{-1}u^{(k)})}$ (2-46)

Let m and M denote the smallest and largest eigenvalues of C; $0 < m < M$. Then, for every vector z, we have $mz'z \leq z'Cz \leq Mz'z$ and $M^{-1}z'z \leq z'C^{-1}z \leq m^{-1}z'z$.

Exercise 2-25   Use an orthogonal transformation to bring C into diagonal form to prove the last-mentioned fact.

It follows that

$$
\frac{1}{u^{(k)'}Cu^{(k)}} \geq \frac{1}{Mu^{(k)'}u^{(k)}} \qquad \text{and} \qquad \frac{1}{u^{(k)'}C^{-1}u^{(k)}} \geq \frac{m}{u^{(k)'}u^{(k)}} \tag{2-47}
$$

and the left side of (2-46) is greater than or equal to m/M.

If we subtract each side of this inequality from unity and simplify, we have

$$f(x^{(k+1)}) - f(\bar{x}) \le \left(1 - \frac{m}{M}\right) [f(x^{(k)}) - f(\bar{x})] \qquad k = 0, 1, 2, \ldots \quad (2\text{-}48)$$

Consequently, $\quad f(x^{(k)}) - f(\bar{x}) \le \left(1 - \frac{m}{M}\right)^k [f(x^{(0)}) - f(\bar{x})] \qquad (2\text{-}49)$

and since $f(x^{(k)}) - f(\bar{x}) = s^{(k)\prime} C s^{(k)}$, as we saw above, and $s^{(k)\prime} C s^{(k)} \ge m s^{(k)\prime} s^{(k)}$, we have

$$s^{(k)\prime} s^{(k)} \le \frac{1}{m} \left(1 - \frac{m}{M}\right)^k [f(x^{(0)}) - f(\bar{x})] \qquad (2\text{-}50)$$

And, since $s^{(k)} = x^{(k)} - \bar{x}$, the proof is complete, for the right side of (2-50) also indicates the rate of convergence as a geometric progression.

Remark 1: The ratio $1 - m/M = (M - m)/M$ can be improved. In fact,

$$s^{(k)\prime} s^{(k)} \le \frac{1}{M} \left(\frac{M - m}{M + m}\right)^{2k} [f(x^{(0)}) - f(\bar{x})]$$

To prove this, reconsider the term

$$(u^{(k)\prime} C u^{(k)})(u^{(k)\prime} C^{-1} u^{(k)})$$

in the proof above. Use the diagonal form which is orthogonally equivalent to C to show that

$$(u^{(k)\prime} C u^{(k)})(u^{(k)\prime} C^{-1} u^{(k)}) \le \frac{1}{4} \left(\sqrt{\frac{M}{m}} + \sqrt{\frac{m}{M}}\right)^2 (u^{(k)\prime} u^{(k)})^2$$

Remark 2: It is clear from the construction in the proof that

$$u^{(k+1)} = (I - \lambda_k C) u^{(k)}$$

and $\qquad x^{(k+1)} - \bar{x} = (I - \lambda_k C)(x^{(k)} - \bar{x})$

The λ_k were chosen so as to make each step in the iteration optimal in the sense that $f(x)$ was decreased as much as possible in each single step. [Indeed, if we put $x^{(k+1)} = x^{(k)} - \lambda u^{(k)}$ with $u^{(k)} = C x^{(k)} + b$, we can calculate $f(x^{(k+1)}) - f(x^{(k)})$ as we did above and find that the value of λ that makes this difference a minimum is precisely our λ_k above.] But the whole iterative process can be improved in general by demanding that each set of r steps be optimal, where r is some integer chosen beforehand (see E. Stiefel[24]). In such a procedure we still go along the path of steepest descent; however, the scalars λ_k are not optimal singly, but rather in sets of r.

Exercise 2-26   Suppose we have a system of linear equations

$$f_i(x) = \sum_{j=1}^{n} a_{ij}x_j - d_i = 0 \qquad i = 1, \ldots, n$$

with the matrix (a_{ij}) nonsingular.   One way to solve this system iteratively is to choose positive weights $\alpha_1, \ldots, \alpha_n$, form the function

$$f(x) = \sum_{i=1}^{n} \alpha_i[f_i(x)]^2$$

and obtain the minimum by means of the previous discussion.   Supply the details. What is the matrix C here?   The vectors, b, u, etc.?   What would be a likely set of weights $\alpha_1, \ldots, \alpha_n$ to use?

When f is not quadratic, a similar procedure can be used.   The following generalization of the steepest-descent method is due to A. A. Goldstein.[13]   We use primes to indicate transposes.

***Theorem* 2-6**   Let $f(x)$ be a continuous real-valued function defined on R_n.   Let $x^{(0)}$ be any point such that the closed set

$$S = \{x : f(x) \leq f(x^{(0)})\}$$

is bounded, and suppose that f is twice continuously differentiable on S. Let the hessian matrix $H(x)$ satisfy

$$|u'H(x)u| \leq Mu'u$$

for every vector u and every x in $S[u'H(x)u$ need not be positive everywhere but must be at the minimum].   Define $x^{(k+1)} = x^{(k)} - \lambda_k \nabla f(x^{(k)})$, and let λ_k satisfy

$$f(x^{(k)} - \lambda_k \nabla f(x^{(k)})) \leq f(x^{(k)} - \lambda \nabla f(x^{(k)})) \qquad \text{for all } \lambda \geq 0$$

(the steepest-descent assumption).   Then

(1) A subsequence $x^{(k_m)}$ converges to a point $\bar{x}$ in S for which $\nabla f(\bar{x}) = 0$.
(2) $f(x^{(k_m)})$ decreases monotonically to $f(\bar{x})$.
(3) If $\bar{x}$ is the only point in S for which $\nabla f(\bar{x}) = 0$, then $x^{(k)}$ converges to $\bar{x}$.

Exercise 2-27   Draw several graphs of functions $f(x)$ of the real variable x, and consider various points $x^{(0)}$.   Verify the hypotheses of the theorem in those graphs and for those points $x^{(0)}$ in which they hold.

Proof: We must first show that the point $x^{(k+1)}$ belongs to S if $x^{(k)}$ does.   Let $<x,y>$ denote the open line segment joining x and y.   Suppose $x^{(k)}$ is in S and $\nabla f(x^{(k)}) \neq 0$.   Let

$$S_k = \{x : f(x) \leq f(x^{(k)})\}$$

Then S_k is a closed bounded subset of S.   Let R denote the half-ray consisting of all points x of the form $x = x^{(k)} - \lambda u_k$ for $\lambda \geq 0$, where $u_k \equiv$

$\nabla f(x^{(k)})$. Define

$$g(\lambda) = f(x^{(k)} - \lambda u_k) - f(x^{(k)}) \qquad \text{for } \lambda \geq 0 \qquad (2\text{-}51)$$

Then, by the mean-value theorem (using primes for transposes),

$$g(\lambda) = -\lambda u_k' u_k + \tfrac{1}{2}\lambda^2 u_k' H(\xi) u_k \qquad (2\text{-}52)$$

in which $\xi = \xi(\lambda)$ belongs to $<x^{(k)}, x^{(k)} - \lambda u_k>$. Since H is continuous, $g(\lambda)$ is continuous, and $dg/d\lambda \big|_{\lambda=0}$ exists and is equal to $-u_k' u_k < 0$. Consequently, there is an open interval $(0, \bar{\lambda})$ such that

$$g(\lambda) < 0 \text{ if } 0 < \lambda < \bar{\lambda} \qquad \text{or} \qquad f(x^{(k)} - \lambda u_k) < f(x^{(k)}) \text{ if } 0 < \lambda < \bar{\lambda}$$

So all points of the form

$$x = x^{(k)} - \lambda u_k$$

with $0 < \lambda < \bar{\lambda}$ belong to R and to S_k. Since $R \cap S_k$ is closed and bounded, and g is continuous, it attains its minimum, say at λ_k, at which $g(\lambda_k) < 0$. Thus $\lambda_k > 0$. So $x^{(k+1)} = x^{(k)} - \lambda_k u_k$ belongs to S_k and, therefore, to S.

Our objective now is to show that $u_k' u_k \to 0$ as $k \to \infty$. This fact will then be used to complete the proof.

Note that we cannot have $g(\lambda) < 0$ for all $\lambda > 0$ because S_k is bounded; as a result $g(\lambda) \geq 0$ for some $\lambda > 0$. Let λ^* be the least positive λ for which $g(\lambda^*) = 0$. [There is such a λ^* because S_k is closed and bounded and g is continuous on it, and there is an interval $(0, \bar{\lambda})$ in which $g(\lambda) < 0$, and hence $\lambda^* \neq 0$.] Since $g(\lambda^*) = 0$, we see by (2-52) that

$$\lambda^* = \frac{2 u_k' u_k}{u_k' H_* u_k}$$

where $H_* = H[\xi(\lambda^*)]$, from which we conclude that $u_k' H_* u_k > 0$. Since

$$0 < u_k' H_* u_k \leq M u_k' u_k$$

we have

$$\lambda^* \geq \frac{2}{M}$$

Also, $\quad g(\lambda) = -\lambda u_k' u_k + \tfrac{1}{2}\lambda^2 u_k' H u_k \leq -\lambda u_k' u_k + \tfrac{1}{2}\lambda^2 M u_k' u_k \qquad (2\text{-}53)$

and the last expression achieves its maximum on any interval $0 < \delta < \lambda < 2/M - \delta$ at either end point, $\lambda = \delta$ or $\lambda = 2/M - \delta$, at which points the value of $-\lambda u_k' u_k + \tfrac{1}{2}\lambda^2 M u_k' u_k$ is $-(\delta - M\delta^2/2) u_k' u_k$. The result is that

$$g(\lambda_k) \leq g(\lambda) \leq -\delta \left(1 - \frac{M\delta}{2}\right) u_k' u_k \qquad (2\text{-}54)$$

for every k.

Since $-g(\lambda_k) > 0$ and $\sum\limits_{k=1}^{\infty} -g(\lambda_k) < \infty$, as the reader may readily prove by substituting for $g(\lambda_k)$ in terms of f and using the boundedness of $f(x)$ in S, it follows from (2-54) that $\sum_{k=1}^{\infty} u_k' u_k < \infty$, since $\delta > 0$, $\delta < 2/M - \delta$ or $\delta < 1/M$, so that $1 - M\delta/2 > \frac{1}{2}$. Consequently, $u_k' u_k \to 0$, or $\nabla f(x^{(k)}) \to 0$ as $k \to \infty$.

To prove (1), note that the sequence $\{x^{(k)}\}$ has a limit point $\bar{x}$ in S since S is closed and bounded, so a subsequence $x^{(k_m)} \to \bar{x}$, and therefore $\nabla f(x^{(k_m)}) \to \nabla f(\bar{x}) = 0$.

To prove (3), if $\bar{x}$ is the only point in S for which $\nabla f(\bar{x}) = 0$, and if $x^{(k)}$ does not converge to $\bar{x}$, then there is another limit point $\bar{\bar{x}}$ of $\{x^{(k)}\}$ for which $\nabla f(\bar{\bar{x}}) = 0$ necessarily, contradicting the uniqueness of $\bar{x}$.

The proof of (2) follows from $g(\lambda_k) < 0$, $k = 0, 1, 2, \ldots$ Thus the proof is complete.

To make use of Theorem 2-6, it is necessary to solve for the point λ_k at which $f(x^{(k)} - \lambda u_k)$ achieves its minimum; this may be difficult if $f(x)$ is not a quadratic function. However, a slightly different iterative procedure which does not require this precise value of λ_k can be used.

Theorem 2-7 Let all the hypotheses of Theorem 2-6 be valid except that pertaining to λ_k. Here we can take any fixed δ such that $0 < \delta \le 1/M$ and let λ_k be any number satisfying $\delta \le \lambda_k \le 2/M - \delta$ (the method-of-gradients assumption). Let $x^{(k+1)} = x^{(k)} - \lambda_k u_k$ again. Then the conclusions of the last theorem are valid.

Proof: The proof is exactly the same as the last proof (except that λ_k has a different meaning here). Equation (2-54) is still valid, but now it is because λ_k satisfies $\delta \le \lambda_k \le 2/M - \delta$.

Exercise 2-28 Suppose $f(x)$ satisfies the hypotheses of Theorem 2-7, and, in addition,

$$0 < mu'u \le u'H(x)u \le Mu'u$$

for every vector $u \ne 0$ and every x in S. Show that we can take $\lambda_k = 2/(m + M)$ for every k, and the conclusion of the theorem will still be valid.

Exercise 2-29 Motzkin and Hart[15] consider the following iterative procedure, which is a combination of the Newton-Raphson and gradient methods for solving $f_j(x_1, \ldots, x_n) = 0$, $j = 1, \ldots, m$. Starting with an initial choice $x^{(0)}$, the iterations are given by

$$x^{(k)} = x^{(k-1)} + \rho^{(k-1)} \Delta x^{(k-1)} \qquad k > 0$$

where $\rho^{(k)} \ne 0$ for $k \ge 0$,

$$\Delta x = \sum_{j=1}^{m} a_j \Delta_j x$$

where $(a_1, \ldots, a_m)$ is an arbitrary set of positive weights,

$$\Delta_j x = (\Delta_j x_1, \cdots, \Delta_j x_n)$$

$$\Delta_j x_i = -f_j(x) \frac{\partial f_j(\xi)/\partial x_i}{w_j^2(\xi)}$$

$$w_j^2(x) = \sum_{i=1}^{n} \left(\frac{\partial f_j(x)}{\partial x_i} \right)^2 \neq 0$$

and ξ is some point in the region not necessarily the same as x, where the latter is some assigned approximation to the solution of the system.

If (1) $\bar{x}$ is a solution of the system [i.e., the hypersurfaces $f_j(x) = 0$ all contain the point $\bar{x}$], (2) the jacobian matrix evaluated at $\bar{x}$, that is, $J(\bar{x})$, has rank n (thus we assume $m = n$), (3) the λ_i are the characteristic values of AA', where $A = J(\bar{x})$ multiplied by the diagonal matrix with coefficients $a_j^{1/2}$, (4) $w_j^2(x) \neq 0$, and (5) $0 < \rho^{(k)} < 2n/\omega$, where $\omega = \sum_{i=1}^{n} \lambda_i$, then there exists a θ, $0 < \theta < 1$ and a $\delta > 0$, such that for arbitrary $x^{(0)}$ and $\xi^{(k)}$ in $\|x - \bar{x}\| < \delta$ we have $\lim_{k \to \infty} x^{(k)} = \bar{x}$ and

$$\|x^{(k)} - \bar{x}\| \leq \theta^k \|x^{(0)} - \bar{x}\|$$

If $n > 1$, it is sufficient to use $\rho^{(k)} = \rho$, a constant less than or equal to $2/\omega$, and if $n = 1$, then $\rho = 2/\omega$. In addition, $\bar{x}$ is the unique solution in $\|x - \bar{x}\| < \delta$.

An important special case occurs when $a_j = 1$ for all j, $\rho^{(k)} = \rho$ is a constant for all k, and $\xi^{(k)} = x^{(k)}$. Then

$$\Delta_j x_i^{(k-1)} = -f_j(x^{(k-1)}) \frac{\partial f_j(x^{(k-1)})/\partial x_i}{w_j^2(x^{(k-1)})}$$

$$\Delta x^{(k-1)} = \sum_{j=1}^{m} \Delta_j x^{(k-1)}$$

Let
$$x_1^2 + x_2^2 - 1 = 0$$
$$x_1^2 - x_2^2 = 0$$
$$(x_1^{(0)}, x_2^{(0)}) = (1,1) \qquad \rho = .1 \qquad a_1 = a_2 = 1$$

Find w_1^2 and w_2^2. Compute all the relevant quantities for the above special case. Show that

$$x^{(1)} = (.925, .960)$$

The next theorem, which is due to Rosenbloom,[22] has been specialized to the finite-dimensional case. Again $H(x)$ is the hessian.

Theorem 2-8 Let $f(x)$ be twice differentiable in the convex domain D. Suppose there is a positive constant A such that $y'H(x)y \geq Ay'y$ for every vector y and every x in D. Let $x^{(0)}$ in D be the initial value, and put $\gamma = \|\nabla f(x^{(0)})\|$. Suppose the sphere $\{\|x - x^{(0)}\| \leq \gamma/A\}$ lies in D. Finally, let $x(t) = (x_1(t), \ldots, x_n(t))$ be the solution of the system of ordinary differential equations

$$\frac{dx_i}{dt} = -f_{x_i}(x_1, \ldots, x_n) \qquad i = 1, \ldots, n$$

for
$$t \geq 0 \qquad x_i(0) = x_i^{(0)}$$

Then

(1) $\displaystyle\lim_{t \to \infty} x(t) = \bar{x}$ exists.

(2) $\displaystyle\lim_{t \to \infty} f(x(t))$ exists.

(3)
$$\|x(t) - \bar{x}\| \leq \frac{\gamma}{A} e^{-At}$$

(4)
$$0 \leq f(x(t)) - f(\bar{x}) \leq \frac{\gamma^2}{2A} e^{-2At}$$

And, for every x in D, we have

$$f(x) \geq f(\bar{x}) + \tfrac{1}{2}A\|x - \bar{x}\|^2$$

Proof: We first show that (3) is true. Since

$$\frac{d}{dt} f(x(t)) = \sum_{i=1}^{n} f_{x_i} \frac{dx_i}{dt} \quad \text{and} \quad \frac{dx_i}{dt} = -f_{x_i}$$

we have
$$\frac{d}{dt} f(x(t)) = - \sum_{i=1}^{n} \left(\frac{dx_i}{dt} \right)^2 \leq 0$$

so $f(x(t))$ is a decreasing function of t. Also,

$$\frac{d^2}{dt^2} f(x(t)) = - \frac{d}{dt} \sum_{1}^{n} \left(\frac{dx_i}{dt} \right)^2 = -2 \sum_{i=1}^{n} \frac{dx_i}{dt} \frac{d^2x_i}{dt^2}$$

But
$$\frac{d^2x_i}{dt^2} = - \frac{d}{dt} f_{x_i} = - \sum_{j=1}^{n} f_{x_i x_i} \frac{dx_j}{dt}$$

The result is that

$$\frac{d^2}{dt^2} f(x(t)) = 2 \sum_{i} \sum_{j} f_{x_i x_i} \frac{dx_i}{dt} \frac{dx_j}{dt} \geq 2A \sum_{1}^{n} \left(\frac{dx_i}{dt} \right)^2$$

or
$$\frac{d^2 f(x(t))}{dt^2} \geq -2A \frac{df(x(t))}{dt} \geq 0$$

From this inequality we find, by transposing and multiplying by e^{2At}, that

$$\frac{d}{dt} \left(e^{2At} \frac{df}{dt} \right) \geq 0$$

Integration between 0 and t yields

$$e^{2At} \frac{df}{dt} - \frac{df}{dt} (x(0)) \geq 0$$

But
$$\frac{df(x(0))}{dt} = - \|\nabla f(x^0)\|^2 = -\gamma^2$$

Hence,
$$0 \geq \frac{df}{dt} \geq -\gamma^2 e^{-2At}$$

Now use
$$\frac{df}{dt} = -\left\| \frac{dx}{dt} \right\|^2$$

to conclude that
$$\left\| \frac{dx}{dt} \right\| \leq \gamma e^{-At}$$

Now
$$x(t) - x(0) = \int_0^t \dot{x}(t)\, dt \qquad \dot{x} = \frac{dx}{dt}$$

so that
$$\|x(t) - x(0)\| \leq \int_0^t \|\dot{x}(t)\|\, dt$$

or
$$\|x(t) - x(0)\| \leq \frac{\gamma}{A}(1 - e^{-At}) < \frac{\gamma}{A}$$

Consequently $x(t)$ is in our sphere contained in D.

To prove (1) we again use
$$\|\dot{x}\| \leq \gamma e^{-At}$$

and integrate between t_1 and t_2; we obtain

$$\|x(t_1) - x(t_2)\| \leq \int_{t_1}^{t_2} \gamma e^{-At}\, dt = \frac{\gamma}{A}(e^{-At_1} - e^{-At_2})$$

$$\leq \frac{\gamma}{A} e^{-At_1} \qquad \text{if} \qquad t_2 > t_1$$

This implies that $\{x(t)\}$ satisfies the Cauchy criterion, and, therefore, there is an $\bar{x}$ in the sphere in D such that $x(t) \to \bar{x}$ as $t \to \infty$.

To prove (2) let $t_2 \to \infty$ in the last inequality, and conclude that

$$\|x(t_1) - \bar{x}\| \leq \frac{\gamma}{A} e^{-At_1}$$

Since $x(t) \to \bar{x}$, we have $f(x(t)) \to f(\bar{x})$.

We showed above that

$$-\frac{df}{dt} \leq \gamma^2 e^{-2At}$$

We integrate between t and ∞ and conclude that

$$0 \leq f(x(t)) - f(\bar{x}) \leq \frac{\gamma^2}{2A} e^{-2At}$$

which yields (4).

For the last part of the theorem, let y belong to D, and form the function

$$g(\lambda, t) = f\{x(t) + \lambda[y - x(t)]\}$$

Since $g(1,t) = g(0,t) + g_\lambda(0,t) + \frac{1}{2}g_{\lambda\lambda}(\bar\lambda,t)$ with

$$0 < \bar\lambda < 1$$

we have
$$\begin{aligned}
f(y) &= f(x(t)) + \nabla f(x(t))'[y - x(t)] \\
&\quad + \frac{1}{2}[y' - x(t)']H(\xi)[y - x(t)]
\end{aligned}$$

where ξ is an intermediate point.

Now we let $t \to \infty$ and recall that $\|\nabla f(x(t))\| = \|\dot x(t)\| \le \gamma e^{-At}$; we obtain

$$\begin{aligned}
f(y) &= f(\bar x) + \frac{1}{2}(y' - \bar x')H(\xi)(y - \bar x) \\
f(y) &\ge f(\bar x) + \frac{1}{2}A(y' - \bar x')(y - \bar x)
\end{aligned}$$

for every y in D; this completes the proof.

The above inequality shows that f has a strict minimum in D at $\bar x$.

Example 1: Suppose we have a system of linear equations with constant coefficients, $Bx = c$, in matrix form, with B symmetric and positive-definite. Then the solution $\bar x$ is the vector that minimizes

$$f(x) = (Bx,x) - 2(x,c)$$

where the parentheses denote inner products.

We have $\nabla f = 2Bx - 2c$ and $H = 2B$; the hypotheses of the theorem are satisfied in the entire vector space. The differential equation is

$$\frac{dx}{dt} = -2Bx + 2c$$

Its solution is given by

$$x(t) = e^{-2Bt}x^{(0)} + B^{-1}(1 - e^{-2Bt})c$$

and evidently $x(t) \to B^{-1}c$ as $t \to \infty$.

Remark: The solution of the differential equation can be carried out on a computer. If B is not positive-definite, we can consider the equivalent system $B'Bx = B'c$, where B' is the transpose of B. Here $B'B$ replaces B and is positive-definite.

Example 2: Consider the problem of minimizing the integral

$$J(x) = \int_0^1 L(x',x,t)\, dt$$

in which L is a smooth function of its three arguments, $x = x(t)$ is a differentiable function, $x' = dx/dt$, and $x(0) = x(1) = 0$. The class of all such functions $x(t)$ is a vector space but is not finite-dimensional. The appropriate theory for a problem of this sort is that of Hilbert space, in which the results given previously carry over with no change once the relevant concepts have been properly defined (gradient, hessian, etc.) However, we can achieve some partial success in such a problem and still stay in the domain of finite-dimensional spaces as follows: Let

$$e_1(t), e_2(t), \ldots, e_n(t), \ldots$$

be a complete orthonormal set of functions, differentiable and such that

$$e_j(0) = e_j(1) = 0$$

Thus
$$\int_0^1 e_i(t)e_j(t)\, dt = 0 \quad \text{if} \quad i \neq j$$

$$\int_0^1 e_i{}^2(t)\, dt = 1 \quad \text{for all } i$$

and every continuous function $g(t)$ [therefore every square-integrable $g(t)$] can be approximated by linear combinations of the $e_j(t)$ in the sense of least squares; i.e.,

$$\int_0^1 \left| g(t) - \sum_1^N \alpha_j e_j(t) \right|^2 dt < \epsilon$$

for appropriate α_j's. An example of such a set of functions is provided by taking

$$e_j(t) = \sqrt{2} \sin \pi j t \quad j = 1, 2, \ldots$$

Let us now decide on some n and ask for that $x(t)$ which is a linear combination of only the first n e_j's and which makes $J(x)$ a minimum. Thus let us write $x(t) = \sum_{j=1}^n \xi_j e_j(t)$. Now we have a problem in the n-dimensional ξ space.

We have

$$\frac{\partial J}{\partial \xi_i} = \int_0^1 [L_{x'} e_i'(t) + L_x e_i(t)]\, dt$$

in which $L_{x'}$ and L_x are functions of $\sum_j \xi_j e_j'$, $\sum_j \xi_j e_j(t)$, and t. Since $e_i(t) = 0$ at $t = 0$ and $t = 1$, we can integrate the first term by parts and get

$$\frac{\partial J}{\partial \xi_i} = \int_0^1 e_i(t)\left[L_x - \frac{d}{dt} L_{x'} \right] dt \quad i = 1, \ldots, n$$

We next examine the matrix $(\partial^2 J/\partial \xi_i\, \partial \xi_j) = H(\xi)$ or, equivalently, the quadratic form $z'Hz$, where z is the arbitrary column vector. Using the expression in $\partial J/\partial \xi_i$, we have

$$\frac{\partial^2 J}{\partial \xi_i \partial \xi_j} = \int_0^1 [L_{x'x'} e_j' e_i' + L_{x'x}(e_i' e_j + e_i e_j') + L_{xx} e_j e_i]\, dt$$

and if z has components $(z_1, \ldots, z_n)$, then

$$z'H(\xi)z = \int_0^1 \left[L_{x'x'} \left(\sum_1^n z_j e_j' \right)^2 + 2L_{x'x} \left(\sum_1^n z_j e_j \right) \left(\sum_1^n z_i e_i' \right) \right.$$
$$\left. + L_{xx} \left(\sum_1^n z_j e_j \right)^2 \right] dt$$

Now if there is an $m > 0$ such that $z'H(\xi)z \geq mz'z$ for all z and all ξ in a convex region D and the other hypotheses of Theorem 2-8 are satisfied, then the best choice of ξ_i is obtained as the limit ξ_i of $\xi_i(\alpha)$, where

$$\frac{d\xi_i}{d\alpha} = -\frac{\partial J}{\partial \xi_i} (\xi_1, \ldots, \xi_n)$$

(We use a different letter α for the curve parameter.) If we can estimate an M such that $z'H(\xi)z \leq Mz'z$ for all z and all ξ in a region D, then the discrete version of the steepest-descent method will produce a sequence of vectors converging to the minimizing ξ.

In either case, the minimizing function $x(t)$ will be

$$x(t) = \sum_{j=1}^n \bar{\xi}_j e_j(t)$$

recalling the fact that $x(t)$ is a linear combination of $e_1(t), \ldots, e_n(t)$.

2-4 SADDLE - POINT METHOD OR STEEPEST - DESCENT METHOD OF COMPLEX INTEGRATION

Before closing this chapter, let us remark that the name "steepest descent" or "saddle point" is often applied to a method quite different from those discussed above. The problem here is to obtain asymptotic expressions for definite integrals containing a real parameter t. It is usually applied to integrals which cannot be evaluated in terms of elementary functions. More specifically, suppose we have a function $f(t)$ expressible as an integral

$$f(t) = \int_C g(z,t) \, dz$$

in which $g(z,t)$ is meromorphic in z for $t \geq t_0$, and C is a contour in the complex z-plane. In many cases we can obtain an approximation $f_1(t)$ to $f(t)$ such that

$$\frac{f_1(t)}{f(t)} \to 1$$

as $t \to \infty$ [$f_1(t)$ is "asymptotically equal" to $f(t)$], so that for large values of t, we can use $f_1(t)$ to approximate $f(t)$.

The idea is to deform the path C into a new path $C'(t)$ (which may depend on t) on which $|g(z,t)|$ is large essentially in one or a few small intervals and small elsewhere; as $t \to \infty$, this behavior often becomes even more pronounced, so that the integral can be evaluated approximately by considering only those parts of the path in which $|g(z,t)|$ is large. [If the new path $C'(t)$ passes through poles of $g(z,t)$, the value of the new integral $\int_{C'} g \, dz$ must be augmented by the addition of appropriate terms, the contributions of the poles.]

If we write

$$g(z,t) = e^{h(z,t)}$$

where $h(z,t)$ is the logarithm of $g(z,t)$, then

$$|g(z,t)| = \exp \{\operatorname{Re} [h(z,t)]\}$$

and, as we want our path to have the property that $|g(z,t)|$ decreases as fast as possible from any maxima on the path, it follows that $\operatorname{Re} h$ must also have this property. Set

$$h(z,t) = u(z,t) + iv(z,t)$$

u, v real, so that $\operatorname{Re} h = u$. We know from previous results that the path of steepest descent for u is the path described by the differential equations

$$\frac{dx}{d\alpha} = -u_x \qquad \frac{dy}{d\alpha} = -u_y$$

in which α is the curve parameter. If there is to be a maximum on the path at z_0, then evidently

$$\frac{d}{d\alpha} u(z_0,t) = -u_x{}^2 - u_y{}^2 = 0$$

and by the Cauchy-Riemann equations we have

$$v_x = v_y = 0 \qquad \text{at } z_0$$

that is,

$$h'(z_0,t) = 0$$

Such points z_0, for which $h'(z_0,t) = 0$, are called *saddle points* of $g(z,t)$. It is also clear that any path of steepest descent of $|g(z,t)|$ must be orthogonal to the level curves of $|g(z,t)|$, that is, the level curves of $u(z,t)$, so that such a path is in fact of the form $v(z,t) = \operatorname{Im} h = \text{const.}$ Note that we have a saddle point because the surface $w = |g(x + iy, \, t)|$ in xyw space near $z = z_0$ has the shape of a saddle (see **Chap. 3** for a discussion of saddle points).

Now we have, in the neighborhood of any saddle point z_0,

$$h(z,t) = h(z_0,t) + \tfrac{1}{2}h''(z_0,t)(z - z_0)^2 + \cdots$$

so that the integral over a neighborhood U of z_0 along the path is approximately

$$e^{h(z_0,t)} \int_U e^{\frac{1}{2}h''(z_0,t)(z-z_0)^2}\, dz$$

This integral will be essentially gaussian (but with a generally complex parameter having a negative real part).

Example (Stirling's formula): Let

$$f(t) = \Gamma(t + 1) = \int_0^\infty e^{-z}z^t\, dz$$

Here

$$g = e^{-z}z^t = e^h$$

and

$$h = -z + t \log z$$

so that

$$h' = -1 + \frac{t}{z} \quad\text{and}\quad z_0 = t$$

is the only saddle point, and for each t, our path already passes through z_0. It is clear that the real axis is the path of steepest descent on either side of $z_0 = t$. We have

$$h(z_0) = -t + t \log t$$

$$h''(z) = -\frac{t}{z_0{}^2} = -\frac{1}{t}$$

so that

$$\Gamma(t + 1) \approx e^{-t}t^t \int_0^\infty \exp\left[-\frac{1}{2t}(z - t)^2 \right] dz$$

$$= e^{-t}t^t \sqrt{2\pi t}$$

which is Stirling's formula.

Note that the success of the method depends on there being peaks along the path through the saddle point in $|g(z,t)|$ which become steeper and steeper as $t \to \infty$. As N. G. de Bruijn ("Asymptotic Methods in Analysis," North Holland Publishing Company, Amsterdam, 1958) expertly points out, "if $g(z,t)$ does not behave violently, the saddle point method has not much of a chance to succeed."

REFERENCES

1.   Booth, A. D.: Non-linear Algebraic Equations, in "Numerical Methods," Butterworth Scientific Publications, London, 1955.

2.   Cauchy, A.: Méthode Générale pour la Résolution des Systémes d'Equations Simultanées, *C. R. Acad. Sci. Paris*, vol. 25, pp. 536–538, 1847.

3.   Cheney, E. W., and A. A. Goldstein: Newton's Method for Convex Programming and Tchebycheff Approximation, *Numerische Math.*, vol. 1, p. 263, 1959.

3a. Collatz, L.: Das Vereinfachte Newtonsche Verfahren bei Algebraischen und Transzendenten Gleichungen, *Z. Angew. Math. Mech.*, vol. 34, pp. 70–71, 1954.

4. Crockett, Jean B., and Herman Chernoff: Gradient Methods of Maximization, *Pacific J. Math.*, vol. 5, pp. 33–59, 1955.

5. Curry, H. B.: The Method of Steepest Descent for Nonlinear Minimization Problems, *Quart. Appl. Math.*, vol. 2, pp. 258–261, 1944.

6. Durand, E.: "Solutions Numeriques des Equations Algébriques," vols. I and II, Masson et Cie, Paris, 1960.

7. Dwyer, P. S.: Application to Non-linear Problems, in "Linear Computations," John Wiley & Sons, Inc., New York, 1951.

8. Fischbach, Joseph W.: Some Applications of Gradient Methods, in "Proceedings of Symposia in Applied Mathematics," vol. VI, pp. 59–72, The American Mathematical Society, 1956.

9. Ford, L. R.; The Solution of Equations by the Method of Successive Approximations, *Am. Math. Monthly*, vol. 32, p. 272, 1925.

10. Forsythe, G. E.: Solving Linear Algebraic Equations Can Be Interesting, *Bull. Am. Math. Soc.*, vol. 59, pp. 299–329, 1953.

11. Forsythe, G. E., and W. R. Wasow: "Finite-difference Methods for Partial Differential Equations," pp. 224–225, John Wiley & Sons, Inc., New York, 1959.

12. Gleyzal, A. N.: Solution of Nonlinear Equations, *Quart. Appl. Math.*, vol. 17, no. 1, p. 95, 1959.

13. Goldstein, A. A.: Cauchy's Method of Minimization, *Numerische Math.*, vol. 4, pp. 146–150, 1962.

14. ———, J. B. Hereshoff, and N. Levine: On the Best and Least qth Approximation of an Over-determined System of Linear Equations, *J. Assoc. Comp. Mach.*, vol. 4, no. 3, pp. 341, 1957.

15. Hart, William L., and Theodore S. Motzkin: A Composite Newton-Raphson Gradient Method for the Solution of Systems of Equations, *Pacific J. Math.*, vol. 6, pp. 691–707, 1956.

16. Haskell, Curry B.: The Method of Steepest Descent for Non-linear Minimization Problems, *Quart. J. Appl. Math.*, pp. 258–261, 1944.

17. Kaczmarz, S.: Angenäherte Auflosung von Systemen Linearer Gleichungen, *Bull. Intern. Acad. Polon. Sci., Cl. Sci. Math. Nat.*, ser. A, pp. 355–357, 1937.

18. Kantorovich, L. V.: Functional Analysis and Applied Mathematics, translated from Russian by Curtis D. Benster, edited by G. E. Forsythe, *Uspehi Mat. Nauk*, vol. 3, no. 6, pp. 89–185, 1948; also *National Bureau of Standards Report*.

18a. Kiss, I.: Uber eine Verallgemeinerung des Newtonschen Näherungsverfahrens, *Z. Angew. Math. Mech.*, vol. 34, pp. 68–69, 1954.

18b. Ludwig, von, Rudolf: Uber Iterationsverfahren fur Gleichungen und Gleichungssysteme, Parts I and II, *Z. Angew. Math. Mech.*, vol. 34, pp. 210–225, 404–416, 1954.

18c. Maehly, von, Hans J.: Zur Iterativen Auflösung Algebraischer Gleichungen *Z. Angew. Math. Phys.* vol. 5, pp. 260–261, 1954.

19. Milnes, H. W.: A Modified Newton-Raphson Process for Approximation of Multiple Roots of Polynomials, *Math. Revs.*, May, 1962, *Ind. Math.*, vol. 9, no. 2, pp. 17–26, 1958.

20. Ostrowski, A. M.: "Solution of Equations and Systems of Equations," Academic Press Inc., New York, 1960.

21. Rheinboldt, Werner: Iterative Methods and Functional Analysis, pt. 1, National Bureau of Standards Training Program in Numerical Analysis, 1959, Lecture Notes.

22. Rosenbloom, P. C.: The Method of Steepest Descent, in "Proceedings of Symposia in Applied Mathematics," vol. 6, pp. 127–176, The American Mathematical Society, 1956.

23. Stiefel, E.: Relaxation Methoden Bester Strategie sur Losung Linearer Gleichungssysteme, *Comm. Math. Helv.*, vol. 29, pp. 157–179, 1955.

24. Stiefel, E.: Kernel Polynomials in Linear Algebra and Their Numerical Applications, *National Bureau of Standards Appl. Math. Ser.* 19, 1949.

24a. Stiefel, E.: "Einführung in die Numerische Mathematik," Teubner Verlag, Stuttgart; "Refined Iterative Methods for Computation of the Solution and the Eigenvalues of Self-adjoint Boundary Value Problems," Mitteilung Nr. 8 aus dem Institut fur Angewandte Mathematik, Birkhauser Verlag, Basel.

24b. Taussky, Olga, and John Todd: Systems of Equations, Matrices and Determinants, *Math. Mag.*, November-December, 1952.

24c. Temple, G.: The General Theory of Relaxation Methods Applied to Linear Systems, *Proc. Roy Soc. London*, ser. A, vol. 169, 1939.

25. Thomas, J. M.: "Systems and Roots," The William Byrd Press, Inc., Richmond, Va., 1962.

25a. Todd, John (ed.): "A Survey of Numerical Analysis," McGraw-Hill Book Company, Inc., New York, 1962.

26. Tompkins, Charles B.: Methods of Steep Descent, in E. F. Beckenbach (ed.), "Modern Mathematics for the Engineer," McGraw-Hill Book Company, Inc., New York, 1956.

27. Vainberg, M. M.: On the Convergence of the Method of Steepest Descent for Non-linear Equations, *Am. Math. Soc. Doklady Transl.* 1, 1960.

28. Wall, H. S.: A Modification of Newton's Method, *Am. Math. Monthly*, vol. 55, p. 90, 1948.

THREE

NONLINEAR OPTIMIZATION; NONLINEAR PROGRAMMING AND SYSTEMS OF INEQUALITIES

3-1 INTRODUCTION

In this chapter systems of inequalities play a more basic role than systems of equations. Maximization or minimization (which we shall often refer to as optimization) of a function $f(x_1, \ldots, x_n)$ subject to equality constraints $g_i(x_1, \ldots, x_n) = 0$, $i = 1, \ldots, m$, is frequently studied through the Lagrange-multiplier method. It is obvious that each of the $g_i(x)$, $x = (x_1, \ldots, x_n)$ defines a surface in n dimensions, and the intersection of n surfaces generally yields a point. Thus it would be trivial to maximize $f(x)$ subject to n constraints. However, if $m < n$, then the surfaces intersect in a region (rather than a point) which is generally a surface of dimension $n - m$, and optimization methods seek a point of this surface which yields the optimum to f. It may be possible to solve for m of the variables x_i in terms of the others in the constraints. Substituting these values in f, one obtains an optimization problem without constraints, and the optimum is found by equating the first partial derivatives to zero and using some of the ideas of the previous chapter to obtain a solution. When Lagrange multipliers are used, it is generally assumed that the intersection surface from which the solution is obtained is determined by all the constraint equalities. Thus utilizing Lagrange multipliers λ_i, one defines the lagrangian function

$$F(x,\lambda) = f(x) + \sum_{i=1}^{m} \lambda_i g_i(x) \tag{3-1}$$

A necessary condition for an optimum is that it be a solution of the system of $m + n$ equations in $m + n$ unknowns $x_1, \ldots, x_n, \lambda_1, \ldots, \lambda_m$,

$$\frac{\partial F}{\partial x_j} = 0 \qquad j = 1, \ldots, n$$
$$\frac{\partial F}{\partial \lambda_i} = 0 \qquad i = 1, \ldots, m \tag{3-2}$$

Thus, frequently, an optimization problem reduces to the solution of a system of equations.

The above necessary condition in the case of two variables and one constraint is obtained as follows: We must show that there is a λ_1 for which

$$\frac{\partial f}{\partial x_1} + \lambda_1 \frac{\partial g_1}{\partial x_1} = 0$$
$$\tag{3-3}$$
and
$$\frac{\partial f}{\partial x_2} + \lambda_1 \frac{\partial g_1}{\partial x_2} = 0$$

At the optimum the differential of f along the constraint curve

$$g_1(x_1,x_2) = 0$$

must vanish, and we have

$$\frac{\partial f}{\partial x_1} dx_1 + \frac{\partial f}{\partial x_2} dx_2 = 0 \tag{3-4}$$

Obviously the analogous differential of $g_1(x_1,x_2) = 0$ vanishes. If we multiply this total derivative by λ_1 and add it to that of f, we obtain

$$\left(\frac{\partial f}{\partial x_1} + \lambda_1 \frac{\partial g_1}{\partial x_1}\right) dx_1 + \left(\frac{\partial f}{\partial x_2} + \lambda_1 \frac{\partial g_1}{\partial x_2}\right) dx_2 = 0 \tag{3-5}$$

Now in $g_1 = 0$ one can solve for dx_2 in terms of dx_1, where dx_1 can be arbitrary (or else solve for dx_1 in terms of dx_2). Thus λ_1 can be chosen such that the coefficient of dx_2 is zero. Since dx_1 is arbitrary, its coefficient must also vanish.

Exercise 3-1   Generalize the above argument to the case of two constraints and three variables.

In the general case we have

$$\nabla f \cdot dx = 0$$

along the constraint surface; that is, $\nabla f \cdot dx = 0$ for all dx for which

$$\nabla g_i \cdot dx = 0 \qquad i = 1, \ldots, m$$

This is possible only if ∇f is a linear combination of $\nabla g_1, \ldots, \nabla g_m$, that is, if there exist $\lambda_1, \ldots, \lambda_m$ for which

$$\nabla f = \lambda_1 \nabla g_1 + \cdots + \lambda_m \nabla g_m$$

Exercise 3-2   Find the dimensions of a rectangular parallelepiped with largest volume [$f(x,y,z) = xyz$] whose sides are parallel to the coordinate planes and which is inscribed in the ellipsoid

$$\frac{x^2}{a^2} + \frac{y^2}{b^2} + \frac{z^2}{c^2} - 1 = 0$$

If the constraints are inequalities

$$g_i(x_1, \ldots, x_n) \leq 0 \qquad i = 1, \ldots, m$$

we no longer need the condition $m < n$, since the intersection of inequalities defines regions whose size merely decreases as m is increased.   Frequently assumptions of convexity (defined below) are imposed on the $g_i(x)$; as a consequence the region of intersection of the inequalities is convex.   Then, if the inequalities are consistent, there is a unique convex region in which $f(x)$ is optimized.   Note that the same statement holds if the $g_i(x)$ are linear.   If, on the other hand, the $g_i(x)$ are arbitrary, then even when the inequalities are consistent, one can obtain more than a single region for the intersection, and the optimum is obtained from one of these regions.   Thus in two dimensions the half-space given by the linear inequality $y \geq 0$ and the region defined by the quartic inequality

$$-(x + 2)(x + 1)(x - 1)(x - 2) \geq y$$

intersect in two regions, as the reader may readily verify (see Fig. 3-1).

As in the previous chapter, in the iterative techniques considered here we assume that a starting point is given; this is then used to construct the next point, which is used as the starting point for subsequent iterations, and so on.   A set of inequalities which represents a practical problem is usually satisfied by the state of the operations it describes at the time, and hence that state can be used as the initial estimate.

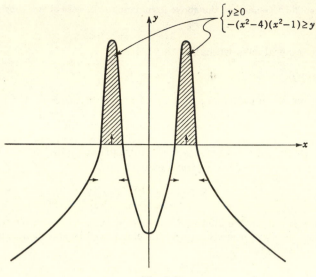

$$\begin{cases} y \geq 0 \\ -(x^2-4)(x^2-1) \geq y \end{cases}$$

Fig. 3-1

3-2 MAXIMA, MINIMA, QUADRATIC FORMS, AND CONVEX FUNCTIONS

This section provides background material for the present chapter. To conserve space we must be brief in our presentation.

The term *extremum* or *extreme value* is used in mathematics to refer to a value of a function (perhaps subject to constraints) which is either a maximum or a minimum. The word *optimum* is used for the particular type of extremum desired in the problem at hand. For instance, if $f(x,y)$ (perhaps with x and y subject to constraints) is a "cost function," then the type of extremum one may seek is a minimum; the optimum is the minimum value of $f(x,y)$, and it is desired to find a pair (x_0,y_0) at which this minimum is attained. But if $f(x,y)$ were an "output function," then the optimum would be a maximum. Note that the term *extremal point* is used to refer to a point in the domain of definition of the function which yields an extremum of the function. It is a well-known fact that a continuous function attains its maxima and minima on a compact domain of definition.

A point $P_0 \equiv (x_1^0, \ldots, x_n^0)$ of a domain on which a function $f(x_1, \ldots, x_n)$ is defined is an *absolute maximum* in D if the following inequality holds for any point of the domain D:

$$f(x_1, \ldots, x_n) \leq f(x_1^0, \ldots, x_n^0) \tag{3-6}$$

It is an *absolute minimum* if, for every point in D,

$$f(x_1, \ldots, x_n) \geq f(x_1^0, \ldots, x_n^0) \qquad (3\text{-}7)$$

A maximum (minimum) is said to be *strict* only if the strict inequalities above hold for every point $(x_1, \ldots, x_n)$ different from P_0.

P_0 is a maximum (minimum) *in the large* (i.e., is an absolute extremum) if there are no points P in D other than P_0 for which $f(P) = f(P_0)$.

The function $f(x_1, \ldots, x_n)$ has a *relative* or *local* maximum (minimum) if $\epsilon > 0$ can be found such that f has an absolute maximum or minimum at P_0 in a subset of points of D which satisfy

$$|x_1 - x_1^0| < \epsilon, \ldots, |x_n - x_n^0| < \epsilon \qquad (3\text{-}8)$$

For example, $\sin^2 x \cos x$ in the domain $-\pi/2 < x < \pi/2$ has a relative minimum at $x = 0$ which is strict, since $\sin^2 x \cos x > 0$ for $\pi/2 > |x| \neq 0$. Note that $x = 0$ is not an absolute minimum when the domain of definition is extended to include the entire real axis.

Taylor's formula gives the following for the series expansion of a function of two variables $f(x,y)$ in the neighborhood of a point (x_0, y_0):

$$f(x_0 + h, y_0 + k) = f(x_0, y_0) + \sum_{i=1}^{n} \frac{1}{i!} \left(h \frac{\partial}{\partial x} + k \frac{\partial}{\partial y} \right)^i f(x_0, y_0)$$

$$+ \frac{1}{(n+1)!} \left(h \frac{\partial}{\partial x} + k \frac{\partial}{\partial y} \right)^{n+1} f(x_0 + \theta h, y_0 + \theta k) \qquad 0 < \theta < 1 \quad (3\text{-}9)$$

If (x_0, y_0) is an extremal point, then, for small h and k, $f(x_0 + h, y_0 + k) - f(x_0, y_0)$ is positive or negative depending on whether the terms involving the second partial derivatives are positive or negative, since for small values of h and k the second-order terms are dominant. The two terms with the first partial derivatives as coefficients vanish as a necessary condition for an extremum.

A similar consideration of the simpler case of a function of a single variable $f(x)$ leads to the well-known sufficient condition which requires that $f''(x) < 0$ for a maximum and $f''(x) > 0$ for a minimum. More generally, if all the derivatives up to but not including the nth derivative vanish at a point x, then if n is even, $f(x)$ has a minimum (maximum) at x_0 if $f^{(n)}(x_0) > 0$ (<0); and if n is odd, $f(x)$ increases (decreases) at x_0 if $f^{(n)}(x_0) > 0$ (<0).

For several variables these conditions generalize to a quadratic form whose coefficients are the second partial derivatives. A quadratic form in n variables is given by

$$\sum_{i=1}^{n} \sum_{j=1}^{n} a_{ij} \lambda_i \lambda_j \qquad \text{where} \qquad a_{ij} = a_{ji}$$

The matrix $[a_{ij}]$ is called the *matrix of the quadratic form*. Note that $a_{ij} = \frac{1}{2}f_{x_i x_j}$ is given by the second derivatives of f. A quadratic form is *positive-definite* if it is positive for all real values of the variables not all zero. It is *positive-semidefinite* if it can also vanish for nonzero values of the variables.

Testing for a Minimum

Let $|D_m|$, $m \leq n$, be the determinant of the matrix consisting of all elements which lie in the first m rows and columns of the matrix whose elements are the second partial derivatives $f_{x_i x_j}$ evaluated at the stationary point (a point at which the first derivatives vanish) under study. It can be shown that the matrix (a_{ij}) is positive-definite if and only if each of the n determinants $|D_m|$ is positive. The matrix is positive-definite if the quadratic form is.

The above condition on the determinants of the matrix is necessary and sufficient for the quadratic form to be positive-definite, which, in turn, is a sufficient condition for a local minimum.

Note that to maximize a function f is the same as to minimize $-f$. Thus, the above argument can also be used as a sufficiency test for a maximum after appropriate adjustment.

In order for a stationary point of a function $f(x,y)$ to be a relative minimum, the following inequalities resulting from the condition on the determinants $|D_m|$ must be simultaneously satisfied:

$$f_{xx} > 0 \qquad f_{xx}f_{yy} - (f_{xy})^2 > 0$$

The subscripts indicate the variables with respect to which the partial derivatives are to be taken and their order.

For a relative maximum these conditions become

$$f_{xx} < 0 \qquad f_{xx}f_{yy} - (f_{xy})^2 > 0$$

In the case of a function of three variables $f(x,y,z)$, the conditions for a minimum become

$$f_{xx} > 0 \qquad f_{xx}f_{yy} - (f_{xy})^2 > 0 \qquad \begin{vmatrix} f_{xx} f_{xy} f_{xz} \\ f_{yx} f_{yy} f_{yz} \\ f_{zx} f_{zy} f_{zz} \end{vmatrix} > 0 \qquad (3\text{-}10)$$

with $f_{xy} = f_{yx}$, $f_{xz} = f_{zx}$, and $f_{yz} = f_{zy}$.

Remark: There are many instances in which an optimization problem with constraints is simpler than the corresponding problem without constraints, in the sense that the constrained-variable problem may have a solution although the unconstrained problems may not.

An example of a function of several variables which has no minimum in all variables at once, but which has one in $n - 1$ variables if x_1 is set equal to zero, is

$$f(x_1, \ldots, x_n) = (2x_1 + x_2)^2 + (2x_1 + x_3)^2 + \cdots + (2x_1 + x_n)^2 - x_1^2$$

It is positive in the remaining variables if not all are zero. It is zero if they are all zero. It is negative if

$$x_1 = \epsilon > 0 \qquad x_2 = x_3 = \cdots = x_{n-1} = x_n = -2\epsilon$$

Thus it does not have a minimum for the values

$$x_1 = x_2 = \cdots = x_n = 0$$

but it does if $x_1 = 0$ is given as a constraint.

Convex Regions, Monotone Functions, and Convex Functions

An n-dimensional region A is said to be *convex* if, whenever $(x_1, \ldots, x_n)$ and $(y_1, \ldots, y_n)$ both belong to A, their convex combination $(\theta x_1 + (1 - \theta)y_1, \ldots, \theta x_n + (1 - \theta)y_n)$, $0 \leq \theta \leq 1$, also belongs to A. Geometrically this means that the straight line connecting any two points of a convex region belongs to the region. In the two-dimensional plane, the area enclosed by a triangle or a circle is a convex region. Similarly the region enclosed by a tetrahedron is convex in three-dimensional space.

Two well-known types of functions which are related and which occur frequently are *monotone* functions and *convex* functions.

A function $f(x)$ is *monotone increasing* on an interval if, for any two values x_1 and x_2 in the interval, with $x_1 < x_2$, the relation $f(x_1) \leq f(x_2)$ holds. It is *strictly increasing* if only the inequality holds. It is *monotone decreasing* if $\geq$ holds. If the derivative of a function is nonnegative (nonpositive) in an interval, then the function is monotone increasing (decreasing) in the interval.

Convex functions occur in an important way in optimization theory. We shall now give conditions for determining when a function is concave or convex. These conditions are related to the foregoing discussion of quadratic forms as applied to determining an extremum of an unconstrained function.

A function of a single variable $f(x)$ is convex (also called *concave upward*) on an interval (see Fig. 3-2) if, for any two points x_1, x_2 on the interval, with $x_1 < x_2$, it satisfies Jensen's inequality,

$$f\left(\frac{x_1 + x_2}{2}\right) \leq \frac{f(x_1) + f(x_2)}{2} \qquad (3\text{-}11)$$

Geometrically, this says that the value of the function at a point which is

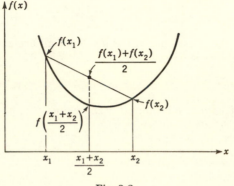

Fig. 3-2

the average of x_1 and x_2 is less than its average value at the two points. This condition may also be stated as

$$f[(1 - \theta)x_1 + \theta x_2] \leq (1 - \theta)f(x_1) + \theta f(x_2) \qquad \text{for} \qquad 0 \leq \theta \leq 1 \quad (3\text{-}12)$$

The function $f(x)$ is strictly convex if only the inequality holds. A twice-differentiable function $f(x)$ on an open interval (not necessarily including the end points of the interval) is convex if and only if $d^2f/dx^2 \geq 0$ on the interval.

This definition may be generalized to a function of several variables. If D is a convex set of the space of points $(x_1, \ldots, x_n)$, a real-valued function defined in D is convex in D if for x and y in D the following holds:

$$f[(1 - \theta)x + \theta y] \leq (1 - \theta)f(x) + \theta f(y) \qquad \text{for} \qquad 0 \leq \theta \leq 1$$

It is strictly convex if $<$ holds for $0 < \theta < 1$ and x and y are distinct points in D.

Theorem **3-1** If $f(x_1, \ldots, x_n)$ is a twice continuously differentiable function in an open convex set D, it is convex in D if and only if the quadratic form

$$\sum_{i=1}^{n} \sum_{j=1}^{n} \frac{\partial^2 f}{\partial x_i \partial x_j} \lambda_i \lambda_j$$

is positive-semidefinite for every point x in D.

Proof:[22a] To prove necessity, we use the extended mean-value theorem and write, for $0 < \theta < 1$,

$$(1 - \theta)f(x) + \theta f(y) - f[(1 - \theta)x + \theta y]$$
$$= (1 - \theta)\{f(x) - f[(1 - \theta)x + \theta y]\}$$
$$+ \theta\{f(y) - f[(1 - \theta)x + \theta y]\}$$
$$= \theta(1 - \theta)(x' - y')f_x[(1 - \theta)x + \theta y]$$
$$+ \tfrac{1}{2}\theta^2(x' - y')f_{xx}(\xi_1)(x - y)$$
$$+ \theta(1 - \theta)(y' - x')f_x[(1 - \theta)x + \theta y]$$
$$+ \frac{(1 - \theta)^2}{2}(y' - x')f_{xx}(\xi_2)(y - x) \tag{3-13}$$

where ξ_1 is between x and $(1 - \theta)x + \theta y$ and ξ_2 is between y and $(1 - \theta)x + \theta y$. The first and third terms cancel. Since f_{xx} is positive-semidefinite, the first expression on the left is greater than or equal to zero, from which $f(x)$ is convex.

To prove sufficiency, we note that if f_{xx} at x_0 is not positive-semidefinite, then there is a vector u such that $u'f_{xx}(x_0)u < 0$. Therefore for all x in a neighborhood of x_0 we also have $u'f_{xx}(x)u < 0$. If we put $x = x_0$ and $y = x_0 + tu$ where t is a small scalar, then the above expression, with this choice of x and y, is negative, and hence the expression on the left violates the convexity condition, a contradiction.

A necessary condition for the convexity of a differentiable function is given by the following theorem, which we shall use later.

Theorem 3-2 If $f(x)$ is differentiable and convex, then

$$f(x) \geq f(x^*) + \nabla f^*(x - x^*) \qquad \text{where} \qquad x^* = (x_1^*, \ldots, x_n^*)$$

and $\nabla f^* = \left(\dfrac{\partial f}{\partial x_1}, \ldots, \dfrac{\partial f}{\partial x_n}\right)_{x = x^*}$

Proof:

$$f(x) - f(x^*) \geq \frac{f[x^* + \theta(x - x^*)] - f(x^*)}{\theta} \qquad 0 < \theta \leq 1 \quad (3\text{-}14)$$

since f is convex. The relation will remain true as $\theta \to 0$, and we obtain the result by calculating the limit as $\theta \to 0$. (In the case in which x lies in a small neighborhood of x^*, the above result follows from the previous theorem.)

We also have the fact that $f(x_1, \ldots, x_n)$ is convex in D if and only if it is convex on every straight segment in D. It is strictly convex if the quadratic form is positive-definite.[24]

A function $f(x_1, \ldots, x_n)$ is concave (strictly concave) if $-f$ is convex (strictly convex).

The above definition of convexity involving the quadratic form is almost identical with the condition previously given to test for a minimum of a function of several variables, except that here all principal minor determinants are required to be nonnegative. These conditions generally provide a set of inequalities in the variables from which the region or regions in which the function is convex may be determined. Let it be required to determine the region of concavity of the function

$$f(x,y) = \frac{1}{2\pi\sigma^2} \exp\left[-\frac{1}{2\sigma^2}(x^2 + y^2) \right]$$

The second partial derivatives are given by

$$\frac{\partial^2 f}{\partial x^2} = \frac{1}{2\pi\sigma^4}\left(\frac{x^2}{\sigma^2} - 1\right) \exp\left[-\frac{(x^2 + y^2)}{2\sigma^2} \right]$$

$$\frac{\partial^2 f}{\partial y^2} = \frac{1}{2\pi\sigma^4}\left(\frac{y^2}{\sigma^2} - 1\right) \exp\left[-\frac{(x^2 + y^2)}{2\sigma^2} \right]$$

$$\frac{\partial^2 f}{\partial y \partial x} = \frac{1}{2\pi\sigma^4}\frac{xy}{\sigma^2} \exp\left[-\frac{(x^2 + y^2)}{2\sigma^2} \right]$$

For concavity the following conditions must be simultaneously satisfied :

$$\frac{y^2}{\sigma^2} - 1 \le 0$$

$$\left(\frac{x^2}{\sigma^2} - 1\right)\left(\frac{y^2}{\sigma^2} - 1\right) - \left(\frac{xy}{\sigma^2}\right)^2 \ge 0$$

which, on simplifying, become

$$y^2 \le \sigma^2 \quad \text{or} \quad -\sigma \le y \le \sigma$$
$$x^2 + y^2 \le \sigma^2 \quad \text{a circle of radius } \sigma$$

Obviously every y which satisfies the second condition also satisfies the first condition, but not conversely. Hence the region of concavity is defined by the region which has the circle $x^2 + y^2 = \sigma^2$ as boundary.

The following two theorems will show that by means of inequality constraints involving convex functions one obtains a convex region; this region will be used as the domain of definition of a function f and will provide what is known as a *feasible* region in which f is maximized or minimized. The convexity assumption on the feasible region enables one to begin the development of a theory for nonlinear programming.

Exercise 3-2a   Prove that a local minimum of a convex function is a global minimum.

Theorem **3-3**   The set of points D which satisfy a constraint $g(x) \le 0$, where $g(x)$ is a convex function, is a convex set.

Proof: Let $(x_1, \ldots, x_n)$ and $(y_1, \ldots, y_n)$ be two points in D. Then $g(x_1, \ldots, x_n) \leq 0$ and $g(y_1, \ldots, y_n) \leq 0$ and

$$g(\theta x_1 + (1 - \theta)y_1, \ldots, \theta x_n + (1 - \theta)y_n) \leq \theta g(x_1, \ldots, x_n) \\ + (1 - \theta)g(y_1, \ldots, y_n) \leq 0$$

Hence $[\theta x_1 + (1 - \theta)y_1, \ldots, \theta x_n + (1 - \theta)y_n]$ belongs to D.

Theorem 3-4   The intersection D of a family F of convex sets is a convex set.

Proof: Let $x = (x_1, \ldots, x_n)$ and $y = (y_1, \ldots, y_n)$ be two points in D; then x and y belong to each member of F. Hence $\theta x + (1 - \theta)y$ belongs to each member of F, since these members are convex. Hence $\theta x + (1 - \theta)y$ belongs to D.

Corollary: The set defined by $g_i(x) \leq 0$ when the $g_i(x)$, $i = 1, \ldots, m$, are convex is a convex set.

Remark: The last-named set is called the *feasible region* for $g_i(x) \leq 0$.

3-3 NONLINEAR PROGRAMMING

The nonlinear-programming problem is concerned with the maximization or minimization of a continuous and differentiable function of n

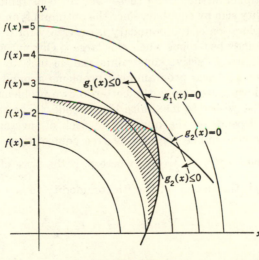

Fig. 3-3

real variables $f(x_1, \ldots, x_n)$, called the *objective function*, subject to m inequality constraints $g_i(x_1, \ldots, x_n) \leq 0$, $i = 1, 2, \ldots, m$, and to $x_j \geq 0$, $j = 1, \ldots, n$ (see Fig. 3-3). It is also frequently referred to as the *general* programming problem. We therefore use the symbol (G) to refer to this problem. We shall impose various restrictions on f and g_i.

The functions g_i, $i = 1, \ldots, m$, will be assumed to be convex. The last corollary asserts that the feasible region is convex. In general, solving (G) requires that a point from the feasible region can be found which optimizes f. The objective function f will be assumed to be convex (concave) if (G) involves the minimization (maximization) of the objective function f. These restrictions on f and g_i together lead to an important property of (G): Any local optimum is a global optimum of the objective function in the feasible region. This follows from the fact that the line segment joining a local optimum attained in the region to a global optimum also attained in the region must itself lie in the region because of convexity. Hence f must be a constant on this segment (the reader should demonstrate why). Thus a local optimum is global.

Ordinary differential-calculus methods for finding the extremum of an unconstrained function, when applied to f, may yield values which are in the region, in which case, one has a solution of (G). One usually begins by testing for an optimum in the region. This process itself may be consuming and costly. On the other hand if the optimum lies on the boundary, then the problem is no longer simple. If it were known which of the constraints determined the optimum, then one could apply the Lagrange-multiplier method using these "active" constraints after replacing the inequality sign by equality, since the optimum is on the boundary. In general, exploration of the boundary is necessary for determining the active set; this may be tedious when m is large. Our purpose is to study methods which lead to active constraints and, in fact, to the optimum.

Examples of nonlinear-programming problems abound in the literature. The following are among the better-known problems.

A monopolist wishes to maximize his revenue. The sale price p_j of a commodity depends on the quantity of each of several commodities which he produces—unlike the case of free competition. Thus, if the revenue is given by $f(x) = \sum_{j=1}^{n} p_j x_j$, where p_j is the price of the jth commodity, and x_j is the amount of the jth commodity to be produced, then

$$p_i = b_i - \sum_{j=1}^{n} a_{ij} x_j \qquad i = 1, \ldots, n \qquad (3\text{-}15)$$

If we substitute these quantities in f, we obtain a quadratic function to be maximized subject to nonnegativity constraints $x_j \geq 0$, $j = 1, \ldots, n$, and subject to capacity constraints which may be prescribed linearly. Thus, since $f(x)$ is nonlinear, we have a nonlinear program.

A similar problem may be formulated in which the cost of producing a commodity depends on the quantity produced. In this case the objective is to find the x_j which minimize the quadratic function $f(x)$ subject to $x_j \geq 0$ and subject to the constraints that the entire capacity should be used. These constraints can be expressed in terms of linear inequalities with nonnegative coefficients.

Another interesting example is to find a best linear approximation to a set of observed data (x_j, y_j), $j = 1, \ldots, n$, subject to constraints on the coefficients of the approximating linear expression. Thus the problem may take the following form: Find the a and b which minimize

$$f(a,b) = \sum_{j=1}^{n} [y_j - (a + bx_j)]^2 \tag{3-16}$$

subject to a, $b \geq 0$. Ordinary least-square methods may not be applicable because of the presence of constraints.

Yet another example is that of portfolio selection, in which an investor wishes to allocate a given sum of money C to different securities. He must maximize the expected return for a fixed variance or minimize the variance for a fixed return. If x_j, $j = 1, \ldots, n$, is the amount to be invested on the jth security, then $\sum_{j=1}^{n} x_j = C$. The total return is $\sum_{j=1}^{n} r_j x_j$, its expected value is $E = \sum_{j=1}^{n} \mu_j x_j$, and its variance is $\sum_{i,j=1}^{n} a_{ij} x_i x_j$, where (a_{ij}) is the covariance matrix of the random variable r_j, the return on the jth security, and μ_j is its expected value. Both (a_{ij}) and μ_j are given. Here the problem is either to maximize $\sum_{j=1}^{n} \mu_j x_j$ subject to $\sum_{i,j} a_{ij} x_i x_j \leq \sigma$, where σ is a given constant, and subject to $\sum_{j=1}^{n} x_j = C$, $x_j \geq 0$, or to minimize $\sum_{i,j} a_{ij} x_i x_j$ subject to $\sum_{j=1}^{n} \mu_j x_j = \mu$, where μ is given, and subject to $\sum_{j=1}^{n} x_j = C$, $x_j \geq 0$. These problems are obviously nonlinear, in the first case because of the presence of a quadratic constraint, and in the second because this constraint is the function to be minimized.

Problems in which profit is described by a nonlinear function of the variables to be maximized subject to constraints can also be nonlinear programs. As an illustration the profit function may be given by $f(x) = \sum_{i=1}^{n} a_i(i - e^{-b_i x_i})$, which must be maximized subject to $\sum_{i=1}^{n} x_i = C$, a_i, $b_i > 0$, $x_i \geq 0$.

As a final illustration we consider an example in chemical equilibrium.[73a] Conservation of mass for m gaseous elements used to form n compounds under constant pressure P, with a_{ij} atoms of the ith element in one molecule of the jth compound, and with b_i moles of the ith element and x_j moles of the jth compound in the mixture, gives

$$\sum_{j=1}^{n} a_{ij}x_j = b_i \qquad i = 1, \ldots, m$$

with
$$x_j \geq 0 \qquad j = 1, \ldots, n$$

with $n \geq m$. Subject to these constraints it is desired to minimize the total Gibbs free energy of the system:

$$\sum_{j=1}^{n} c_j x_j + \sum_{j=1}^{n} x_j \log \frac{x_j}{\sum_{i=1}^{n} x_i} \qquad (3\text{-}17)$$

where $c_j = F_j/RT + \log P$

F_j = Gibbs energy per mole of jth gas at temperature T and unit atmospheric pressure

R = universal gas constant

3-4 LINEAR PROGRAMMING

A special case of (G) is the linear-programming problem in which f and g_i, $i = 1, \ldots, m$, are linear, that is, $f(x) = c_1x_1 + \cdots + c_nx_n$, and

$$g_i(x) \leq 0 \qquad \text{or} \qquad a_{i1}x_1 + \cdots + a_{in}x_n - b_i \leq 0$$
$$i = 1, \ldots, m \qquad (3\text{-}18)$$

We shall not discuss this well-known problem except for a brief summary of the *simplex* process, the algorithm most frequently used for solving such a problem, sometimes also used as an auxiliary tool for solving a nonlinear-programming problem.

In the linear-programming problem, the feasible region is a polyhedron (since the g_i's are linear and hence are hyperplanes in n-dimensional space), the objective function defines a hyperplane which must touch the region, and the solution lies on at least one of its vertices, a theorem proved by Weyl[73] and intuitively obvious in three dimensions. Here it is conceivable that all the vertices can be tested for the solution, since the number of vertices is finite; however, shorter methods are used in practice. As we shall see later, in the case of (G), the solution does not necessarily lie on the intersection of the surfaces corresponding to the constraints; consequently, one does not have available such finite and exhaustive procedures.

In matrix notation a linear-programming problem requires that a column vector $x \geq 0$, i.e., with components $x_j \geq 0, j = 1, \ldots, n$, be found which satisfies the constraints $Ax \leq b$ and maximizes the linear function cx, where $A = (a_{ij})$, $i = 1, \ldots, m, j = 1, \ldots, n, c = (c_1, \ldots, c_n)$, and b is a column vector with components $b_1, \ldots, b_m$. With the original problem, known as the *primal*, is associated a *dual* problem $yA \geq c$, $y \geq 0$; yb is minimum where $y = (y_1, \ldots, y_m)$.

A well-known duality theorem of linear programming asserts that if either the primal or the dual has a solution, then the values of the objective functions of both problems at the optimum are the same. The solution of either problem can be obtained from the solution of the other. It is sometimes computationally convenient to solve the dual problem. More will be said about this theorem in the next section. For a proof see Ref. 61.

The simplex process currently used to solve linear programs begins by choosing basis vectors in m dimensions, where m is the number of inequalities. Successive bases are chosen; one is finally obtained which solves the problem. With each iteration, i.e., new basis, the value of the objective function improves towards the optimum value or, at worst, remains the same, for iterations in the feasible region. Since the number of possible bases is finite, it is clear that, barring pathological cases (e.g., a process known as *cycling* in which bases may recur), the optimum is reached in a finite number of steps. There are methods devised to avoid these difficulties. We shall now give some theory and illustrate the simplex process. The following existence theorems are well known.

Theorem If one feasible solution exists, then there exists a feasible solution (called a *basic feasible* solution) with, at most, m points P_i, with positive weights x_i and $n - m$ or more points P_i with $x_i = 0$.

Theorem If the values of the objective function for the class of feasible solutions have a finite upper bound, then there exists a maximum feasible solution which is a basic feasible solution.

The simplex method will be applied to an illustrative problem. The long method of calculation is used here to clarify the ideas. A tableau method will be illustrated in solving a quadratic programming problem in a later section. The constraint set for the problem is

$$\begin{aligned}
x_1 + 4x_2 + 2x_3 &\geq 5 \\
3x_1 + x_2 + 2x_3 &\geq 4 \\
x_i &\geq 0 \qquad i = 1, 2, 3
\end{aligned} \qquad (3\text{-}19)$$

and the form (cost function) to be minimized is

$$2x_1 + 9x_2 + x_3 \qquad (3\text{-}20)$$

Step 1: Change to equalities by introducing slack variables, non-negative variables used to reduce the inequalities to equalities.

$$\begin{aligned}
x_1 + 4x_2 + 2x_3 - x_4 &= 5 \\
3x_1 + x_2 + 2x_3 - x_5 &= 4 \\
-2x_1 - 9x_2 - x_3 &= \max
\end{aligned} \qquad \left.\begin{aligned} x_4 &\geq 0 \\ x_5 &\geq 0 \end{aligned}\right\} \text{ slack variables} \qquad (3\text{-}21)$$

Step 2: Write out the matrix

$$[P_1, P_2, P_3, P_4, P_5; P_0] = \begin{bmatrix} 1 & 4 & 2 & -1 & 0 & 5 \\ 3 & 1 & 2 & 0 & -1 & 4 \end{bmatrix} \qquad (3\text{-}22)$$

and write $c_1 = -2$ $c_2 = -9$ $c_3 = -1$ $c_4 = 0$ $c_5 = 0$

Step 3: Start by selecting a basis, which is a set of linearly independent vectors (i.e., the determinant of their matrix does not vanish) every other vector can be expressed as a linear combination of the basis vectors. A linear combination of vectors P_1 and P_2, for example, may be given as $aP_1 + bP_2$, where a and b are real numbers. In this case two vectors are needed to form a basis, as can be seen from the matrix

Remark: Using a set of artificial vectors as a basis avoids an initial choice of nonfeasible vectors, i.e., a set of vectors for which the corresponding variables are negative. A simple example of an artificial vector basis is (1,0) and (0,1).

Suppose that the initial choice is P_1 and P_2 as a basis. Express P_0 as a linear combination of them. Obtain $P_0 = P_1 + P_2$ here. In this problem the cost function will be negative for positive x_i. Hence, if a positive cost is obtained for P_0 as a combination of the basis vectors chosen, then at least one basis vector must be changed. Also express the remaining vectors as a linear combination of P_1 and P_2, obtaining

	P_1	P_2
P_1	1	0
P_2	0	1
P_3	$6/11$	$4/11$
P_4	$1/11$	$-3/11$
P_5	$-4/11$	$1/11$

(3-23)

Let β_{ij} be the coefficient in the above array whose first subscript is the same as that of the vector on top and whose second subscript is the same as that of the vector to its left. For example, to obtain P_4 as a linear combination of P_1 and P_2, write $P_4 = aP_1 + bP_2$. Note that $\beta_{14} = a$, $\beta_{24} = b$. To show how a and b are obtained, the relation among the vectors may be written as

$$\overset{P_4}{\begin{bmatrix} -1 \\ 0 \end{bmatrix}} = a \overset{P_1}{\begin{bmatrix} 1 \\ 3 \end{bmatrix}} + b \overset{P_2}{\begin{bmatrix} 4 \\ 1 \end{bmatrix}} = \begin{bmatrix} a \\ 3a \end{bmatrix} + \begin{bmatrix} 4b \\ b \end{bmatrix} = \begin{bmatrix} a + 4b \\ 3a + b \end{bmatrix} \quad (3\text{-}24)$$

or, by equating sides: $-1 = a + 4b$, $0 = 3a + b$. On solving these two simultaneous equations in a and b, one has $a = \frac{1}{11}$, $b = -\frac{3}{11}$. Similarly, all other vectors are expressed as linear combinations of the basis P_1 and P_2.

Step 4: Consider $z_j = \beta_{1j}c_1 + \beta_{2j}c_2$ $(j = 1, \ldots, 5)$.

Remark: In general, if the basis vectors have subscripts p, q, r, etc., write

$$z_j = \beta_{pj}c_p + \beta_{qj}c_q + \beta_{rj}c_r + \cdots \quad (3\text{-}25)$$

$z_1 = c_1 \qquad\qquad\quad = -2$	Compare with $c_1 = -2$
$z_2 = c_2 \qquad\qquad\quad = -9$	Compare with $c_2 = -9$
$z_3 = \quad \frac{6}{11}c_1 + \frac{4}{11}c_2 = -\frac{48}{11}$	Compare with $c_3 = -1$
$z_4 = \quad \frac{1}{11}c_1 - \frac{3}{11}c_2 = \frac{25}{11}$	Compare with $c_4 = 0$
$z_5 = -\frac{4}{11}c_1 + \frac{1}{11}c_2 = -\frac{1}{11}$	Compare with $c_5 = 0$

Compare z_j with c_j as indicated.

If $z_j \geq c_j$ for all j, the process is finished; that is, $P_0 = P_1 + P_2$, and the cost would be (when maximizing, negative) given as follows: If $P_0 = aP_1 + bP_2$, then the cost is $ac_1 + bc_2$. In this case one has

$$-2 - 9 = -11 \text{ units}$$

that is, 11 units for the minimum. If one were directly minimizing, then the criterion would be $z_j \leq c_j$.

Remark: It is clear that $z_j < c_j$ for some j. Then consider max $(c_j - z_j)$. In this case $c_3 - z_3 = $ max. Hence for a next choice of basis one uses P_3 along with either P_1 or P_2. One decides which of P_1 or P_2 to use by the following method (if $c_j \geq z_j$ for all j, and $\beta_{ij} \leq 0$ for all i, the maximum feasible solution is infinite):

Step 5: Express P_0 as a linear combination of P_1 and P_2:

$$P_0 = P_1 + P_2$$

Express P_3 as a linear combination of P_1 and P_2, and multiply by θ:

$$\theta P_3 = \theta \tfrac{6}{11} P_1 + \theta \tfrac{4}{11} P_2$$

Subtract the second equation from the first:

$$P_0 = \theta P_3 + (1 - \tfrac{6}{11}\theta)P_1 + (1 - \tfrac{4}{11}\theta)P_2$$

Choose
$$\theta = \min\left(\frac{1}{\frac{6}{11}}, \frac{1}{\frac{4}{11}}\right) = \frac{11}{6}$$

Hence obtain

$$P_0 = 1\tfrac{1}{6}P_3 + \tfrac{2}{6}P_2$$

and the new basis will consist of P_2 and P_3. The cost in this case is given by

$$1\tfrac{1}{6}c_3 + \tfrac{2}{6}c_2 = 1\tfrac{1}{6}(-1) + \tfrac{2}{6}(-9) = -2\tfrac{9}{6}$$

which is clearly an improvement on the previous cost, since one is maximizing.

Remark: In general, if $P_0 = \alpha_1 P_1 + \cdots + \alpha_q P_q$, $P_1, \ldots, P_q$ being the basis vectors, and if $c_j - z_j = \max$ yields P_j for the new vector, and

$$P_j = \beta_{1j}P_1 + \cdots + \beta_{qj}P_q \tag{3-26}$$

then choose

$$\theta = \min_i \left(\frac{\alpha_i}{\beta_{ij}}\right) \qquad \beta_{ij} > 0 \tag{3-27}$$

In this manner one of the vectors in

$$P_0 = \theta P_j + (\alpha_1 - \theta\beta_{1j})P_1 + \cdots + (\alpha_q - \theta\beta_{qj})P_q \tag{3-28}$$

is eliminated.

Step 6: Again express the remaining vectors as a linear combination of P_2 and P_3:

	P_2	P_3
P_1	$-\tfrac{2}{3}$	$1\tfrac{1}{6}$
P_2	1	0
P_3	0	1
P_4	$-\tfrac{1}{3}$	$\tfrac{1}{6}$
P_5	$\tfrac{1}{3}$	$-\tfrac{2}{3}$

Consider now $z_j = \beta_{2j}c_2 + \beta_{3j}c_3$.

$$z_1 = (-\tfrac{2}{3})(-9) + 1\tfrac{1}{6}(-1) = 2\tfrac{5}{6} \qquad \text{Compare with } c_1 = -2$$
$$z_2 = c_2 \qquad\qquad\qquad\qquad = -9 \qquad \text{Compare with } c_2 = -9$$
$$z_3 = c_3 \qquad\qquad\qquad\qquad = -1 \qquad \text{Compare with } c_3 = -1$$
$$z_4 = (-\tfrac{1}{3})(-9) + \tfrac{1}{6}(-1) = 1\tfrac{7}{6} \qquad \text{Compare with } c_4 = 0$$
$$z_5 = \tfrac{1}{3}(-9) - \tfrac{2}{3}(-1) = -\tfrac{7}{3} \qquad \text{Compare with } c_5 = 0$$

Here one has $c_5 - z_5 = \max$. Hence P_5 is to replace either P_2 or P_3 in the next choice of basis. To decide which, write

$$P_0 = 1\tfrac{1}{6}P_3 + \tfrac{2}{6}P_2$$
$$\theta P_5 = -\theta\tfrac{2}{3}P_3 + \theta\tfrac{1}{3}P_2$$

Subtracting the second from the first, one has

$$P_0 = \theta P_5 + (1\tfrac{1}{6} + \tfrac{2}{3}\theta)P_3 + (\tfrac{2}{6} - \tfrac{1}{3}\theta)P_2$$

Thus $\theta = 1$, since one considers only those values of β_{ij} that are greater than zero. This choice gives

$$P_0 = P_5 + 1\tfrac{5}{6}P_3$$

and the cost is given by

$$1(0) + 1\tfrac{5}{6}(-1) = -\tfrac{5}{2}$$

which is greater than the preceding costs.

Step 7: Again, express the remaining vectors as a linear combination of P_3 and P_5:

	P_3	P_5	
P_1	$\tfrac{1}{2}$	-2	
P_2	2	3	(3-29)
P_3	1	0	
P_4	$-\tfrac{1}{2}$	-1	
P_5	0	1	

Consider once more $z_j = \beta_{3j}c_3 + \beta_{5j}c_5$.

$$z_1 = -\tfrac{1}{2} \qquad \text{Compare with } c_1 = -2$$
$$z_2 = -2 \qquad \text{Compare with } c_2 = -9$$
$$z_3 = -1 \qquad \text{Compare with } c_3 = -1$$
$$z_4 = \tfrac{1}{2} \qquad \text{Compare with } c_4 = 0$$
$$z_5 = 0 \qquad \text{Compare with } c_5 = 0$$

It is clear that all $z_j \geq c_j$, and the solution is complete. In other words, since $P_0 = 1\tfrac{5}{6}P_3 + P_5$, one has $(x_1,x_2,x_3,x_4,x_5) = (0,0,1\tfrac{5}{6},0,1)$. The total cost is $\tfrac{5}{2}$ (the sign having been changed to obtain the minimum). The vector P_5, which has zero cost, contributes nothing. There is no item corresponding to it.

Sometimes an integer solution is required. Note that the simplex process guarantees no integer values for the solution vector. There is no a priori reason why any vertex of the polyhedron of feasible solutions should be a lattice point, i.e., have integer coordinates. Approximating the coordinate values with integers does not necessarily yield a feasible vector, since the vertex yielding a solution may be nearest to a lattice point outside the region of feasible solutions, and the nearest feasible lattice point may not be near enough to justify any approximation.

Rapid progress has been made in providing methods of solving such problems.

Solution of the Dual

If the basis leading to a maximum solution of the primal has been obtained by the simplex process, then the solution of the dual is obtained as follows:

Let B be the matrix whose column vectors constitute the basis vectors leading to the solution with inverse B^{-1}, and let C^* be the cost vector corresponding to the solution basis. Then the solution of the dual is $y = C^*B^{-1}$, where the components of the vector y comprise the solution $(y_1^0, \ldots, y_m^0)$.

Proof: Note that $z_j = C^*(B^{-1}A)$. Since $z_j - c_j \geq 0$, $C^*(B^{-1}A) - c \geq 0$, which becomes (on substituting $y = C^*B^{-1}$) $yA - c \geq 0$ or $yA \geq c$, which is the dual problem.

Given a linear-programming problem with three constraints and n variables, it is possible to obtain the simplex solution by forming the dual and solving geometrically. To show this, suppose for simplicity that (y_1, y_2, y_3) are the values constituting the solution. Then P_1, P_2, and P_3 are the basis vectors in the simplex solution of the dual; when the submatrix comprised of these vectors is taken from the matrix of the problem, inverted, and multiplied by the vector whose components are the cost coefficients corresponding to P_1, P_2, P_3, the solution (x_1, x_2, x_3) of the primal is obtained; that is,

$$(x_1, x_2, x_3) = (\text{cost vector})(B^{-1})$$

Once the values of x_1, x_2, and x_3 are known, the corresponding vectors from the matrix constitute the basis vectors leading to the solution of the primal. A similar argument applies for the simpler case of two constraints in the primal.

Exercise 3-3 Following the above description of the simplex process, show that

$$2x_1 + 3x_2 + x_3$$

is minimized at $(0, \frac{7}{20}, \frac{51}{40})$ subject to the constraints

$$4x_1 + x_2 + 6x_3 \geq 8$$
$$3x_1 + 7x_2 + 2x_3 \geq 5$$
$$x_1, x_2, x_3 \geq 0$$

Exercise 3-4 Obtain the solution to Exercise 3-3 by geometric means; i.e., draw a picture of the problem.

Exercise 3-5 Find the dual of the foregoing problem, and obtain the solution geometrically and algebraically using the solution of the primal.

3-5 CHARACTERIZATION OF THE OPTIMUM ON THE BOUNDARY; SADDLE POINTS; DUALITY

In this section we shall give necessary and sufficient conditions for an optimum on the boundary. These conditions occasionally enable one to obtain this optimum. However, they provide basic ideas used in developing iterative algorithms. This characterization is obtained through concepts involving gradients and saddle points and through differential equations. The greater the variety of characterization theorems, the greater is the possibility of constructing ingenious algorithms, perhaps by borrowing ideas from related fields.

We shall be using the lagrangian function $F(x,\lambda)$ defined early in the chapter for a maximization problem (G), except that instead of a plus sign we use a minus sign for reasons given below. We have

$$F(x,\lambda) = f(x) - \lambda G(x) \tag{3-30}$$

where
$$\lambda = (\lambda_1, \ldots, \lambda_m)$$

and where
$$G(x) = \begin{pmatrix} g_1(x) \\ \cdots \\ g_m(x) \end{pmatrix} \tag{3-31}$$

Note that $G(x) \leq 0$ implies that $g_i(x) \leq 0$, $i = 1, \ldots, m$.

Remark: As to the reason for the negative sign, note that in the case of equality constraints it makes no difference whether one defines $F(x,\lambda)$ with a plus sign or a minus sign, since in any case the $g_i(x)$ are zero. In a programming problem, however, the constraints may be given as ≥ 0 or ≤ 0. It turns out that one takes the positive sign when all the inequalities are greater than or equal to zero, and hence $g_i(x) \leq 0$ must be put in the form $-g_i(x) \geq 0$. The choice of the sign is not a simple matter of convention. Our object is to define F in such a way as to be able to obtain useful characterization theorems for a maximum of (G) in terms of F.

The next theorem, which does not use $F(x,\lambda)$, proves a result which confirms this choice of sign. Since the theorem will be used in proving later theorems involving F, it is natural that F must conform to the requirements of this theorem. Subsequent concepts involving saddle points of $F(x,\lambda)$ also indicate that a point x which yields an absolute maximum to (G) yields an absolute maximum to F for some $\bar\lambda \geq 0$ provided the $g_i(x)$ are convex. Since the sum of concave functions is concave, $F(x,\lambda)$ is concave with respect to x for $\lambda = \bar\lambda$. This fact provides an easy check for the sign of λ necessary to produce concavity.

We now define

$$\nabla G(x) = \begin{pmatrix} \dfrac{\partial g_1}{\partial x_1} & \cdots & \dfrac{\partial g_1}{\partial x_n} \\ \cdots & \cdots & \cdots \\ \dfrac{\partial g_m}{\partial x_1} & \cdots & \dfrac{\partial g_m}{\partial x_n} \end{pmatrix} \tag{3-32}$$

We shall use primes to indicate the transposes of vectors and matrices.

In forming F, we have the additional condition $x \geq 0$. This can be included in the definition of $F(x,\lambda)$ by treating each component of x as an additional constraint. For convenience let us denote x_1 by $-g_{m+1}$, x_2 by $-g_{m+2}, \ldots, x_n$ by $-g_{m+n}$. Then we have the lagrangian function without any additional constraints,

$$F(x,\lambda) = f(x) - \sum_{i=1}^{m+n} \lambda_i g_i(x) \tag{3-33}$$

For convenience we define $\lambda_{m+1} \equiv \mu_1, \ldots, \lambda_{m+n} \equiv \mu_n$, and hence $\mu \equiv (\mu_1, \ldots, \mu_n)$ and $\lambda = (\lambda_1, \ldots, \lambda_m)$.

Note that if $\bar{x}$ is a point on the boundary, then it need not yield equality for all the constraints. Since it is a feasible point, the constraints on whose boundary it does not lie are satisfied strictly; that is, $g_i(\bar{x}) < 0$. The remaining constraints give $g_i(\bar{x}) = 0$.

We shall, hereafter, restrict our attention to those feasible regions R which satisfy the following Kuhn-Tucker constraint qualification:

For every boundary point x^0 and for every sufficiently small displacement dx from x^0 such that

$$\nabla g_i(x^0) \, dx < 0$$

whenever $g_i(x^0) = 0$, and $dx_j \geq 0$ whenever $x_j^0 = 0$, the vector dx lies in R. (By "lying in R" we mean that the vector dx is a positive multiple of a tangent vector to some differentiable curve in R with x^0 as the initial point.) We now prove:

Theorem 3-5 A necessary condition that $f(x)$ attains its maximum at a boundary point $\bar{x}$ of $G(x) \leq 0$, with $x \geq 0$, is that there exist $\lambda \geq 0$ and $\mu \geq 0$ such that

$$\nabla f(\bar{x}) = \lambda \cdot \nabla G(\bar{x}) - \mu \tag{3-34}$$

where, if $\bar{x}_j > 0$, then $\mu_j = 0$, and if $g_i(\bar{x}) < 0$, then $\lambda_i = 0$.

Exercise 3-6 Verify that our definition of F is in harmony with this theorem and hence that the theorem requires the vanishing of the derivatives of F with respect to the x_i as a necessary condition for a maximum.

Remark: The theorem obviously also holds if $\bar{x}$ is an interior point, for in this case $\bar{x} > 0$, $G(\bar{x}) < 0$, so that $\mu = 0$ and $\lambda = 0$.

Proof: Let $\bar{x}$ maximize $f(x)$ on the boundary, and let $\alpha(\bar{x})$ be the set of subscripts of those $g_i(\bar{x})$ which satisfy $g_i(\bar{x}) = 0$.

Define the region

$$R = \{x : G(x) \leq 0, \, x \geq 0\} \tag{3-35}$$

and note that if x is any point of R in a neighborhood of $\bar{x}$ not on the boundary, then the vector $x - \bar{x}$ points inward in R. However, since $\nabla f(x)$ points in the direction of maximum increase, and since we are at the maximum, it follows that $\nabla f(\bar{x})$ projected along $x - \bar{x}$ points in a direction of decrease. Thus

$$\nabla f(\bar{x}) \cdot (x - \bar{x})' \leq 0 \tag{3-36}$$

Similarly, the gradient of each $g_i(x)$, where i is a member of α, points in the direction of maximum increase of $g_i(x)$ which is in the region $g_i(x) \geq 0$ away from R [note that $g_i(x)$ decreases from the boundary into R], and we have

$$\nabla g_i(\bar{x}) \cdot (x - \bar{x})' \leq 0 \qquad \text{for } i \text{ in } \alpha \tag{3-37}$$

Both the last two inequalities are valid if $\nabla f(\bar{x})$ or $\nabla g_i(\bar{x})$ is zero.

Thus we have

$$\nabla f(\bar{x}) \cdot x' \leq \nabla f(\bar{x}) \cdot \bar{x}' \tag{3-38}$$

and
$$\nabla g_i(\bar{x}) \cdot x' \leq \nabla g_i(\bar{x}) \cdot \bar{x}' \qquad i \, \epsilon \, \alpha \tag{3-39}$$

The first relation implies that $\bar{x}$ is a maximum of $\nabla f(\bar{x}) \cdot x'$. Hence consider the problem of finding $\bar{x} \geq 0$ which maximizes

$$\nabla f(\bar{x}) \cdot x' \tag{3-40}$$

subject to the constraints

$$\nabla g_i(\bar{x}) \cdot x' \leq \nabla g_i(\bar{x}) \cdot \bar{x}' \qquad i \, \epsilon \, \alpha \tag{3-41}$$

This is a linear-programming problem, and its solution is evidently $x = \bar{x} = \bar{x}$. It has a dual in which, for our purposes, we use the new variables $\lambda_i \geq 0$ instead of the variables $y_i \geq 0$ which we used in the previous section. The dual requires that $\lambda_i \geq 0$ be found which minimize

$$\sum_{i \, \epsilon \, \alpha} \lambda_i [\nabla g_i(\bar{x}) \cdot \bar{x}'] \tag{3-42}$$

subject to the linear constraints

$$\sum_{i \, \epsilon \, \alpha} \lambda_i \nabla g_i(\bar{x}) \geq \nabla f(\bar{x}) \tag{3-43}$$

Consequently the problem of finding $\bar{x} = (\bar{x}_1, \ldots, \bar{x}_n)$ which maximizes the initial programming problem has been reduced to the problem of finding $\lambda_i \geq 0$ which solve the above linear-programming problem.

Now if we take $\lambda_i = 0$ for i not in α, inequalities (3-43) may be written in the form

$$\nabla f(\bar{x}) = \lambda \cdot \nabla G(\bar{x}) - \mu \tag{3-44}$$

where $\mu = (\mu_1, \ldots, \mu_n) \geq 0$. If we take the scalar product of this expression with $\bar{x}'$, we obtain

$$\mu \cdot \bar{x}' = (\lambda \cdot \nabla G) \cdot \bar{x}' - \nabla f(\bar{x}) \cdot \bar{x}' = 0 \tag{3-45}$$

By the duality theorem the two objective functions appearing in the middle expression must be equal, and hence their difference is zero. Thus $\mu_j = 0$ whenever $\bar{x}_j > 0$.

As we shall see in Theorem 3-8, the conditions of this theorem are also sufficient if $f(x)$ is concave and $G(x)$ is convex.

Remark: The Kuhn-Tucker condition is aimed at removing degeneracies in the gradients of the constraints by allowing us the use of the statement immediately after (3-35). For example, two constraints may meet in such a way as to form a cusp at a point x^0, and hence their gradients coincide. In that case (if they are the only constraints meeting at x^0) we do not have a basis; hence ∇f cannot be obtained as a linear combination of the constraints. An example of this is given by

$$g_1(x_1, x_2) \equiv x_2 + (x_1 - 1)^3 \leq 0$$
and $$x_1 \geq 0 \qquad x_2 \geq 0$$

at the point $(1,0)$. The gradient of g_1 at $(1,0)$ is $(0,1)$, which coincides (with opposite direction) with the gradient of $g_2(x_1, x_2) = -x_2$ at $(1,0)$.

Corollary (Gibbs): A necessary condition that $(x_1^0, \ldots, x_n^0)$ be a maximum of $\sum_{i=1}^{n} f_i(x_i)$ subject to $\sum_{i=1}^{n} x_i = 1$ and $x_i \geq 0$, $i = 1, \ldots, n$, is that there exists a real number λ such that

$$\begin{aligned} f_i'(x_i^0) &= \lambda & \text{if} \quad x_i^0 > 0 \\ f_i'(x_i^0) &\leq \lambda & \text{if} \quad x_i^0 = 0 \end{aligned} \tag{3-46}$$

If the $f_i(x_i)$ are concave, and hence $f_i''(x_i) \leq 0$ for $0 \leq x_i \leq 1$, the condition is also sufficient.

Remark: If the functions are concave, a sufficient condition for $(x_1^0, \ldots, x_n^0)$ to be a solution is that the above λ satisfy

$$\begin{aligned} f_i'(0) &> \lambda & \text{if and only if} \quad x_i^0 > 0 \\ f_i'(0) &\leq \lambda & \text{if and only if} \quad x_i^0 = 0 \end{aligned} \tag{3-47}$$

Our objective now is to characterize the maximum of (G) through the use of the lagrangian function and the concept of a saddle point.

Saddle Point

The point $(\bar{x},\bar{\lambda}) \equiv (\bar{x}_1, \ldots ,\bar{x}_n,\bar{\lambda}_1, \ldots ,\bar{\lambda}_n)$ is a saddle point of $F(x,\lambda)$ if

$$F(x,\bar{\lambda}) \leq F(\bar{x},\bar{\lambda}) \leq F(\bar{x},\lambda) \tag{3-48}$$

for all x and λ.

To see why the term *saddle point* is used, note that a saddle rises away from its center in either direction parallel to the horse's spine and falls away from its center in either direction perpendicular to the spine. If the saddle surface is given by the equation

$$z = F(x,\lambda)$$

(where the λ axis is horizontal and parallel to the spine, the x axis is horizontal and perpendicular to the spine, and the positive z axis is directed vertically upward), then F will have a saddle point at the pair (x,λ) yielding its center. The function $F(x,\lambda) = \lambda^2 - x^2$ has a saddle point at $(0,0)$. Of course, in the general case x and λ are vectors.

In studying a maximization problem (G) we use the lagrangian function F, and we require that the saddle condition be satisfied for nonnegative values of x and λ. This condition can also be written as

$$F(x,\lambda) = \max_{x \geq 0} \min_{\lambda \geq 0} F(x,\lambda) = \min_{\lambda \geq 0} \max_{x \geq 0} F(x,\lambda) \tag{3-49}$$

The object is to show that the component $\bar{x}$ of a saddle point solves the maximization problem (G). Thus if one were able to obtain a saddle point, one would have a solution to (G). The saddle point, at least in theory, may be obtained as follows: We define

$$\nabla_x F(x,\lambda) = \left(\frac{\partial F}{\partial x_1}, \ldots , \frac{\partial F}{\partial x_n} \right)$$

$$\nabla_\lambda F(x,\lambda) = \left(\frac{\partial F}{\partial \lambda_1}, \ldots , \frac{\partial F}{\partial \lambda_m} \right)$$

$$\nabla_x F(\bar{x},\lambda) = \nabla_x F(x,\lambda) \Big|_{x = \bar{x}}$$

$$\nabla_\lambda F(x,\bar{\lambda}) = \nabla_\lambda F(x,\lambda) \Big|_{\lambda = \bar{\lambda}}$$

Theorem 3-6   A necessary condition for $(\bar{x},\bar{\lambda})$ to be a saddle point of F is that $\bar{x}$ and $\bar{\lambda}$ satisfy

$$\begin{array}{lll} \nabla_x F(\bar{x},\bar{\lambda}) \leq 0 & \nabla_x F(\bar{x},\bar{\lambda}) \cdot \bar{x}' = 0 & \bar{x} \geq 0 \\ \nabla_\lambda F(\bar{x},\bar{\lambda}) \geq 0 & \nabla_\lambda F(\bar{x},\bar{\lambda}) \cdot \bar{\lambda}' = 0 & \bar{\lambda} \geq 0 \end{array} \tag{3-50}$$

The foregoing two sets of conditions, together with the two conditions [analogous to series expansions in the neighborhood of $(\bar{x},\bar{\lambda})$ up to linear terms]

$$F(x,\bar{\lambda}) \leq F(\bar{x},\bar{\lambda}) + \nabla_x F(\bar{x},\bar{\lambda}) \cdot (x - \bar{x})'$$
$$F(\bar{x},\lambda) \geq F(\bar{x},\bar{\lambda}) + \nabla_\lambda F(\bar{x},\bar{\lambda}) \cdot (\lambda - \bar{\lambda})' \tag{3-51}$$

are sufficient for $(\bar{x},\bar{\lambda})$ to be a saddle point with $x \geq 0$, $\lambda \geq 0$.

Proof: The first two conditions are necessary. To see this, note that

$$\nabla_x F(\bar{x},\bar{\lambda}) \cdot (x - \bar{x})' \leq 0 \tag{3-52}$$

which is true since $\nabla_x F(\bar{x},\bar{\lambda})$ is in the direction of maximum increase along the x values and is the component of the gradient of F with respect to x at $\bar{x}$, and $\bar{x}$ yields the maximum. Hence along any vector $(x - \bar{x})$, where $x \geq 0$, in the feasible region, $\nabla_x F(\bar{x},\lambda)$ points in a direction of decrease. To elucidate further, the total derivative with respect to x, that is,

$$dF = \sum_{i=1}^{n} \frac{\partial F}{\partial x_i}\bigg|_{x_i = \bar{x}_i} dx_i$$

with $dx_i = x_i - \bar{x}_i$, $x_i \geq 0$, must be nonpositive since F decreases, and must be the same as (3-52). It is certainly possible to choose all but one of the x_i to be equal to $\bar{x}_i$. Let the exception be x_k. Then the expression for the total derivative gives

$$\frac{\partial F}{\partial x_k}\bigg|_{x_k = \bar{x}_k,\lambda} \cdot (x_k - \bar{x}_k) \leq 0$$

If $\quad \bar{x}_k = 0 \quad$ then $\quad \dfrac{\partial F}{\partial x_k}\bigg|_{x_k = \bar{x}_k,\lambda} \leq 0 \quad$ since $\quad x_k \geq 0$

On the other hand, if $\bar{x}_k > 0$, then x_k can be chosen such that $x_k \geq \bar{x}_k$ or $\bar{x}_k > x_k \geq 0$. The only way in which the above inequality can be satisfied for all choices of $x_k \geq 0$ is that

$$\frac{\partial F}{\partial x_k}\bigg|_{x_k = \bar{x}_k} = 0$$

Consequently we have

$$\nabla_x F(\bar{x},\bar{\lambda}) \leq 0 \tag{3-53}$$

since $\bar{x} > 0$ yields the maximum, and

$$\nabla_x F(\bar{x},\bar{\lambda}) \cdot \bar{x}' = 0 \tag{3-54}$$

as we have just shown. One can similarly prove the remaining part of the condition, since $\bar{\lambda}$ yields the minimum value with respect to λ for $x = \bar{x}$.

Exercise 3-7   Finish the proof of the necessary condition, and prove sufficiency.

Theorem 3-7   A necessary (sufficient) condition that $\bar{x}$ be a solution of the maximum nonlinear-programming problem is that there exists some $\bar{\lambda}$ such that $\bar{x}$ and $\bar{\lambda}$ satisfy the necessary (first three sufficiency) conditions of Theorem 3-6.

Remark: This theorem reduces the problem of maximizing a function subject to constraints to that of minimizing a single function, i.e., the lagrangian, with respect to λ after taking its maximum with respect to x.   Thus one is able to apply iterative techniques to minimize the lagrangian with respect to λ [Ref. 68].

Proof: To prove necessity, we note from Theorem 3-5 that at $\bar{x}$ we have

$$\nabla f(\bar{x}) \equiv \lambda \cdot \nabla G(\bar{x}) - \mu \tag{3-55}$$

Hence
$$\nabla_x F(\bar{x},\lambda) \equiv \nabla f(\bar{x}) - \lambda \cdot \nabla G(\bar{x}) = -\mu \leq 0 \tag{3-56}$$

and
$$\nabla_x F(\bar{x},\lambda) \cdot \bar{x}' = 0 \tag{3-57}$$

since
$$\mu \cdot \bar{x}' = 0$$

But
$$\nabla_\lambda F(x,\bar{\lambda}) = -G(\bar{x}) \geq 0$$

and from Theorem 3-6

$$\nabla_\lambda F(\bar{x},\bar{\lambda}) \cdot \bar{\lambda} = -G(\bar{x}) \cdot \bar{\lambda}' = 0 \tag{3-58}$$

Thus we have necessity.

To prove sufficiency, note from the third sufficient condition that for $\bar{\lambda} \geq 0$ we have

$$F(x,\bar{\lambda}) = f(x) - \bar{\lambda} \cdot G(x) \leq F(\bar{x},\bar{\lambda}) + \nabla_x F(\bar{x},\lambda) \cdot (x - \bar{x})'$$
$$\leq F(\bar{x},\bar{\lambda}) = f(\bar{x}) - \bar{\lambda} \cdot G(\bar{x}) = f(\bar{x}) \tag{3-59}$$

and, since $-\bar{\lambda} \cdot G(x) \geq 0$, we have $f(x) \leq f(\bar{x})$, and thus $\bar{x}$ is the maximum.

Theorem 3-8 (*Equivalence Theorem*)   Let $f(x)$ be concave and $g_i(x)$, $i = 1, \ldots, m$, be convex (both differentiable).   A necessary and sufficient condition that $\bar{x}$ maximizes f subject to

$$g_i(x) \leq 0 \qquad x_j \geq 0 \qquad i = 1, \ldots, m; j = 1, \ldots, n$$

is that $\bar{x}$ and some $\bar{\lambda}$ comprise a saddle point for the lagrangian function $F(x,\lambda)$.

Proof: Since $G(x)$ is convex, we have, from Theorem 3-2,

$$G(x) \geq G(\bar{x}) + \nabla G(\bar{x}) \cdot (x - \bar{x})'$$
$$f(x) \leq f(\bar{x}) + \nabla f(\bar{x}) \cdot (x - \bar{x})' \tag{3-60}$$

for all x, $\bar{x} \geq 0$.   For any $\bar{\lambda} \geq 0$,

$$F(x,\bar{\lambda}) = f(x) - \bar{\lambda} \cdot G(x) \leq F(\bar{x},\bar{\lambda}) + \nabla_x F(\bar{x},\bar{\lambda}) \cdot (x - \bar{x})' \tag{3-61}$$

after the appropriate substitutions. Thus the third sufficiency condition of Theorem 3-6 is satisfied in this case. Now the fourth sufficiency condition always holds because of linearity in λ; that is,

$$F(\bar{x},\lambda) = F(\bar{x},\bar{\lambda}) + \nabla_\lambda F(\bar{x},\bar{\lambda}) \cdot (\lambda - \bar{\lambda})' \tag{3-62}$$

holds identically. Because of Theorem 3-7, all the conditions of Theorem 3-6 are necessary and sufficient in this case. Thus we have the equivalence between concave programming and the saddle-value problem.

Remark: Note that the duality theorem of linear programming, which was used to prove Theorem 3-5, can be deduced from Theorem 3-8 by defining, for the primal,

$$F_p(x,\lambda) = \sum_{j=1}^{n} c_j x_j + \sum_{i=1}^{m} b_i \lambda_i - \sum_{j=1}^{n} \sum_{i=1}^{m} a_{ij} \lambda_i x_j \tag{3-63}$$

and, for the dual,

$$F_d(\lambda,x) = - \sum_{i=1}^{m} b_i \lambda_i - \sum_{j=1}^{n} c_j x_j + \sum_{i=1}^{m} \sum_{j=1}^{n} a_{ij} \lambda_i x_j \tag{3-64}$$

Then $F_p = -F_d$, and if either one has a saddle point, then so does the other, and the duality theorem follows from Theorem 3-8. But of course the duality theorem was used to prove Theorem 3-8.

The following theorem[45] enables numerical solution of the saddle-value problem by solving a system of differential equations.

Theorem 3-9 Let $F(x,\lambda)$ be strictly concave in $x \geq 0$ and convex in $\lambda \geq 0$, and suppose it has a saddle point. Then as $t \to \infty$, the solution $[x(t),\lambda(t)]$ of the system

$$\frac{dx_j}{dt} = \begin{cases} 0 & \text{if } x_j = 0 \text{ and } F_{x_j} < 0 \\ F_{x_j}[x(t),\lambda(t)] & \text{otherwise} \end{cases}$$

$$\frac{d\lambda_i}{dt} = \begin{cases} 0 & \text{if } \lambda_i = 0 \text{ and } F_{\lambda_i} > 0 \\ -F_{\lambda_i}[x(t),\lambda(t)] & \text{otherwise} \end{cases}$$

$$j = 1, \ldots, n; i = 1, \ldots, m \tag{3-65}$$

with an arbitrary nonnegative initial value $x^{(0)} \equiv x(t_0)$, $\lambda^{(0)} \equiv \lambda(t_0)$ converges to a saddle point of F.

Exercise 3-8 Show that the component $\bar{x}$ of a saddle point of F is unique by assuming the opposite and using strict concavity and the definition of a saddle point.

Exercise 3-9 Show that if $(\bar{x},\bar{\lambda})$ is a singular solution of (3-65), that is, if $dx_j/dt = 0$, $d\lambda_i/dt = 0$ at this point, then $(\bar{x},\bar{\lambda})$ is a saddle point of F; that is, verify the two necessary conditions of Theorem 3-6.

Exercise 3-10   To prove convergence of solutions of the differential equations to $(\bar{x},\bar{\lambda})$, let

$$r^2(t) = \sum_{j=1}^{n} (x_i(t) - \bar{x}_i)^2 + \sum_{i=1}^{m} (\lambda_i(t) - \bar{\lambda}_i)^2 \tag{3-66}$$

be the euclidean distance in the space $x \geq 0$, $\lambda \geq 0$ from $(\bar{x},\bar{\lambda})$ to any other point, and show that

$$\frac{dr}{dt} < 0 \qquad \text{for all } t \text{ for which} \qquad x(t) \neq \bar{x} \tag{3-67}$$

then show that $dr/dt < 0$ for all t by considering the sums over indices for which $\dot{x}_j$ and $\dot{\lambda}_i$ are different from zero for $t \geq 0$. Now use the fact that F satisfies the last two sufficiency conditions of Theorem 3-6, and combine these conditions to obtain

$$\sum_{j=1}^{n} (x_j - \bar{x}_j)F_{x_j} - \sum_{i=1}^{m} (\lambda_i - \bar{\lambda}_i)F_{\lambda_i} < F(x,\bar{\lambda}) - F(\bar{x},\lambda) < 0 \tag{3-68}$$

The strict inequality (3-68) holds because of the saddle-point property with strict concavity. If the sums on the left are taken with respect to the same indices at which nonvanishing occurs for $t \geq 0$ as described above, one obtains the same result expressed by (3-68) by substituting in both members of the inequality $\bar{x}_j$ and $\bar{\lambda}_i$ for x_j and λ_i with these indices. Let x^* and λ^* be the totality of these $\bar{x}_j$ and $\bar{\lambda}_i$ to be substituted, and use $x = x^*$, $\lambda = \lambda^*$ in the middle expression above. Conclude that $dr/dt < 0$ for $0 \leq t < \infty$ whenever $(x,\lambda) \neq (\bar{x},\bar{\lambda})$. Thus r is monotone decreasing, and hence, as $t \to \infty$, all solutions starting from a nonnegative initial point converge to the unique saddle point $(\bar{x},\bar{\lambda})$.

The following are known as *duality* theorems. They apply to the dual problems:

1. *The primal problem:* Maximize the concave differentiable function $f(x)$ subject to the convex differentiable inequality constraints $G(x) \leq 0$ and $x \geq 0$.

2. *The dual problem:* Minimize $F(x,\lambda) = f(x) - \lambda G(x)$ or, equivalently, maximize $-F(x,\lambda)$ subject to

$$\nabla_x[f(x) - \lambda G(x)] = 0 \qquad \text{and} \qquad \lambda \geq 0 \cdot$$

Note that there is no explicit mention that $x \geq 0$ be true.

Theorem 3-10 (*Wolfe, Huard-Mangasarian*) If $\bar{x}$ maximizes $f(x)$ in the primal problem, then there exists a $\bar{\lambda} \geq 0$ such that $(\bar{x},\bar{\lambda})$ is a solution of the dual problem. Conversely, if $(\bar{x},\bar{\lambda})$ minimize $F(x,\lambda)$ in the dual problem, and if the matrix of second partial derivatives $\{\partial^2 F/\partial x_i\partial x_j\}$ evaluated at $(\bar{x},\bar{\lambda})$ has an inverse, then $\bar{x}$ maximizes $f(x)$ in the primal problem. In both cases $\max\limits_{x} f(x) = \min\limits_{x,\lambda} F(x,\lambda)$.

Proof: If $\bar{x}$ solves the primal problem, and $\bar{\lambda}$ exists satisfying Theorem 3-5, then since f is concave, G is convex, $-\bar{\lambda} \cdot G(\bar{x}) = 0$, $\lambda \geq 0$, and $G(x) \leq 0$,

$$
\begin{aligned}
F(\bar{x},\bar{\lambda}) - F(x,\lambda) &= f(\bar{x}) - f(x) - \bar{\lambda} \cdot G(\bar{x}) + \lambda \cdot G(x) \\
&\leq \nabla f(x) \cdot (\bar{x} - x)' + \lambda \cdot G(x) \\
&\leq \nabla f(x) \cdot (\bar{x} - x)' + \lambda \cdot G(x) - \lambda \cdot G(\bar{x}) + \lambda \cdot G(\bar{x}) \\
&\leq [\nabla f(x) - \lambda \cdot \nabla G(x)] \cdot (\bar{x} - x)' + \lambda \cdot G(\bar{x}) \\
&= \lambda \cdot G(\bar{x}) \leq 0
\end{aligned}
\tag{3-69}
$$

hence $F(\bar{x},\bar{\lambda}) \leq F(\bar{x},\lambda)$ for all $x \geq 0$, $\lambda \geq 0$, a minimum at (x,λ), and $F(\bar{x},\bar{\lambda}) = f(\bar{x}) - \bar{\lambda} \cdot G(\bar{x}) = f(\bar{x})$.

Conversely,[†] suppose that $(x,\bar{\lambda})$ give the maximum to the dual problem; then $(\bar{x},\bar{\lambda})$ must satisfy the conditions of Theorem 3-5 with respect to both x and λ. Therefore the condition applied to x gives

$$
-\nabla f(\bar{x}) + \bar{\lambda}\nabla G(\bar{x}) = \kappa \left[H_f(\bar{x}) - \sum_{k=1}^{m} \lambda_k H_{g_k}(\bar{x}) \right]
\tag{3-70}
$$

where

$$
H_f(\bar{x}) = \left\{ \frac{\partial^2 f}{\partial x_i \partial x_j} \right\}
$$

is the hessian matrix of f evaluated at $\bar{x}$; similarly the matrix

$$
H_{g_k}(\bar{x}) = \left\{ \frac{\partial^2 g_k}{\partial x_i \partial x_j} \right\}
$$

is evaluated at $\bar{x}$. Note that since $x \geq 0$ is not stated explicitly (but may be included in G), we need not subtract a constant vector.

The condition with respect to λ gives

$$
G'(\bar{x}) = -\kappa \nabla G'(\bar{x}) - \mu \qquad \mu \geq 0
\tag{3-71}
$$

Also from Theorem 3-5 we must have the third condition,

$$
\mu \bar{\lambda}' = 0
\tag{3-72}
$$

Now from the hypotheses of the theorem, the left side of (3-70) is zero, and because the quantity in brackets [which is $\{\partial^2 F/\partial x_i \partial x_j\}$ evaluated at $(\bar{x},\bar{\lambda})$] has an inverse, we conclude that $\kappa = 0$. Hence (3-71) gives $G'(\bar{x}) \leq 0$. Thus $\bar{x}$ is in the feasible region of the primal problem. If we now multiply (3-71) by $\bar{\lambda}$ and use (3-72), we obtain $\bar{\lambda}G(\bar{x}) = 0$. Thus all the conditions of Theorem 3-5 are satisfied for the primal problem. Since $f(x)$ is concave and $G(x)$ is convex, these conditions are also sufficient (see Theorem 3-8).

---

† Mangasarian states the converse theorem requiring twice-continuous differentiability of $f(x)$ and $g_i(x)$, and his condition is that either f be strictly concave in the neighborhood of $\bar{x}$ or at least one of the active $g_i(x)$ be strictly convex in this neighborhood, or both. If $f(x)$ is quadratic, and the $g_i(x)$ are linear, the only requirements are that $f(x)$ be strictly concave and twice differentiable.

Theorem 3-11 (*a duality theorem of Dorn*[20]† *for quadratic programs*)   Given the following two problems:

Problem I: Maximize $f(x) = \frac{1}{2}x'Cx + p'x$ subject to the constraints $Ax \leq b$, $x \geq 0$, where x and p are n vectors, C is an $n \times n$ symmetric negative-semidefinite matrix, A is an $m \times n$ matrix, and b is an m vector.

Problem II: Minimize $g(u,\lambda) = \frac{1}{2}u'u + b'\lambda$ subject to $A'\lambda - Cu \geq p$, $\lambda \geq 0$.

If $x = \bar{x}$ is a solution to problem I, then a solution $\{\bar{u},\bar{\lambda}\}$ exists to problem II, and conversely.   In both cases $\max_{x} f(x) = \min_{u,\lambda} g(u,\lambda)$.

Proof: Suppose $x = \bar{x}$ is the maximizing solution to problem I Then by Theorem 3-5 there exists a column vector $\bar{\lambda}$ such that

$$\bar{\lambda} \geq 0 \tag{3-73}$$
$$A'\bar{\lambda} \geq C'\bar{x} + p' \tag{3-74}$$
$$\bar{\lambda}'(A\bar{x} - b) = 0 \tag{3-75}$$
$$\bar{x}'(A'\bar{\lambda} - C\bar{x} - p) = 0 \tag{3-76}$$

From (3-73) and (3-74), $(\bar{x},\bar{\lambda})$ is a feasible solution of problem II.   If (u,λ) is any other feasible solution, then

$$g(\bar{x},\bar{\lambda}) - g(u,\lambda) = \frac{1}{2}(u - \bar{x})'C(u - \bar{x}) + \bar{x}'C(u - \bar{x}) + b'\bar{\lambda} - b'\lambda$$

Since C is negative-semidefinite,

$$g(\bar{x},\bar{\lambda}) - g(u,\lambda) \leq \bar{x}'C(u - \bar{x}) + b'\bar{\lambda} - b'\lambda$$

From (3-75) and (3-76),

$$b'\bar{\lambda} = \bar{\lambda}'A\bar{x} = \bar{x}'C\bar{x} + p'\bar{x} \tag{3-77}$$

so that
$$g(\bar{x},\bar{\lambda}) - g(u,\lambda) \leq \bar{x}'Cu + p'\bar{x} - b'\lambda \tag{3-78}$$

Since u satisfies the constraints of problem II and $\bar{x} \geq 0$, it follows that

$$x'Cu \leq \bar{x}'(A'\lambda - p) \tag{3-79}$$

Similarly, since $\bar{x}$ satisfies the constraints of problem I and $\bar{x} \geq 0$, we have

$$-b'\lambda \leq \bar{x}'A\lambda \tag{3-80}$$

Substituting (3-79) and (3-80) in (3-78), we obtain

$$g(\bar{x},\bar{\lambda}) - g(u,\lambda) \leq 0$$

which shows that $(\bar{x},\bar{\lambda})$ is the minimum solution of problem II.   Using (3-77), we have

$$g(\bar{x},\bar{\lambda}) = -\frac{1}{2}\bar{x}'C\bar{x} + b'\bar{\lambda} = \frac{1}{2}\bar{x}'C\bar{x} + p'\bar{x} = f(\bar{x})$$

and the first half of the theorem is proved.

† We are indebted to Dr. Dorn, who in a private communication indicated this new proof of his theorem.

To verify the second half, let $(\bar{u}, \bar{\lambda})$ be the minimizing solution to problem II. Then by Theorem (3-5) we have, for some $\bar{x}$,

$$\bar{x} \geq 0 \tag{3-81}$$
$$C\bar{x} = C\bar{u} \tag{3-82}$$
$$A\bar{x} \leq b \tag{3-83}$$
$$\bar{\lambda}'(A\bar{x} - b) = 0 \tag{3-84}$$
$$\bar{x}'(-C\bar{u} - A'\bar{\lambda} - p) = 0 \tag{3-85}$$

From (3-81) and (3-83), $\bar{x}$ is a feasible solution of problem I. Since $\bar{\lambda}$ satisfies the constraints of problem II, (3-82) implies that $(\bar{x}, \bar{\lambda})$ satisfies (3-73) and (3-74). Moreover, (3-84) is exactly (3-75), and (3-85) with (3-82) imply (3-76). Since conditions (3-73) to (3-76) are sufficient for $\bar{x}$ as a maximizing solution of problem I, the proof is complete.

The following is an application, given by Dorn, of Theorem 3-11. It is well known (in applied mechanics) that in the analysis of an elastic-perfectly plastic structure, of all the stress distributions which satisfy the equilibrium and yield conditions, those which actually occur are those which minimize the elastic strain energy.

For a pin-jointed truss this can be expressed as

$$\text{Minimize} \quad U = \frac{1}{2} \sum_{j=1}^{n} \frac{L_j}{A_j E_j} S_j{}^2$$

where

$$\sum_{j=1}^{n} a_{ij} S_j = F_i \quad i = 1, \ldots, m$$

$$|S_j| \leq Y_j$$

Here L_j, A_j, and E_j are the length, cross-sectional area, and Young's modulus of the jth bar; S_j is the force (or stress) in the jth bar, and Y_j is the maximum force the bar can withstand. The equations above are the equilibrium equations, where the F_i are applied loads. These equations define the *Haar-Karman principle* and date back to 1909. The dual to this problem, quite formally, is

$$\text{Minimize} \quad V = \frac{1}{2} \sum_{j=1}^{n} \frac{A_j E_j}{L_j} (e_j')^2 + \sum_{j=1}^{n} Y_j |e_j''| - \sum_{i=1}^{m} F_i u_i$$

subject to

$$\sum_{i=1}^{m} a_{ij} u_i = e_j' + e_j''$$

The variables here may be interpreted as follows: e_j' is the elastic or recoverable elongation of the jth bar, e_j'' is the plastic or permanent elongation of the jth bar, and u_i is the displacement of the ith joint or node.

The problem says that of all the elongations and displacements which are compatible (leave no holes in the structure), the actual ones are those which minimize the potential energy. This is a generalization of the principle of minimum potential energy and, as far as is known, was not discovered until the duality theorem produced it.

One could easily construct a diet-type problem in which the prices are a linear function of the amount of food purchased. Let

a_{ij} = amount of ith nutrient in jth food

d_i = minimum requirement of ith nutrient

$\frac{1}{2}c_j x_j + b_j$ = price of one additional unit of jth food when x_j units have been purchased

and   Minimize $\frac{1}{2} \sum_j c_j x_j^2 + \sum_j b_j x_j$

where $\sum_j a_{ij} x_j \geq d_i \qquad x_j \geq 0$

If the $c_j \geq 0$, then the problem is convex. This implies that the more one buys, the more it costs, because the material begins to become scarce (a monopolistic-type market).

The dual is

Maximize $-\frac{1}{2} \sum_j c_j u_j^2 + \sum_i d_i \lambda_i$

subject to $\sum_i a_{ij} \lambda_i \leq c_j u_j + b_j \qquad \lambda_i \geq 0$

One can interpret this in terms of a vitamin-pill salesman who says to the dietician, "I will sell you pills at prices v_i so that the cost of the pills to synthesize any food does not exceed the marginal cost of that food." The dietician must, of course, agree to buy under these conditions, and subject to this, the salesman maximizes his income.

Let $f(x) = [f_1(x), \ldots ,f_n(x)]$, $x' = (x_1, \ldots ,x_n)$. Then the following is a direct result of Theorem 3-5:

Theorem   If min $x'f$ exists for $x \geq 0$, $f \geq 0$, and if $\lambda'\{\partial f/\partial x\}\lambda \geq 0$, where $\{\partial f/\partial x\}$ is the jacobian matrix of f, and $\lambda' = (\lambda_1, \ldots ,\lambda_n)$, then min $x'f = 0$.

Using this theorem, Dantzig, Eisenberg, and Cottle have recently proved that the following general duality theorem holds:

Theorem   Let $F(x,y)$ be convex in x and concave in y, and consider the primal problem,

$$\min \left[F(x,y) - y' \frac{\partial F}{\partial y} \right]$$

$$\frac{\partial F}{\partial y} \leq 0 \qquad x \geq 0 \qquad y \geq 0$$

and the dual problem,

$$\max \left[F(x,y) - x' \frac{\partial F}{\partial x} \right]$$

$$\frac{\partial F}{\partial x} \geq 0 \qquad x \geq 0 \qquad y \geq 0$$

If either problem has a solution, then so does the other, and the extreme values are the same. The proof assumes that the Kuhn-Tucker constraint qualifications hold.

3-6 CONSTRUCTION OF SOLUTIONS

An iterative procedure that converges to the optimum may be employed in solving (G). We select an arbitrary point in the feasible region as an initial point; if it does not yield the optimum, we obviously would like to shift to another point at which the objective function has an improved value. Often the gradient determines the direction of our movement. In a maximization problem we move along the gradient; otherwise we move in a direction opposite to that of the gradient. The actual distance to be traversed may be determined in a manner analogous to that used in the method of steepest descent. By repetition of this process, the optimum may be reached in a finite or an infinite number of steps. Several of the algorithms discussed here actually employ variations of the procedure just described. We illustrate with examples.

For convenience we consider the problem of minimizing a convex function. Hence at each step we calculate $-\nabla f$ instead of ∇f. The initial step towards the solution of (G) involves a routine check by differentiation to determine whether the maximum lies in the interior of the feasible region R. If not, we know it lies on the boundary.

We start with any point $x^{(0)}$ in the feasible region and calculate $-\nabla f(x^{(0)})$. If $x^{(0)}$ is an interior point, we move along the vector $-\nabla f(x^{(0)})$ a distance equal to its length and arrive at a point whose coordinates are

$$x^{(1)} = x^{(0)} - \nabla f(x^{(0)}) \tag{3-86}$$

If $x^{(1)}$ belongs to the region (i.e., satisfies the constraints) and $f(x^{(1)}) < f(x^{(0)})$, then $x^{(1)}$ is our new point. Otherwise we obtain a new point by applying the gradient method on the segment $x^{(0)}$, $x^{(1)}$ described in the last chapter and obtain $x^{(1)}$. If, in either case, $x^{(1)}$ does not belong to the region, we take those constraints of (G) which are not satisfied by $x^{(1)}$ and calculate their intersection with the line passing through $x^{(0)}$ and $x^{(1)}$. This line in n space is given as the intersection of $n - 1$ hyperplanes. Hence for each constraint one must solve for n unknowns from n equations, of which $n - 1$ are linear and the remaining one is a constraint.

Among all these points of intersection, the one nearest to $x^{(0)}$ must satisfy all the constraints and, hence, would belong to R. We denote this point by $x^{(1)}$. If $f(x^{(1)}) < f(x^{(0)})$, then $x^{(1)}$ is the new starting point; otherwise we apply the gradient criterion to pick a point on $x^{(0)}$, $x^{(1)}$ at which the value of f would be smaller than at $x^{(0)}$.

If $x^{(0)}$ lies on the boundary of R and is the required optimum, the gradients of the objective function and the constraints (evaluated at $x^{(0)}$) satisfy Theorem 3-5. Otherwise we compute the projection of $-\nabla f$ on the intersection of the tangent planes to all the surfaces $g_i(x) = 0$ at $x^{(0)}$. The equation of such a tangent plane is given by

$$\nabla g_i(x^0) \cdot (x - x^0) = 0 \tag{3-87}$$

The intersection of these planes is a plane of dimension at least equal to $n - m$. The projection of $-\nabla f$ on this intersection is a line whose length is

$$\cos \theta \left[\left(\frac{\partial f}{\partial x_1} \right)^2 + \cdots + \left(\frac{\partial f}{\partial x_n} \right)^2 \right]^{1/2} \Bigg|_{x = x^{(0)}} \tag{3-88}$$

where θ is the angle between the direction $-\nabla f$ and the hyperplane of intersection. [Note that if $g_i(x)$ is a hyperplane, the tangent coincides with the plane $g_i(x) = 0$.] This projection determines a ray along which we must move to our next point. The magnitude of the step is determined by the gradient criterion. If this new point does not lie in R [if the $g_i(x)$ are convex, the point cannot lie in R], then a new appropriate point may be chosen in R such that f shows an improvement for the new value. This procedure of choosing the point in R is involved and is discussed by Rosen.[60] In the example considered below, we shall only show how to obtain the length of the projection of $-\nabla f$ on the tangent planes. If the projection of $-\nabla f$ on the intersection of these hyperplanes is zero, we locate the surfaces which, if dropped from the set of $g_i(x)$ containing $x^{(0)}$, will yield a nonzero projection of the gradient on the intersection of the tangent planes to the remaining surfaces. Once this is accomplished, the procedure is the same as above.

To illustrate, let

$$f(x,y) = (x - 3)^2 + (y - \tfrac{1}{2})^2$$
$$g_1(x,y) = x - 2$$
$$g_2(x,y) = x^2 - 2x - 3 + 4y$$

and suppose it is desired to minimize f subject to $g_i \leq 0$, $i = 1, 2$.

Since f is convex, a local minimum is a global one.

Let $x^{(0)} = (1,1)$; then $-\nabla f(x^{(0)}) = (4,-1)$. The farthest we can go along the direction determined by $-\nabla f(x^{(0)})$ is $x^{(1)} = (2,\tfrac{3}{4})$, which is on the boundary of both constraints; i.e., the line pierces the intersection of both constraints at this point. Then $-\nabla f = (2,-\tfrac{1}{2})$. The reader should show that the projection of $(2,-\frac{1}{2})$ on $x - 2 = 0$ is $(0,-\frac{1}{2})$. This furnishes the direction. Using the gradient method, we substitute in f the quantity $x^{(1)} + \lambda$ multiplied by the projection of $\nabla f(x^{(1)})$, that is, we put

$$x = 2 + \lambda \cdot 0 \qquad y = \frac{3}{4} - \frac{\lambda}{2}$$

and minimize with respect to λ for $0 \leq \lambda \leq 1$. We obtain $\lambda = \frac{1}{2}$. Thus our next point is $(2,\frac{1}{2})$ (obtained by putting $\lambda = \frac{1}{2}$ in x and y), which gives the required minimum by Theorem 3-5. The minimum value of f is 1.

As another example, let

$$f(x,y) = 3x^2 + 3y^2 + 2xy - 28x - 20y + 76$$
$$g(x,y) = x^2 - 2x + 4y - 3$$

The problem is to minimize f subject to $g \leq 0$. In the absence of the constraints, f takes its minimum value at the point $(4,2)$, which does not lie in the feasible region. Let $x^{(0)} = (2,\frac{1}{2})$; then $-\nabla f = (15,13)$. The farthest we can move in the feasible region along this direction is to the point on the boundary whose coordinates are $(2.2,.64)$. Let this point be denoted by $x^{(1)}$. Then $-\nabla f(x^{(1)}) = (13.53,11.76)$. The equation of the tangent to $g(x,y) = 0$ is $y - .64 = -.6 \ (x - 2.2)$. The equation of the line passing through $(2.2,0.64)$ and $(14.72,12.40)$ is $11.76(x - 2.2) = 13.52 \ (y - .64)$.

The angle θ between these two lines is approximately $71.58°$, for which $\cos \theta = .309$. Hence the length of the projection of $-\nabla f$ is calculated to be 5.53, etc. The discussion is continued in Rosen's method, given below.

We shall present here a collection of algorithms for solving (G) bearing the names of their respective inventors. In consideration of the conflicting requirements of brevity and clarity, we are forced to handle them all in a somewhat arbitrarily uniform fashion. We shall, for each algorithm give first the rationale or motivation of the process, with a formal justification, then the explicit details of the algorithm itself, and finally an example. We shall not give complete consideration to convergence. While the explanation of these algorithms is necessarily an unwieldly business, their presence in this chapter is essential to the spirit in which it is written, namely, with the aim of optimizing the accessibility of this theory as an analytic tool for the reader. The variety of these methods might inspire one to devise new ones.†

---

† As mentioned in the preface, the authors are indebted to Mr. Dhirendra Nath Ghosh for his help in studying some of these methods, which he included in an M.A. thesis.

Frank and Wolfe's Method[29]

Frank and Wolfe consider the problem of maximizing a concave quadratic function $f = \alpha x' - xAx'$ subject to linear constraints $Bx' \leq b'$ and $x \geq 0$, where $A = (a_{ij})$ is an $n \times n$ positive-semidefinite symmetric matrix, that is, $A = A'$, $B = (b_{ij})$ is an $m \times n$ matrix, $b = (b_1, \ldots , b_m)$, $x = (x_1, \ldots , x_n)$, and $\alpha = (a_1, \ldots , a_n)$.

The algorithm is based on the Equivalence Theorem proved in the previous section and uses the simplex method of linear programming. Geometrically, the Equivalence Theorem implies that f has a local extremum at x if and only if the normal hyperplane to the gradient vector ∇f at x is a locally supporting hyperplane of the constraint set. A hyperplane touching the constraint set at a point x satisfies the above condition if and only if x has the maximum projection along the outward normal to the hyperplane. Since f has been taken to be a quadratic function, the expression for an extremum on the boundary can be written as a linear function of x, and the simplex method can be used to arrive at the optimum.

It follows from the above that x^* is a solution if and only if

$$\nabla f(x^*) \cdot x^{*\prime} = \max \{\nabla f(x^*) \cdot w' : w \geq 0; Bw' \leq b'\}$$

where $w = (w_1, \ldots , w_n)$.

By the duality theorem for linear programming, the right side of the above equation is equal to $\min \{ub' : u \geq 0; uB \geq \nabla f(x^*)\}$, where $u = (u_1, \ldots , u_m)$. Hence the necessary and sufficient condition for x^* to be a solution is that $\max \{\nabla f(x^*) \cdot x^{*\prime} - ub' : u \geq 0, uB \geq \nabla f(x^*)\} = 0$, since

$$\nabla f(x^*) \cdot x^{*\prime} - \min ub' = 0 \qquad \text{and} \qquad \nabla f(x^*) \, x^{*\prime} - \min ub'$$
$$= \max [\nabla f(x^*) \, x^* - ub']$$

Let $F(x,u) = \nabla f(x) \cdot x' - ub' = \alpha x' - ub' - 2xAx'$ (since $\nabla f = \alpha - 2xA$). Hence x^* is the optimum solution of our problem if and only if, for some u, the pair (x,u) satisfies

$$x \geq 0 \qquad u \geq 0 \qquad Bx' \leq b' \qquad \nabla f(x^*) \leq uB \qquad F(x,u) = 0 \qquad (3\text{-}89)$$

Thus we maximize $F(x,u)$ (whose value at the maximum is known to be zero) subject to the constraints. The algorithm of Wolfe and Frank applies the simplex method of linear programming. By introducing slack variables $v = (v_1, \ldots , v_n)$ and $y = (y_1, \ldots , y_m)$, we can restate the problem as follows: Given the constraints $x \geq 0$, $u \geq 0$, $v \geq 0$, $y \geq 0$,

$$Bx' + y' = b'$$
$$2Ax' + B'u' - v' = \alpha'$$
$$vx' + uy' = 0$$

$$\text{Maximize} \qquad F(x,u) = \alpha x' - ub' - 2xAx' = -vx' - uy'$$

The value of this maximum is zero. The last expression is obtained by using the constraints.

The algorithm proceeds as follow:

1. Form the equation

$$Dw' = \begin{pmatrix} b' \\ \alpha' \end{pmatrix} \equiv d'$$

where

$$D = \begin{pmatrix} B & 0 & I & 0 \\ 2A & B' & 0 & -I \end{pmatrix}$$

and

$$w = (x, u, y, v)$$

2. An initial feasible solution is reached just as in the simplex method. First choose an initial basis as in the simplex method.

3. Then express d' and all the columns of D in terms of the chosen basis.

4. Let $w_1^{*\prime}$ be the coefficients of the linear combination of d in terms of the new vectors. Using the coefficients of w_1^* in the positions indicated by the subscripts of the basis vectors, and zeros elsewhere, obtain $w_1 = [x, u, y, v]$.

5. Calculate $(vx' + uy')$.

   (a) If $vx' + uy' = 0$, the algorithm terminates.

   (b) If $vx' + uy' \neq 0$, the algorithm proceeds as follows:

6. Let $\nabla F(w)$ be the gradient obtained by taking the partials of F with respect to the components of w, in the order in which they appear in w.

7. Evaluate this vector at $w = w_1$, and use $\nabla F(w_1)$ as the cost-coefficient vector of the simplex method to obtain a new basis.

8. Obtain w_2^*, which is the vector d' expressed as a linear combination of the new basis.

9. Calculate $F(w_2) = vx' + uy'$.

   (a) If $vx' + uy' = 0$, the algorithm terminates.

   (b) If $vx' + uy' \neq 0$, and

   (c) If $vx' + uy'$ has not decreased from the previous step, find λ such that $F(w) = F[w_1 + \lambda(w_2 - w_1)]$, $0 \leq \lambda \leq 1$, is minimized. This gives the new w for which ∇F is calculated, yielding the cost coefficients. When (c) is completed, return to (5).

Example: Maximize $f = 6x_1 - 2x_1^2 + 2x_1x_2 - 2x_2^2$, subject to the constraints $x_1 + x_2 \leq 2$, $x_i \geq 0$, $i = 1, 2$.

$$A = \begin{pmatrix} 2 & -1 \\ -1 & 2 \end{pmatrix} \qquad b = (2) \qquad B = (1,1) \qquad \alpha = (6,0)$$

$$D = \begin{pmatrix} D_1 & D_2 & D_3 & D_4 & D_5 & D_6 \\ 1 & 1 & 0 & 1 & 0 & 0 \\ 4 & -2 & 1 & 0 & -1 & 0 \\ -2 & 4 & 1 & 0 & 0 & -1 \end{pmatrix} \begin{pmatrix} w' \\ x_1 \\ x_2 \\ u \\ y \\ v_1 \\ v_2 \end{pmatrix} = \begin{pmatrix} d' \\ 2 \\ 6 \\ 0 \end{pmatrix}$$

Let the initial basis be D_2, D_3, D_4, and express the remaining vectors of D as a linear combination of these. Obtain

	w_1^*	1	2	3	4	5	6
D_2	-1	-1	1	0	0	$\frac{1}{6}$	$-\frac{1}{6}$
D_3	4	2	0	1	0	$-\frac{4}{6}$	$-\frac{1}{3}$
D_4	3	2	0	0	1	$-\frac{1}{6}$	$\frac{1}{6}$

$$w_1 = [\,0 \quad \underbrace{-1}_{x} \quad \underbrace{4}_{u} \quad \underbrace{3}_{y} \quad \underbrace{0 \quad 0}_{v}\,]$$

Note that w_1^* is obtained by expressing d' as a linear combination of the basis vectors. The vector w_1 is formed by inserting zeros everywhere except where the corresponding component of w_1^* is nonzero. Note that $F(w) = vx' + uy' = 12$.

	c^*	w_1^*	$\nabla F(w_1)\ c_1 = 0$	$c_2 = 0$	$c_3 = 3$	$c_4 = 4$	$c_5 = 0$	$c_6 = -1$
D_2	0	-1	-1	1	0	0	$\frac{1}{6}$	$-\frac{1}{6}$
D_3	3	4	2	0	1	0	$-\frac{4}{6}$	$-\frac{1}{3}$
D_4	4	3	2	0	0	1	$-\frac{1}{6}$	$\frac{1}{6}$

$$z_1 = 14 \quad z_2 = 0 \quad z_3 = 3 \quad z_4 = 4 \quad z_5 = -1\frac{6}{6} \quad z_6 = -\frac{1}{3}$$

where, for example, z_4 is the product of the column above $z_4 = 4$ and the vector c^*, the cost vector corresponding to D_2, D_3, and D_4 is obtained from $\nabla F(w_1)$. Calculate $c_j - z_j$ for $j = 1, \ldots, 6$. From among the vectors for which $c_j - z_j$ is negative, choose the one (in this case D_1) with the largest absolute value. This will be the new basis vector. To determine the vector which is to be eliminated, compute

$$\min \frac{\text{component of basic solution } w_1^*}{\text{corresponding positive component of } D_1} = \min(\tfrac{4}{2}, \tfrac{3}{2}) = \tfrac{3}{2}$$

(note that no ratio is included in which the component of D_1 is nonposi-tive), thereby finding that D_4 is to be eliminated. The new format is as follows:

	w_2^*	1	2	3	4	5	6
D_1	$\frac{3}{2}$	1	0	0	$\frac{1}{2}$	$-\frac{1}{12}$	$\frac{1}{12}$
D_2	$\frac{1}{2}$	0	1	0	$\frac{1}{2}$	$\frac{1}{12}$	$-\frac{1}{12}$
D_3	1	0	0	1	-1	$-\frac{1}{2}$	$-\frac{1}{2}$

The components of w_2^* give $vx' + uy' = 0$. Hence the optimum solu-tion is $x_1 = \frac{3}{2}$, $x_2 = \frac{1}{2}$. Substituting these values, obtain the optimum value of f.

Exercise 3-11 Using the above method, show that $x_1 = 2$, $x_2 = 1$ minimizes $f(x) = x_1^2 + 3x_2^2 - 4x_1 - 6x_2$ subject to $x_1 \geq 0$, $x_2 \geq 0$, $x_1 + 2x_2 \leq 4$.

Wolfe's Variation of the Algorithm[81] (*the Simplex Method of Quadratic Programming*)

Wolfe furnishes a variation of the above technique. To indicate it, we must first prove the following theorem, after which we shall give a sketch of the algorithm.

Theorem 3-12 If $x = (x_1, \ldots, x_n) \geq 0$, $Bx' = b'$, and there exist

$$v = (v_1, \ldots, v_n) \geq 0 \qquad u = (u_1, \ldots, u_m)$$

such that

(1) $$vx' = 0$$
(2) $$Ax' - v' + B'u' + \lambda p' = 0$$

then x yields the minimum of

$$f(\lambda, y) \equiv \lambda py' + \tfrac{1}{2} yAy'$$

subject to $$\lambda \geq 0 \qquad y \geq 0 \qquad By' = b$$

where A is an $n \times n$ positive-semidefinite matrix, $p = (p_1, \ldots, p_n)$ and $\lambda \geq 0$ is a scalar.

Proof: Let $y = (y_1, \ldots, y_n)$ be any other vector. We shall show that y yields a value for f greater than x does. Let $By' = b'$ be given. From the positive-semidefiniteness of A we have

$$(y - x)A(y - x)' \geq 0$$

Hence $$yAy' + xAx' \geq 2xAy'$$

or

$$f(\lambda, y) - f(\lambda, x) = \lambda p'(y - x)' + \tfrac{1}{2} yAy' - \tfrac{1}{2} xAx' \geq (\lambda p + xA)(y' - x')$$

From (2), however,

$$\lambda p + xA = v - uB$$
$$f(\lambda, y) - f(\lambda, x) \geq vy' - vx' - uBy' + uBx'$$
$$= vy' - 0 - ub' + ub' = vy' \geq 0$$

(since $v, y \geq 0$), and the proof is complete.

Wolfe's variation is outlined as follows:

Step 1:   Begin with the set of relations

$$Bx' + w' = b'$$
$$Ax' - v' + Bu' + z^{(1)'} - z^{(2)'} = -px'$$
$$v, z^{(1)}, z^{(2)}, w \geq 0$$

where
$$z^{(1)} = (z_1^{(1)}, \ldots, z_n^{(1)})$$
$$z^{(2)} = (z_1^{(2)}, \ldots, z_n^{(2)})$$

are two successive iterations, and

$$w = (w_1, \ldots, w_n)$$

Step 2:   Use the simplex method to minimize Σw_i to zero, keeping v and u nonnegative.   Discard w and the unused components of $z^{(1)}$ and $z^{(2)}$.   Let the remaining n components be denoted by z, and their coefficients by E.   A solution of the system is

$$Ax' = b'$$
$$Ax' - v' + B'u' + Ez' = -\lambda p'$$
$$x, v, z \geq 0$$

Now we wish to minimize $\sum_{i=1}^{n} z_i$ subject to the constraints under the following side conditions: For $k = 1, \ldots, m$, if x_k is in the basis, we do not admit v_k; if v_k is in the basis, we do not admit x_k.   It will take at most $\binom{3n}{n}$ iterations to reach the solution.   The x component of the terminal basic solution is a solution of the quadratic problem.

Exercise 3-12   Using this method, minimize $\frac{1}{2}(x_1^2 + x_2^2 + x_3^2) + x_1 - 2x_3$, subject to $x_1, x_2, x_3 \geq 0$ and $x_1 - x_2 + x_3 = 1$.   The solution is given by $(0, \frac{1}{2}, \frac{3}{2})$.

The Gradient-projection Method of Rosen[60]†

This method is applicable to the linear-constraint case $Ax' \leq b'$ as well as to the general case in which the components of $G(x) \leq 0$ are convex.

† We are indebted to Commander D. W. Sencenbaugh for help in preparing this section.

Case 1: *Nonlinear objective function with linear constraints:* The main problem in the linear case is to iterate by projecting the gradient of the concave function $f(x)$ (to be maximized) on the boundary and, after traveling a certain length on this projection, to evaluate the gradient again at the terminal point, and so forth. In the nonlinear case, at each new iteration point the gradient is projected on the tangent planes, and, after the determination of a legitimate length of the projection on the tangent planes, a nearby point on the constraint surfaces is obtained by an iteration procedure.

We need the following facts:

If A_q is a matrix whose rows are the normals to the boundary planes in the linear case (or to the tangent planes in the nonlinear case), the matrix

$$P_q = I - A_q'(A_q A_q')^{-1} A_q \tag{3-90}$$

projects any vector into the intersection of the tangent planes. Thus, if P_q is multiplied on the right by $\nabla f(x^{(n)})$, we obtain the projection on this intersection of the gradient vector at the nth iteration. The reader should have no difficulty in verifying this fact by showing that the projected vector is orthogonal to any one of the normals (multiply on the left by a row of the matrix A_q and show that the result must be zero).

In the linear case, if we start with $x^{(0)}$ in the feasible region, our new iteration point is given by

$$x^{(1)} = x^{(0)} + \lambda_0 r^{(0)} \tag{3-91}$$

where

$$r^{(0)} = \frac{\nabla f(x^{(0)})}{|\nabla f(x^{(0)})|} \tag{3-92}$$

is the unit projected gradient, and

$$\lambda_0 = \min_i \left[\frac{A_i \cdot x^{(0)} - b_i}{A_i \cdot r^{(0)}} \right], \qquad A_i r^{(0)} > 0 \tag{3-93}$$

where the A_i are the rows of A. The projection matrix P_q is selected so that $A_i \cdot P_q \nabla f(x^{(0)}) \geq 0$ for all i such that $A_i x^{(0)} - b_i = 0$. The numerator of the quantity in brackets is the distance from $x^{(0)}$ to each constraint plane, while the denominator is the projection of the unit gradient at $x^{(0)}$ onto the line joining them. Our next problem is to project the gradient of f at $x^{(1)}$ onto the hyperplane or intersection of hyperplanes on which $x^{(1)}$ is located. If their matrix is A_q, then our next point $x^{(2)}$ is determined as follows:

Compute $P_q \cdot \nabla f(x^{(1)})$. If this is zero, and if

$$(A_q A_q')^{-1} A_q \nabla f(x^{(1)}) \geq 0 \tag{3-94}$$

then $x^{(1)}$ is a global maximum. If at least one component of the last inequality, say the ith, is negative, then the corresponding row A_i is eliminated from A_q, leaving A_{q-1} with its corresponding P_{q-1}.

Our next iteration is

$$x^{(2)} = x^{(1)} + \lambda_1 r^{(1)}$$

where λ_1 and $r^{(1)}$ are computed exactly as λ_0 and $r^{(0)}$, except that $x^{(1)}$ is used in place of $x^{(0)}$. If

$$r^{(1)} \cdot \nabla f(x^{(2)}) \geq 0 \tag{3-95}$$

one forms the new projection matrix A_q and its projection P_q by adjoining the row corresponding to $\lambda^{(1)}$ and then proceeds to $x^{(3)}$.

If, on the other hand, $r^{(1)} \cdot \nabla f(x^{(2)}) < 0$ holds, then the maximum of f has been passed, and to retrace our steps we define the interpolation value

$$\theta = \left[\frac{r^{(1)} \cdot \nabla f(x^{(2)})}{r^{(1)} \cdot \nabla f(x^{(1)}) - r^{(1)} \cdot \nabla f(x^{(2)})} \right] \tag{3-96}$$

The new iteration point is

$$x^{(3)} = \theta x^{(2)} + (1 - \theta) x^{(1)} \tag{3-97}$$

and the projection matrix remains the same. The process is repeated, and one obtains convergence according to a theorem of Rosen.

Example: Suppose it is desired to maximize the function

$$f(x) = (x_1 - 3)^2 (x_2 - 4)$$

subject to $x \geq 0$ and

$$b - Ax = \begin{pmatrix} 2 \\ 3 \\ 2 \end{pmatrix} - \begin{pmatrix} 0 & 1 \\ 1 & 1 \\ 1 & 0 \end{pmatrix} \begin{pmatrix} x_1 \\ x_2 \end{pmatrix} \geq 0$$

The rows of A are

$$A_1 = (0,1), \qquad A_2 = (1,1), \qquad \text{and} \qquad A_3 = (1,0)$$

Let
$$x^{(0)} = (.2, 1.8)$$

$$\nabla f(x) = \begin{pmatrix} 2(x_2 - 4)(x_1 - 3) \\ (x_1 - 3)^2 \end{pmatrix}$$

$$r^{(0)} = (.844, .54)$$

$$\lambda_0 = \min\ (.371, .723, 2.24) = .371$$

$$x^{(1)} = (.2, 1.8) + .371(.844, .54) = (.512, 2.0)$$

which is on the boundary of A_1 corresponding to the row yielding λ_0. We now project $\nabla f(x^{(1)})$ onto A_1. If it projects onto a point, we may have the maximum; otherwise we continue:

$$P_1 = \begin{pmatrix} 1 & 0 \\ 0 & 1 \end{pmatrix} - \begin{pmatrix} 0 & 0 \\ 0 & 1 \end{pmatrix} = \begin{pmatrix} 1 & 0 \\ 0 & 0 \end{pmatrix}$$

where the subscript on P indicates that the first row of A is used in its formation;

$$\nabla f(x^{(1)}) = (9.95, 6.18)$$

$$r^{(1)} = \begin{pmatrix} 1 \\ 0 \end{pmatrix}$$

$$\lambda_1 = \min\ (.488, 1.488) = .488$$

$$x^{(2)} = \begin{pmatrix} .512 \\ 2.00 \end{pmatrix} + .488 \begin{pmatrix} 1 \\ 0 \end{pmatrix} = \begin{pmatrix} 1 \\ 2 \end{pmatrix}$$

and $r^{(1)} \cdot \nabla f(x^{(2)}) > 0$; hence $x^{(2)}$ is the next feasible point. Substituting in the inequalities, we find that we are in the intersection of A_1 and A_2 (see Fig. 3-4). Next

$$\nabla f(x^{(2)}) = \begin{pmatrix} .895 \\ .446 \end{pmatrix}$$

$$P_{1,2} = \begin{pmatrix} 1 & 0 \\ 0 & 1 \end{pmatrix} - \begin{pmatrix} 0 & 1 \\ 1 & 1 \end{pmatrix}\begin{pmatrix} -1 & 1 \\ 1 & 0 \end{pmatrix} = \begin{pmatrix} 0 & 0 \\ 0 & 0 \end{pmatrix}$$

$$(A_{1,2}A'_{1,2})^{-1}\,A_{1,2}\nabla f(x^{(2)}) = \begin{pmatrix} -1 & 1 \\ 1 & 0 \end{pmatrix}\begin{pmatrix} .895 \\ .446 \end{pmatrix} = \begin{pmatrix} -.446 \\ .865 \end{pmatrix}$$

The subscripts indicate which rows of A are used.

Since we have a negative value for the first entry, we eliminate A_1 and project ∇f down on the plane corresponding to A_2 in search of a maximum or the next intersection. Thus we must compute

$$P_2 = \begin{pmatrix} 1 & 0 \\ 0 & 1 \end{pmatrix} - \begin{pmatrix} \frac{1}{2} & \frac{1}{2} \\ \frac{1}{2} & \frac{1}{2} \end{pmatrix} = \begin{pmatrix} \frac{1}{2} & -\frac{1}{2} \\ -\frac{1}{2} & \frac{1}{2} \end{pmatrix}$$

and

$$P_2 \cdot \nabla f(x^{(2)}) = (.223, -.223)$$

from which

$$r^{(2)} = \begin{pmatrix} .707 \\ -.707 \end{pmatrix}$$

$$\lambda_2 = 1.41$$

Thus

$$x^{(3)} = (1,2) + 1.41(.707, -.707) = (2,1)$$

which is at the intersection of A_2 and A_3. Again

$$\nabla f(x^{(3)}) = \begin{pmatrix} 6 \\ 1 \end{pmatrix}$$

and we examine

$$P_{2,3} = \begin{pmatrix} 0 & 0 \\ 0 & 0 \end{pmatrix}$$

$$(A_{2,3}A'_{2,3})^{-1}A_{2,3}\nabla f(x^{(3)}) = \begin{pmatrix} 0 & 1 \\ 1 & -1 \end{pmatrix}\begin{pmatrix} 6 \\ 1 \end{pmatrix} = \begin{pmatrix} 1 \\ 5 \end{pmatrix}$$

which has no negative component; hence we have reached a global maximum at $(2,1)$.

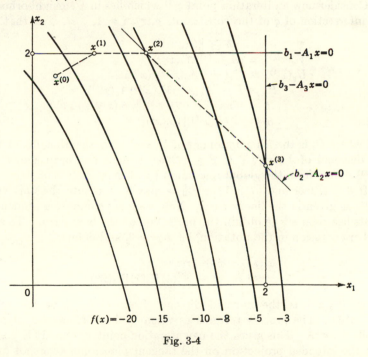

Fig. 3-4

Case 2: Nonlinear objective function with nonlinear constraints: One begins here by specifying tolerances to the location of points. The tolerance of x is a certain δ neighborhood of x where

$$\delta = \frac{\epsilon^2}{3a}(8nbc)^2$$

where n is the number of variables,

$$a = \max(1, \gamma c^2 n^{3/2}) \geq 1$$
$$\gamma \geq \|H\|$$

where H is the hessian of $g_i(x)$ evaluated at the point, and if $A_q(x)$ is the set of hyperplanes tangent to the intersection of the set of q non-linear hypersurfaces on which x lies, then

$$c^2 \geq \|[A_q(x) A'_q(x)]^{-1}\|$$

is a measure of the linear independence of vectors, and

$$b = \sup \{\|x - y\|; \, x \, \epsilon \, R, \, y \, \epsilon \, R\}$$

where R is the constraint region, and hence b is its diameter. The quantity ϵ is merely a specified tolerance.

Consider now an iteration point $x^{(k)}$ which lies in a δ neighborhood of the intersection of q of the constraints $g_i(x)$, $i = 1, \ldots, m$, that is,

$$\delta \geq \|G(x^{(k)})\|$$

Let
$$V_q(x^{(k)}) \equiv [A_q(x^{(k)})A'_q(x^{(k)})]^{-1}$$
$$U_q(x^{(k)}) \equiv V_q(x^{(k)})A_q(x^{(k)})f(x^{(k)})$$
$$P_q \equiv I - A'_q(x^{(k)})V_q(x^{(k)})A_q(x^{(k)}) \tag{3-98}$$
$$\beta = \max [\|P \cdot \nabla f(x^{(k)})\|, \, \beta_j(x^{(k)}), \, j \, \epsilon \, q]$$
where
$$\beta_j = \tfrac{1}{2} U_j(x^{(k)})[V_{jj}(x^k)]^{-1/2}$$

and where U_j is the jth component of U, and V_{jj} is the jth component on the diagonal of $V_q(x^{(k)})$. If $\beta \leq \epsilon/2nbc$, then it has been shown that $f(x^{(k)})$ differs from the global maximum by at most ϵ.

If $\beta > \epsilon/2nbc$ and $\|P_q \cdot \nabla f(x^{(k)})\| > \max \beta_j$, compute the unit vector $r(x^{(k)})$ as given in the linear case. Now we must compute a sequence of points based on $x^{(k)}$ to obtain the projection on the boundary. To avoid further confusion in the notation, let $\bar{x} = x^{(k)}$, and define

$$\bar{x}^0(\lambda) = \bar{x} + \lambda r(x^{(k)}) \tag{3-99}$$
and
$$\bar{x}^{m+1} = \bar{x}^m - A'_q(\bar{x})V_q(\bar{x})W_q(\bar{x}^m) \tag{3-100}$$

where $W_q(\bar{x}^m)$ is the normed distance from $\bar{x}^m$ back to the region, $\lambda = c\beta/6a$. The process terminates when we have reached a point for which $W_q < \delta$. This gives the new iteration point $x^{(k+1)}$. It is possible that the intended projection on the tangent planes has overshot one of the boundary constraints. In this case one must interpolate using a value $\bar{\lambda} < \lambda$ until the projection is in the intersection involving that constraint and the remaining ones.

If $\qquad \beta > \dfrac{\epsilon}{2nbc} \qquad$ and $\qquad \|P_q \cdot \nabla f(x^{(k)})\| < \max_j \beta_j$

then the row with subscript corresponding to the value of j which maximizes β_j must be eliminated from $A_q(x^{(k)})$, yielding corresponding P_{q-1}, V_{q-1}, U_{q-1}, and r at $x^{(k)}$. And again the projection on the boundary involves the determination of a sequence based on $x^{(k)}$. The process is repeated until a maximum is reached.

Example: Maximize

$$f(x) = (x_1 - 3)^2(x_2 - 4)$$

subject to $x \geq 0$ and $G(x) \geq 0$ defined by

$$4 - x_1^2 - x_2^2 + 2x_1 \geq 0$$
$$5 - x_1^2 - x_2^2 \geq 0$$
$$4 - x_1^2 - x_2^2 + 2x_2 \geq 0$$

Let $x^{(0)} = (\tfrac{1}{2}, 1)$. Then

$$\nabla f(x^{(0)}) = (15, 6.25)$$
$$r^{(0)} = (.923, .385)$$
$$\lambda_0 = \min_i \; [t_i : g_i(x^{(0)} + t_i r^{(0)}) = 0; \, i = 1, 2, 3]$$

Thus we must solve a set of equations in t_i for which $g_i = 0$ holds. Hence

$$x^{(1)} = \binom{\tfrac{1}{2}}{1} + 1.25 \binom{.923}{.385} = \binom{1.652}{1.482}$$

Since 1.25 corresponds to t_2, we concentrate on g_2. We compute $\nabla g_2(x^{(1)})$ from which the tangent plane may be obtained (see Fig. 3-5). Thus

$$A_2 = (.745, .666)$$
$$V_2 = 1$$
$$U = 6.26$$
$$P_2 \cdot \nabla f(x^{(1)}) = (2.12, -2.34)$$
$$\beta = \max \, (3.16, 3.13) = 3.16$$
$$r(x^{(1)}) = (.671, -.74)$$
$$\lambda_1 = 3.16/6 = .523$$

We now compute the sequence corresponding to $\bar{x} = x^{(1)}$. We have

$$\bar{x}^{(0)} = \binom{1.652}{1.482} + .523 \binom{.671}{-.74} = \binom{2}{1.1}$$

and

$$\bar{x}^{(1)} = \binom{2}{1.1} - \binom{.745}{.666}(.21)^2 = \binom{1.97}{1.07}$$

where we have defined the normal error as

$$W(\bar{x}^{(0)}) = [g_2(\bar{x}^{(0)})]^2 = (.21)^2$$

Now $W(\bar{x}^{(1)}) = .0001$, which is small and less than the tolerance δ with $a = 1$, $n = 2$, $b^2 = 5$, $c = 1$. Therefore, we can put $x^{(2)} = \bar{x}^{(1)}$.

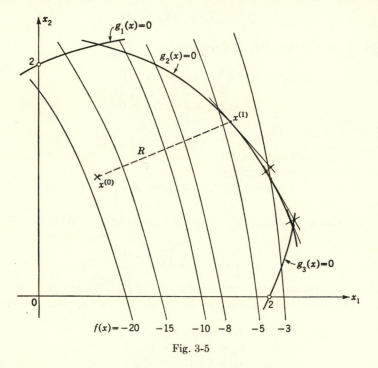

Fig. 3-5

Note that $x^{(2)}$ is on $g_2(x)$. Now, for the next iteration,

$$A_2 = (.88,.478)$$
$$V_2 = 1$$
$$\nabla f(x^{(2)}) = (6.72,1.14)$$
$$U(x^{(2)}) = 6.07$$
$$P_2 \cdot \nabla f(x^{(2)}) = (.92,-1.77)$$
$$\beta = \max (2,3.13) = 3.13$$

Hence
$$\lambda_2 = 3.13/6 = .505$$
$$r(x^{(2)}) = (.46,-.89).$$

Let
$$\bar{x} = x^{(2)}$$

Then
$$\bar{x}^{(0)} = (2.2,.62)$$
$$\bar{x}^{(1)} = (2.16,.6) \equiv x^{(3)}$$

which is on $g_2(x)$. Again

$$A_2 = (.963,.276)$$
$$V_2 = 1$$
$$\nabla f(x^{(3)}) = (5.71,.701)$$
$$U(x^{(3)}) = 5.7$$
$$P_2 \cdot \nabla f(x^{(3)}) = (.22,-.85)$$
$$\beta = \max (.88,3.13) = 3.13$$

Hence
$$\lambda_3 = .505$$
$$r(x^{(3)}) = (.25,-.965)$$

Let $\qquad\qquad \bar{\bar{x}} = x^{(3)}$

Then
$$\bar{\bar{x}}^{(0)} = (2.29,.123)$$
$$\bar{\bar{x}}^{(1)} = (2.25,.112)$$
$$g_1(\bar{\bar{x}}^{(1)}) > 0$$
$$g_2(\bar{\bar{x}}^{(1)}) = 0$$
$$g_3(\bar{\bar{x}}^{(1)}) < 0$$

and we have gone beyond the intersection of g_3 with g_2 and must interpolate in the multiplier λ. Thus we have

$$\bar{\lambda}_3 = (.18/.936)\lambda_3 = .097$$

We reconstruct

$$\bar{\bar{x}}^{(0)} = (2.16,.6) + .097(.25,-.965) = (2.18,.507)$$

which, on substituting in g_2 and g_3, is within our tolerance limit and hence is our new iteration $x^{(4)}$. In this case we have, for tangent-plane directions,

$$A_{2,3} = \begin{pmatrix} .976 & .226 \\ .97 & -.219 \end{pmatrix}$$
$$V_{2,3}A_{2,3} = \begin{pmatrix} .56 & 2.13 \\ .49 & -2.13 \end{pmatrix}$$

and from these values we can check the projection on the hyperplanes tangent at $x^{(4)}$. We find that

$$V_{2,3}A_{2,3} \cdot \nabla f(x^{(4)}) = \begin{pmatrix} 4.64 \\ 1.38 \end{pmatrix} > 0$$

and we have arrived at a global maximum within the tolerance limit.

The gradient-projection program is available for the IBM 7090.

Carroll's Response-surface Technique with Development by Fiacco and McCormick[12,26]

The problem is to minimize $f(x)$, where $x = (x_1, \ldots, x_n)$, subject to $q_i(x) \geq 0$, $i = 1, \ldots, m + n$, which include the nonnegativity constraints $x \geq 0$. The iterative procedure revolves about the expression

$$P(x,r) \equiv f(x) + r \sum_{j=1}^{m+n} \frac{1}{g_i(x)} \tag{3-101}$$

where $r > 0$ is a perturbation parameter. The second term on the right is considered a penalty for requiring a minimum of f subject to the constraints. The reason for this form is that we may, by appropriate choices of a decreasing (to zero) sequence of values of r, minimizing P for each r (thus solving a set of simultaneous equations), obtain a sequence of values of x which comes arbitrarily close to the minimum of f as $r \to 0$. Thus one converges to the actual minimum of f on a path of minima of P. The function P is infinite on the boundary $g_i(x) = 0$, and, by imposing appropriate restrictions on f and g_i, P is made to be well behaved in the region. The successive iterations of the problem are all within the region of constraints ultimately leading towards the optimum (often on the boundary). Thus the method avoids a direct boundary search and its associated difficulties. One starts by choosing an initial point $x^{(0)}$ in the region and computing a value r_0 which minimizes the length of the gradient or $(\nabla P)^2$. This procedure yields, as the reader may show without difficulty,

$$r_0 = \left. \frac{\left| \nabla f \cdot \nabla \sum_{i=1}^{m+n} 1/g_i \right|}{\left(\nabla \sum_{i=1}^{m+n} 1/g_i \right)^2} \right|_{x = x^{(0)}} \tag{3-102}$$

One then minimizes $P(x,r_0)$ with respect to x, obtaining $x^{(1)}$. By the choice of any value $r_1 < r_0$, $P(x,r_1)$ is minimized again, yielding $x^{(2)}$. In this manner one continues to choose a decreasing sequence of values of r and obtains the corresponding sequence of values of x. One cannot use the same formula for computing r, since the minimum of $(\nabla P)^2$ at $x^{(1)}$ is already zero. However, in practice additional computational information yields better guesses at values of r. As the values of r become smaller, the minimum of P (under favorable circumstances mentioned below) tends to the minimum of f, subject to the constraints.

Let the interior of the feasible region be nonempty, and let the convex function f have a minimum on the boundary of the convex region $g_i(x) \geq 0$ determined by the concave functions $g_i(x)$. Let $\{r_k\}$, $k = 0, 1, \ldots$, be a sequence which decreases to zero; then:

Theorem 3-13

$$\lim_{r_k \to 0} \min_{x} P(x,r_k) = \min_{x} f(x) \equiv v_0 \tag{3-103}$$

Proof: Note first that if $g(x)$ is concave, then $1/g(x)$ is convex for all x which satisfy $g(x) > 0$. Thus, for fixed r, $P(x,r)$ has a unique minimum in x. Also, for any two values of r, for example, r_a and r_b with $r_a < r_b$, we have

$$\min_x P(x,r_a) < \min_x P(x,r_b)$$

proved by evaluating $P(x,r_b)$ at that x which minimizes $P(x,r_a)$. To prove the theorem, it suffices to show that for $\epsilon > 0$ there exists k_0 such that for $k > k_0$,

$$\left| \min_x P(x,r_k) - v_0 \right| < \epsilon$$

Since $\min f(x) = v_0$, we have, for x sufficiently close to the minimum,

$$f(x) < v_0 + \frac{\epsilon}{2}$$

Let x^* satisfy this condition, and choose r_{k_0} such that

$$\frac{r_{k_0}}{\min_i g_i(x^*)} < \frac{\epsilon}{2(m+n)}$$

is satisfied. Then for $k > k_0$,

$$\min_x P(x,r_k) < \min_x P(x,r_{k_0}) \leq v_0 + \frac{\epsilon}{2} + \frac{\epsilon}{2} = v_0 + \epsilon$$

Also
$$\min_x P(x,r_k) > v_0 - \epsilon$$

since the penalty factor is always positive. This completes the proof.

Example: Suppose that it is required to minimize $f(x) = x_1 + x_2$ subject to $x_1 \geq 0$, $x_2 \geq 0$. Then

$$P(x,r) = x_1 + x_2 + \frac{r}{x_1} + \frac{r}{x_2} \qquad \nabla P = \left(1 - \frac{r}{x_1{}^2}, 1 - \frac{r}{x_2{}^2}\right)$$

and, by the method of steepest descent given in Chap. 2, the minimum of P is attained at $(r^{1/2}, r^{1/2})$, at which $f(x) = 2r^{1/2}$ and $P = 4r^{1/2}$. Instead of the criterion given for choosing r_0, suppose one arbitrarily starts with $r_0 = 1$, $(x_1{}^{(0)}, x_2{}^{(0)}) = (1.5, .75)$. Then the iterations may be represented by

Step	P	x_1	x_2	$-\partial P/\partial x_1$	$-\partial P/\partial x_2$	r
0	4.250	1.5	.75	$-.555$	.777	1.0
1	4.059	1.25	1.1	$-.360$	$-.173$	1.0
2	4.002	1.050	1.003	$-.092$	$-.0059$	1.0
2	2.54	1.050	1.003	$-.773$	$-.751$	.25
3	2.0011	.5218	.4898	$-.0818$	.0421	.25

The dual of the minimization problem stated above is

$$\max_{x,\lambda} F(x,\lambda) = f(x) - \sum_{j=1}^{m+n} \lambda_j g_j(x)$$

subject to $\nabla f(x) = \sum_{j=1}^{m+n} \lambda_j \nabla g_j(x)$ $\lambda_j \geq 0$ $j = 1, \ldots, m$

The solution of the dual problem is obtained very simply. We have $\bar{x}$ by the above method, and $\bar{\lambda}$ is given by $\bar{\lambda}_j = \lim_{\bar{r} \to 0} \bar{r}/g_j{}^2(\bar{x})$, where $\bar{x}$ lies on the trajectory of the minimum of the function P.

When the convexity requirement is removed, the method can be used to obtain a local solution.

Example: Minimize

$$f(x) = x_1{}^3 - 6x_1{}^2 + 11x_1 + x_3$$

subject to

$$\begin{aligned} -x_1{}^2 - x_2{}^2 + x_3{}^2 &\geq 0 \\ x_1{}^2 + x_2{}^2 + x_3{}^2 - 4 &\geq 0 \\ -x_3 + 5 &\geq 0 \\ x_1, x_2, x_3 &\geq 0 \end{aligned}$$

The global solution of the problem is at $(0,\sqrt{2},\sqrt{2})$, with the first two constraints and $x_1 \geq 0$ binding. The following table describes the iterations:

r	x_1	x_2	x_3	F
1.0	.37896	1.68076	2.34720	5.70854
.1	.10088	1.41984	1.68317	2.73285
.0001	.00302	1.41421	1.42266	1.45581
10^{-9}	.00001	1.41421	1.41424	1.41434
Actual theoretical solution	0	$\sqrt{2}$	$\sqrt{2}$	$\sqrt{2}$

This problem actually has another local solution.

Fiacco and McCormick have been developing computational algorithms particularly suited to solving nonlinear problems by this method. They have coded for a computer and solved several moderately difficult problems, including linear problems, nonconvex problems, and problems characterized by nonlinearities in both the constraints and objective function. It appears that the method is computationally feasible and is an important addition to the construction of solutions of nonlinear-programming problems.

Zoutendijk's Method of Feasible Directions[84,85]

Zoutendijk considers the problem of maximizing a quadratic function subject to linear constraints using the method of feasible directions. He also indicates how this method can be extended to the case of non-linear constraints. We shall discuss only the case of linear constraints and refer the reader to Ref. 85 for an extension of the method to the nonlinear-constraint case.

The problem may be given as follows: Maximize the concave function $f(x) = a + \alpha x' + xAx'$ subject to the linear constraints

$$b_{i1}x_1 + \cdots + b_{in}x_n \leq b_i \qquad i = 1, 2, \ldots, m$$

and $x \geq 0$, where A is an $n \times n$ negative-semidefinite matrix, a is a constant, $\alpha = (a_1, \ldots, a_n)$, and $f(x)$ is continuous and has continuous partial derivatives.

In this method we start from a point in the feasible region and determine a feasible direction of movement such that a small step in that direction violates none of the constraints. The feasible direction which makes the smallest possible angle with the gradient evaluated at the point is the best possible direction, since the gradient points towards the maximum increase of the objective function. Thus, in small steps in the region, the process converges.

To formalize this, we let $x^{(0)} = (x_1^{(0)}, \ldots, x_n^{(0)})$, where $x_i^{(0)} \geq 0$, be a feasible point, that is, $b_{i1}x_1^{(0)} + \cdots + b_{in}x_n^{(0)} \leq b_i, i = 1, \ldots, m$. Let

$$P_i(x^{(0)}) \equiv b_i - b_{i1}x_1^{(0)} - \cdots - b_{in}x_n^{(0)} \geq 0$$

Let s denote a vector in n space at the initial point $x^{(0)}$. In order for the direction of s to be the best feasible direction, it must satisfy the following conditions:

$$\begin{aligned}
B_i s' &\leq 0 \qquad \text{if} \qquad P_i(x^{(0)}) = 0 \qquad \text{where } B_i = (b_{i1}, \ldots, b_{in}) \\
ss' &= 1 \\
\nabla f(x^{(0)}) &\cdot s' \text{ is maximum}
\end{aligned} \qquad (3\text{-}104)$$

It is assumed that $\nabla f(x^{(0)}) \neq 0$; if the opposite holds, we shall take a feasible point $y^{(0)}$ in the neighborhood of $x^{(0)}$ such that $\nabla f(y^{(0)}) \neq 0$. If such a point does not exist, obviously $x^{(0)}$ is the optimum. Since we intend to determine the direction such that the objective function increases if we move from $x^{(0)}$ in the direction of s, we are only interested in a vector s with $\nabla f(x^{(0)}) \cdot s' \geq 0$. The second condition in (3-104) can be replaced by an inequality $ss' \leq 1$. However, a further simplification can be achieved by considering a slightly different problem.   Consider

$$B_i s' \leq 0 \quad \text{if} \quad P_i(x^{(0)}) = 0$$
$$-1 \leq s \leq 1 \tag{3-105}$$
$$\nabla f(x^{(0)}) \cdot s' \text{ is maximum}$$

Although the direction of s determined from (3-105) is not the best feasible direction, there is an obvious advantage to (3-105); it becomes a linear-programming problem if we make a trivial transformation.   Thus at every step we have only to solve a linear-programming problem for determining the direction of the next step.   Once s is determined, we must determine the length of the step in that direction.

If λ denotes the length of the step, then λ must be such that (1) none of the constraints will be violated by the next solution

$$x^{(1)} = x^{(0)} + \lambda s$$

and (2) $f(x^{(0)} + \lambda s)$ is maximized as a function of λ (Cauchy gradient criterion).   Hence, λ must be min (λ_1, λ_2), where

$$\lambda_1 = \min_i \frac{P_i(x^{(0)})}{B_i s'} \text{ for those } i \text{ for which } B_i s' > 0 \tag{3-106}$$

$$\lambda_2 = \frac{\nabla f(x^{(0)}) \cdot s'}{s A s'} \tag{3-107}$$

The method proceeds as follows:

Step 1 (*not repeated*):   Obtain a feasible solution $x^{(0)}$ for the problem. Either this can be obtained by inspection or we can introduce an extra variable ξ and first solve the problem

$$\begin{aligned} B_i x' &\leq b_i & i = 1, \ldots, m \\ B_i x' - \xi &\leq b_i & i = i_1, \ldots, i_k \\ \xi &\geq 0 \\ \xi &\text{ is a minimum} \end{aligned} \tag{3-108}$$

Starting with $x = \bar{x} = -\min P_i(\bar{x})$, $i = i_1, \ldots, i_k$, the value of x for which $\xi = 0$ is a feasible solution.

Step 2:   Make the transformation

$$t_j = 1 + s_j$$

and let

$$\beta_i = \sum_j b_{ij}$$

then $B_i t' \leq \beta_i$ for those i such that $P_i(x^{(0)}) = 0$.

$$0 \leq t \leq 2$$
$$\nabla f(x^{(0)}) \cdot t' \text{ is maximum}$$

Sometimes the above problem can be solved by inspection; otherwise we use linear-programming techniques. Let $\bar{l}$ be a solution of the above. Then $\bar{s} = \bar{l} - 1$.

Step 3: Calculate $\lambda_1 = \min_i [P_i(x^{(0)})/B_i\bar{s}']$ for those i for which $B_i\bar{s}' > 0$.

Step 4: Calculate

$$\lambda_2 = \frac{f(x^{(0)}) \cdot \bar{s}'}{\bar{s}A\bar{s}'}$$

Let $\lambda = \min(\lambda_1, \lambda_2)$.

Step 5: The new feasible point is $x^{(0)} + \lambda\bar{s}$.

Step 6: Proceed from step 2 again, and repeat the process until the optimum is reached.

Example: Maximize

$$f = -5 + 4x_1 + 2x_2 - x_1{}^2 - x_2{}^2$$

subject to $x_1 + x_2 \leq 2 \qquad x_1 \geq 0 \qquad x_2 \geq 0$

Let $x^{(0)} = (1, \tfrac{1}{4})$ be a feasible point

$$\nabla f(x^{(0)}) = (4 - 2x_1,\ 2 - 2x_1) = (2, \tfrac{3}{2})$$

Hence $2s_1 + \tfrac{3}{2}s_2$ must be maximized subject to the constraints

$$-1 \leq s_1 \leq 1 \qquad -1 \leq s_2 \leq 1$$

Clearly $s_1 = 1 = s_2$ is a solution of the above problem: $\lambda_1 = \tfrac{3}{8}$, $\lambda_2 = \tfrac{7}{8}$. Hence $\lambda = \tfrac{3}{8}$. Therefore, the new feasible point is

$$(1 + \tfrac{3}{8},\ \tfrac{1}{4} + \tfrac{3}{8}) = (1\tfrac{3}{8}, \tfrac{5}{8}) = x^{(1)}$$

and $\nabla f(x^{(1)}) = (\tfrac{5}{4}, \tfrac{3}{4})$.

Hence $\tfrac{5}{4}s_1 + \tfrac{3}{4}s_2$ must be maximized subject to

$$-1 \leq s_1 \leq 1 \qquad -1 \leq s_2 \leq 1$$
and $$s_1 + s_2 \leq 0$$

The solution is given by $s_1 = 1$, $s_2 = -1$. Thus $\lambda_1 = \tfrac{1}{8}$, $\lambda_2 = \tfrac{1}{8}$; hence $\lambda = \tfrac{1}{8}$. The new feasible point is $(\tfrac{3}{2}, \tfrac{1}{2})$. It is readily verified that this is the maximum.

We shall now prove the convergence of the algorithm. If the feasible region is bounded, the sequence of points $x^{(k)}$ is also bounded; hence there will be at least one point of accumulation. Let us assume that $x^{(k)}$ has only one point of accumulation (denoted by x^*) which is not a stationary point; i.e., there exist a vector $\bar{s}$ and a real number $\gamma > 0$ such that

$$\nabla f \cdot \bar{s}' > \gamma \qquad B_i \cdot \bar{s}' \leq 0$$

for all i for which $P_i(x^*) = 0$.

It can be proved that the above assumption leads to a contradiction. Let I denote the set of i for which $P_i(x^*) = 0$. Let $R(\epsilon)$ denote the set of points x within a distance ϵ of x^* inside the feasible region. Let $\gamma(\epsilon)$ denote the set of directions that would be chosen for the next move. Starting at any point x with $R(\epsilon)$, the constraints $B_i s' \leq 0$ are imposed for any subset of I but not for $i \notin I$. Now the components of ∇f are continuous functions of x; hence the set $\gamma(\epsilon)$ increases continuously as ϵ increases from 0. Further, $-\nabla f(x^*) \cdot s' > \gamma > 0$ for all γ in $\gamma(\epsilon)$. Hence there exists $\epsilon > 0$ such that $\nabla f \cdot s' > 0$ for all x within $R(\epsilon)$ and all γ within $\gamma(\epsilon)$. We can suppose that ϵ is chosen so small that for all x in $R(\epsilon)$, $P_i(x) > 0$ for all $i \epsilon I$. Now our convergence assumption implies that there exists some $k_0(\epsilon)$ such that $x^{(k)}$ is within $R(\epsilon)$ for all $k > k_0$. Thus, after a finite number of steps, we drop all restrictions

$$B_i \cdot \bar{s}' \equiv 0 \qquad \text{for} \qquad i \epsilon I$$

Hence we have arrived at a contradiction. If, however, the sequence of points has more than one point of accumulation, it can be proved that they are all stationary points. As this proof is lengthy, we refer the reader to Zoutendijk.

Kelley's Method of Cutting Planes[44,76]

Kelley considers the problem of minimizing a linear function

$$f = \sum_{i=1}^{n} a_i x_i \equiv \alpha x'$$

subject to the nonlinear constraint $g(x) \leq 0$, where

$$\alpha = (a_1, \ldots, a_n) \qquad x = (x_1, \ldots, x_n)$$

and $g(x)$ satisfies the following conditions:

1. g is continuous and convex.
2. g is defined in $S = \{x : Bx' \leq b'\}$, where $B = (b_{ij})$ is an $n \times n$ matrix; $b = (b_1, \ldots, b_n)$.
3. At every t of S, there is a supporting hyperplane $p(x,t)$ for $g(x)$ such that $\|\nabla p(x,t)\|$ (the square root of the sum of squares, or any norm, for that matter) is bounded for all t.

This algorithm employs the theorem which states that any compact convex set can be approximated (circumscribed) by a compact convex polyhedron to any degree of accuracy.   We choose a compact polyhedron which includes the feasible region and attempt to minimize the linear objective function over this polyhedron.   As this is a linear-programming problem, some vertex of this polyhedron yields a solution (however, this point does not belong to the feasible region).   We then pass a hyperplane (cutting plane) between the point which yielded a minimum for the objective function and the feasible region, thus obtaining a new polyhedron which is contained in the previous one.   This process of making cuts and minimizing the linear form is continued until a solution is reached which is sufficiently near the optimum.   The iterations converge to a limit.   A method for accelerating the algorithm has been given by Wolfe,[76] to whom we owe the general idea illustrated in Fig. 3-6.

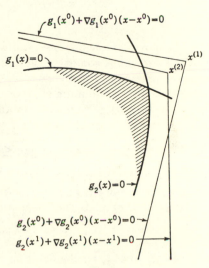

Fig. 3-6

Let $R = \{x : g(x) \leq 0\}$ be the feasible region of our problem. Let t be any point in $S - R$.   An extreme support to the graph of $g(x)$ at this point may be written in the form

$$p(x,t) = g(t) + \nabla p(x,t) \cdot (x - t)' = C \qquad (3\text{-}109)$$

and when $g(x)$ is smooth, $C = p(x,t)$ is the tangent plane to the graph of $g(x)$ at $x = t$.   Since $g(x)$ is convex, $p(x,t) \leq g(x)$ for all $x \, \epsilon \, S$.   Thus if $x \, \epsilon \, R \equiv \{g(x) \leq 0\}$, then $p(x,t) \leq 0$.   If $y \, \epsilon \, R$, then $p(y,t) = g(y) > 0$. Thus the set R and the point y lie on opposite sides of $p(x,t) = 0$.

Let $S_0 \equiv S$, let t_0 be the point which yields a minimum to the linear objective function, and let $S_1 = S_0 \cap \{x : p(x,t_0) \leq 0\}$ (t_0 lies in $S - R$; otherwise t_0 is the optimum).   In general let $f = \alpha t_k'$ (note that t_k is a vector).   In this way a sequence $\{S_k\}$ of convex sets is generated such that $S_k \subset S_{k-1}$, and hence $g(t_{k-1}) \leq g(t_k)$.   In general t_0 does not belong to S_1 .   Thus there exists a point t_1 which minimizes f in S_1 .   It is proved below that $\{t_k\}$ contains a subsequence converging to $\bar{t} \, \epsilon \, R$, which is the optimum.

The problem thus reduces to determining $p(x,t_k)$, and hence $\nabla p(x,t_k)$; since $g(x)$ is differentiable, $\nabla p(x,t_k) = \nabla g(t_k)$.   The last equation is correct only if we assume that $g(x)$ is differentiable, since the right- and left-hand derivatives at a point may not be equal for a general convex function. In such cases $\nabla g(t)$ is not defined at that point. It is for this reason that $p(x,t)$ is introduced. Now we must solve the following problem:

$$Bx' \leq b'$$
$$-\nabla p(x,t_i)x' \geq g(t_i) - \nabla p(x,t_i)t_i$$
$$\alpha x' = \text{minimum} \qquad 0 \leq i \leq k - 1 \qquad (3\text{-}110)$$

The above problem is essentially a linear-programming problem. If, however, R is not compact, the problem can be solved if there exists a finite minimum and if it is possible to select S containing R^* (which is in R) such that the minimum of f over R is contained in R^*. However, if it is known that it is impossible for any coordinate of the solution vector x to be numerically greater than a fixed upper bound, then the hypercube

$$s = \{x : |x_j| \leq \lambda,\ 1 \leq j \leq k\}$$

may be a suitable choice.

We now detail the steps of the algorithm.

Step 1: Formulate the linear-programming problem which minimizes

$$f = \alpha x' \qquad \text{subject to} \qquad Bx' \leq b'$$

Step 2: Let t_0 be the solution of the problem in step I. Calculate

$$\nabla g(t_0) = \nabla p(x,t_0)$$

Step 3: Calculate $p(x,t_0) = g(t_0) - \nabla p(x,t_0) \cdot (x - t_0)'$.

Step 4: Formulate the linear-programming problem minimizing f subject to

$$Bx' \leq b' \qquad \text{and} \qquad p(x,t_0) \leq 0$$

Repeat this process until the solution is sufficiently near the optimum. Although a foolproof criterion for terminating the process is not presently known, stopping when $g(t_k) \leq \epsilon$ (ϵ small and positive) seems to be a useful condition.

Example: Minimize

$$f = x_1 + x_2$$
subject to $\quad g(x) = x_1{}^2 + x_2{}^2 - 4x_1 - 4x_2 + 7 \leq 0$
$$R = \{x : g(x) \leq 0\}$$
$$S_0 = \{(x_1,x_2) : 1 \leq x_1,\ x_2 \leq 3\}$$

Note that R is a disk of radius 1 centered at $(2,2)$, and that the solution is $x = (2 - 1/\sqrt{2},\ 2 - 1/\sqrt{2})$, $f = 4 - \sqrt{2}$. The corresponding linear-programming problem is: Minimize

$$x_1 + x_2$$

subject to

$$1 \leq x_1 \leq 3$$
$$1 \leq x_2 \leq 3$$

Solution: $x = (1,1)$ with $f = 2$. Set $t_0 = (1,1)$. The new constraint required for the second step is

$$p(x,t_0) = g(t_0) - \nabla p(x,t_0) \cdot (x - t_0)'$$

If we write

$$t_0 = (t_{10}, t_{20})$$

then $\nabla p(x,t_0) = (2t_{10} - 4, 2t_{20} - 4)$

$$p(x,t_0) = 1 + (x_1 - 1, x_2 - 1) \cdot (-2,-2)' = 5 - 2x_1 - 2x_2$$
$$S_1 = S_0 \cap \{x : p(x,t_0) \leq 0\}$$

Next solve the following: Minimize

$$x_1 + x_2$$

subject to

$$1 \leq x_1 \leq 3$$
$$1 \leq x_2 \leq 3$$
$$5 - 2x_1 - 2x_2 \leq 0$$

Solution $\quad x = (\tfrac{3}{2},1) \quad f = \tfrac{5}{2} \quad g(x) = \tfrac{1}{4}$

Set

$$t_1 = (\tfrac{3}{2},1)$$
$$p(x,t_1) = 15 - 4x_1 - 8x_2$$
$$S_2 = S_1 \cap \left\{ \frac{x}{p(x,t_1)} \leq 0 \right\}$$
$$x = (1,1\tfrac{1}{8}) \quad f = 1\tfrac{9}{8}$$

Note that this solution does not satisfy the constraint of the previous cutting plane. The true solution at this point is $x = (\tfrac{5}{4},\tfrac{5}{4})$, $f = \tfrac{5}{2}$, $g(x) = \tfrac{1}{8}$.

Exercise 3-13 Carry out the example through two additional iterations.

Proof of Convergence

The proof of convergence of the algorithm is given by the following theorem:

Theorem 3-14 Let $g(x)$ be a continuous convex function defined on the n-dimensional compact convex set S such that at every point $t \epsilon S$ there exists an extreme support $C = p(x,t)$ to the graph of $g(x)$ with the property that, for some constant K, $\|\nabla p(x,t)\| \leq K$ for all $x \epsilon S$. Further, let $f(x) = \alpha x'$, and let $R = \{x : g(x) \leq 0\}$ with R nonempty. If $t_k \epsilon S_k$ is such that $f(t_k)$ is minimum for $\{f(x) : x \epsilon S_k\}$, $k = 0, 1, \ldots,$

where $S_0 = S$, and where

$$S_k = S_{k-1} \cap \{x : p(x, t_{k-1}) \le 0\}$$

then the sequence $\{t_k\}$ contains a subsequence that converges to a point $\bar{t} \in R$ with $f(\bar{t}) \le f(x)$ for all $x \in R$.

Proof: Since t_k minimizes $f(x)$ on S_k,

$$g(t_i) + \nabla p(t_k, t_i) \cdot (t_k - t_i)' \le 0 \qquad 0 \le i \le k - 1$$

Suppose $\{t_k\}$ does not have a subsequence which converges to a point in R; then there exists an $r > 0$ independent of k such that

$$r \le g(t_i) \le -\nabla p(t_k, t_i) \cdot (t_k - t_i)' \le k\|t_i - t_i\| \qquad 0 \le i \le k - 1$$

It follows that for every subsequence $\{k_p\}$ of indices,

$$\|t_{k_q} - t_{k_p}\| \ge \frac{r}{k} \qquad q < p$$

Thus $\{t_k\}$ does not contain a Cauchy subsequence. But this is impossible, since S is compact. Hence $\{t_k\}$ contains a subsequence which converges to a point $\bar{t} \in S$. Since it also follows that the corresponding subsequence $\{f(t_k)\}$ converges to zero, $\bar{t}$ must be in R.

Beale's Method[9]

Beale considers the problem of minimizing a convex quadratic function

$$f(x) = a + \alpha x' + xAx' \tag{3-111}$$

subject to linear constraints $Bx' \le b'$ and $x \ge 0$, where primes indicate transposes, $x = (x_1, \ldots, x_n)$, $A = (a_{ij})$ is an $n \times n$ matrix, a is a constant, $\alpha = (a_1, \ldots, a_n)$, $B = (b_{ij})$ is an $m \times n$ matrix, and $b = (b_1, \ldots, b_m)$.

We shall first describe the process intuitively and then follow it by a formal presentation. The inequality constraints are reduced to equations by introducing nonnegative "slack variables," thus increasing the number of variables to $m + n$. It is then possible to express m of the variables (now called the *basic* variables) using our m equations, in terms of the remaining n variables (called the *nonbasic* variables). Assigning an arbitrary nonnegative value to the basic variables yields a value for the nonbasic variables which may not be nonnegative. A feasible point consists of a choice of nonnegative values of the basic variables which also yields nonnegative values of the nonbasic variables. Without loss of generality, we may assume that a feasible point has been found in which all nonbasic variables have the value zero. At the initial step a nonbasic variable is some x_j, but later we introduce new nonbasic variables u_k which need not be any of the x's; for this reason z_k will be used to denote a nonbasic variable in the general discussion. In actual calculation we shall use x's and u's. We note that a nonbasic variable need not represent a direction parallel to the coordinate axes, but instead may represent a direction passing through the feasible region; hence the use of z_k.

The process begins by increasing from zero (which is taken as the initial feasible point) the value of one of the nonbasic variables (to determine which one, see below) and is continued until one of three cases obtains:

1. The function f begins increasing. [This happens if $\partial f/\partial z_k$ becomes zero and then positive, where z_k is the nonbasic variable for which $|\partial f/\partial z_k|$ among all $\partial f/\partial z_k \leq 0$ has its largest value at the initial point. Note that in the direction of this z_k, $f(x)$ decreases most rapidly. Hence variation along this z_k yields the greatest improvement towards the minimum of f.]

2. One of the constraints is violated as this z_k increases.

3. As this z_k increases, one of the basic variables which originally assumed a positive value becomes negative, thus violating the condition $x_j \geq 0$. This condition is essentially the same as (2).

A new set of nonbasic variables is introduced, according to which of these cases occurs, by a method described below. One then proceeds with the new set of nonbasic variables. Continuing in this manner, one reaches a point where no further change in the value of a nonbasic variable produces any decrease in the value of f. Since f is a convex function, a global maximum is reached. The justification for these statements is given below.

Let x_i denote a basic variable, and z_j a nonbasic variable. Then, using the linear constraints, we write, for any m of the n variables,

$$x_i = a_{i0} + \sum_{j=1}^{n-m} a_{ij}z_j \qquad i = 1, \ldots, m \qquad (3\text{-}112)$$

where z_j denotes x_{m+j}. Similarly, f can be expressed in terms of the nonbasic variables by substituting z_j's for the basic variables x_i:

$$f = \sum_{i=0}^{n-m} \sum_{j=0}^{n-m} a_{ij}z_iz_j \qquad (3\text{-}113)$$

Here $z_0 = 1$, and $z_1, \ldots, z_{n-m}$ are the nonbasic variables. If $\partial f/\partial z_k < 0$ for any k, $k = 1, \ldots, n - m$, then a small increase in z_k, with the other nonbasic variables (z_i, $i \neq k$) held equal to zero, will reduce f. Hence it is profitable to go on increasing z_k until either (1) some x_k becomes zero and decreases to negative values or (2) $\partial f/\partial z_k$ becomes zero and increases to positive values. In (1), the set of nonbasic variables is changed by replacing z_k by x_k, and f is expressed in terms of the new nonbasic variables. In (2), as $\partial f / \partial z_k$ is a linear function of the z_j's since f is a quadratic function, we introduce $u_k = \frac{1}{2} \partial f / \partial z_k$ as a nonbasic variable. This u_k is similar to any other nonbasic variable, except that u_k may assume both positive and negative values; hence it is called a *free variable*, as distinguished from the other nonbasic variables which cannot be negative. The justification for choosing this particular form as a nonbasic variable is that after substitution for z_k in terms of u_k and the remaining variables in f, f will contain only one term in u_k involving u_k^2. After this is accomplished, we proceed by repeating the steps until the optimum is reached.

By following the foregoing treatment, we can derive the steps of the algorithm as follows:

Step 1:   Find a basic feasible solution (in general, zero is convenient).

Step 2:   Express f in terms of the nonbasic variables.

Step 3:   Calculate $\partial f / \partial z_k$, where z_k is nonbasic for all k, and choose z_k such that $|\partial f / \partial z_k|$ is largest among all k for which $\partial f / \partial z_k$ is negative.

Step 4:   We have given three conditions on the choice of z_k which we now apply. Solve for z_k in $\partial f / \partial z_k = 0$, which is linear in z_k and hence yields a value for z_k in which all nonbasic variables other than z_k are zero. Also solve for z_k for each i in the equations

$$x_i = a_{i_0} + \sum_{j=1}^{n-m} a_{ij} z_k = 0 \qquad (3\text{-}114)$$

where $z_j = 0$ if $j \neq k$. Note that here we have put all x_i equal to zero.

Now compare the different values obtained from these two computations. Let z_{k_0} be the minimum of all values of z_k. We follow step 5a if z_{k_0} is obtained from Eqs. (3-114); otherwise we follow step 5b.

Step 5a: Put $u_k = \frac{1}{2} \partial f / \partial z_k$, and solve for z_k. (This expression is linear in z_k.) Then replace z_k by the result throughout. Note that $u_k = 0$ at this stage, since $\partial f / \partial z_k = 0$.

Step 5b: Having determined $z_k = z_{k_0}$ to be the minimum corresponding to a certain equation in (3-114), replace z_k in f by the x_i in that equation, and obtain a new expression for f in terms of the nonbasic variables. Note that z_k at this stage is basic and equal to $-a_{i_0} / a_{i_k}$.

Step 6:   Calculate $\partial f / \partial z_k$, where z_k is any nonbasic variable. Calculate $|\partial f / \partial z_k|$ for all z_k such that either $\partial f / \partial z_k$ is negative or $z_k = u_k$ of step 5a. We distinguish the two cases.

Case 1:   If $\partial f / \partial z_k$ is maximum for $z_k = u_k$, then u_k must be decreased or increased according as $\partial f / \partial u_k$ is positive or negative, and one follows **step** 5 using the new feasible point with the minimum value of z_k just computed. (Note that f is different now, and the original z_k has been eliminated.)

Case 2: If z_k is not u_k, then the procedure is similar to step 4. The algorithm terminates when it is not profitable to change any of the nonbasic variables, and the minimum is thereby reached.

Example: Minimize

$$f = x_1{}^2 + x_2{}^2 - 4x_1 - 2x_2 + 5$$

subject to
$$x_1 + x_2 \leq 4 \qquad (3\text{-}115)$$
$$x_1, \, x_2 \geq 0 \qquad (3\text{-}116)$$

The constraint (3-115) can be reduced to $x_1 + x_2 + x_3 = 4$, where x_3 is a nonnegative slack variable. Let an initial feasible solution be $x_1 = 0 = x_2$, which yields $x_3 = 4$. Next we note that f is already given in terms of the nonbasic variables x_1 and x_2. We have

$$\frac{\partial f}{\partial x_1} = -4 + 2x_1$$

Putting
$$\frac{\partial f}{\partial x_1} = 0 \qquad \text{yields} \qquad x_1 = 2 \qquad (3\text{-}117)$$

On solving for x_1 in $x_3 = 4 - x_1 - x_2 = 0$ with $x_2 = 0$, we obtain

$$u_1 \equiv x_1 = 4 \qquad (3\text{-}118)$$

The smaller of the two values for x_1 is given in (3-117); that is, $x_1 = 2$. Hence

$$u_1 = \frac{1}{2}\frac{\partial f}{\partial x_1} = -2 + x_1$$
$$x_1 = u_1 + 2$$

Expressing f in terms of x_2 and u_1, we obtain $f = 1 - 2x_2 + x_2{}^2 + u_1{}^2$. To decide which of the variables must be increased, we again take $u_1 = 0$, $x_2 = 0$ and evaluate $\partial f / \partial u_1 = 2u_1 = 0$, and

$$\frac{\partial f}{\partial x_2} = -2 + 2x_2 = -2$$

Hence x_2 is the variable to be increased.

Putting $\partial f / \partial x_2 = 0$ gives $x_2 = 1$ and $x_3 = 4 - x_1 - x_2 = 0$, or $4 - 2 - x_2 = 0$; that is, $x_2 = 2$, using the value of x_1 from (3-117).

Since $x_2 = 1$ has decreased, introduce $u_2 = -1 + x_2$.

Expressing f in terms of u_2 and u_1, we obtain $f = u_1{}^2 + u_2{}^2$, which is minimum at $u_1 = u_2 = 0$; any change in u_1 or u_2 would increase f. Hence the minimum value of f is zero. [Note that f represents a family of concentric circles centered at $(2,1)$. This point falls in the region $x_1 + x_2 \leq 4$. Hence the problem could also be solved by elementary calculus. One would then have to solve two simultaneous linear equations. And if one does this by first eliminating x_1, one essentially follows the steps of the quadratic programming algorithm just outlined.] The solution now is given by $x_1 = 2$ and $u_2 = 1$.

Below, a slight variation of the above problem is discussed. We replace the constraint (3-115) by a new constraint

$$x_1 + x_2 \leq 2 \tag{3-119}$$

To reduce the inequality to an equation, we introduce the slack variable x_3. We have

$$x_3 = 2 - x_1 - x_2 \tag{3-120}$$

Now $\dfrac{\partial f}{\partial x_1} = -4 + 2x_1 = 0$ gives $x_1 = 2$

Also $x_3 = 2 - x_1 = 0$ gives $x_1 = 2$.

Let us take Eq. (3-120) and solve for x_1. This gives

$$x_1 = 2 - x_3 - x_2$$

Substituting in f, we have

$$f = 2x_2{}^2 + x_3{}^2 + 1 - 2x_2 + 2x_2x_3$$

Again $\dfrac{\partial f}{\partial x_2} = 4x_2 - 2 - 2x_3 = -2$ since $x_2 = 0$

$$\dfrac{\partial f}{\partial x_3} = 2x_3 + 2x_2 = 0$$

Hence x_2 is to be increased.

Solving for x_2 in $\partial f/\partial x_2 = 0$ yields $x_2 = \frac{1}{2}$, and solving for x_2 in $x_1 = 2 - x_3 - x_2 = 0$ gives $x_2 = 2$. Hence x_2 can only be increased up to $\frac{1}{2}$. We replace x_2 by $\frac{1}{2}\partial f/\partial x_2 = u_1$:

$$\frac{1}{2}\frac{\partial f}{\partial x_2} = 2x_2 - 1 - x_3 = u_1 \quad\text{or}\quad x_2 = \frac{u_1 + 1 + x_3}{2}$$

Hence $f = 2\,\dfrac{(u_1 + 1 + x_3)^2}{2} + x_3{}^2 + 1 - 2\,\dfrac{u_1 + 1 + x_3}{2}$

$$+ 2x_3\,\frac{u_2 + 1 + u_2}{2}$$

$$= \frac{u_1{}^2}{2} + \frac{5}{2}x_3{}^2 + \frac{1}{2} + x_3 + 2u_1x_3$$

$\dfrac{\partial f}{\partial u_1} = 2u_1 + 2x_3 = 0$ when $u_1 = 0,\ x_2 = 0$

$\dfrac{\partial f}{\partial x_3} = 5x_3 + 2u_1 = 0$ when $u_1 = 0,\ x_3 = 0$

Hence no change in x_3 or u_1 produces any decrease in f, and we have reached the minimum. The solution is given by $x_1 = \frac{3}{2}$, $x_2 = \frac{1}{2}$, at which the minimum is given by $f = \frac{1}{2}$.

Convergence

We give here a feasibility argument for the convergence of Beale's method in a finite number of steps. We shall say that the objective function is in standard form if the linear term contains no free variable.

For convergence we need the following procedure: If, at any stage, there is a nonbasic free variable u for which $\partial f/\partial u = 0$, then a free variable must be removed from the nonbasic set. Now when f is in standard form, its value in the present trial solution is a stationary value of f subject to the restriction that all the present nonbasic restricted variables take the value zero. There can be only one value for which any set of nonbasic restricted variables take the value zero, so there can be only one such value for any set of nonbasic restricted variables. There are only a finite number of possible sets of nonbasic restricted variables, so in order to show that the iterations, if not initially in standard form, must terminate, it suffices to show that they invariably reach a standard form in a finite number of steps.

We prove this as follows: Our procedure ensures that when f is not in standard form a free variable will be removed from the nonbasic set. Let p be the number of nonbasic free variables. It is easy to show that, in the new expression for f, the off-diagonal elements in the row associated with the new nonbasic variable must vanish. It follows that f does not contain a linear term in this variable, and furthermore f can never contain a linear term in the variables unless some other restricted variable becomes nonbasic, thereby decreasing p.

Therefore, if f is not in standard form, and $p = p_0$, then p cannot increase and must decrease after at most p_0 steps, unless p meanwhile achieves a standard form. Since f is always in standard form when $p = 0$, the required result follows.

Wegner's Method[72] (a Nonlinear Extension of the Simplex Method)

This problem is concerned with maximizing

$$f = \sum_{j=1}^{n} c_j x_j$$

subject to m nonlinear separable† constraints $\sum_{j=1}^{n} a_{ij}(x_j) = b_i$, $m \leq n$. It is assumed that it is possible to obtain a basic feasible solution in terms of the first m variables so that we can write the constraints in the form

$$b_i = x_i + \sum_{j=m+1}^{n} g_{ij}(x_j) \qquad (3\text{-}121)$$

where $\qquad g_{ij}(0) = 0 \qquad i = 1, \ldots, m$

for $\qquad\qquad j = m + 1, \ldots, n$

This method, which is an extension of the simplex method of linear programming, uses two quantities, namely the marginal substitution coefficient and the average substitution coefficient of one variable with respect to another, to get a new basic feasible solution from a previous one. In the simplex method both quantities are equal to the constraint coefficients a_{ij}.

Here we define

$$a_{ij}{}^M(x_j) = \frac{\partial a_{ij}(x_j)}{\partial x_j} \qquad (3\text{-}122)$$

as the marginal substitution coefficient of x_i with respect to x_j, and

$$a_{ij}{}^A(x_j) = \frac{a_{ij}(x_j)}{x_j} \qquad (3\text{-}123)$$

as the average substitution coefficient of x_i with respect to x_j. The quantity

$$F_j \equiv \sum_{i=1}^{m} a_{ij}{}^M c_i - c_j$$

---

† The method given in Ref. 72 was formulated for general nonlinear constraint equations. However, Wegner has indicated in a private communication that the method does not, in its present form, work for nonseparable constraints, since an equation of the form $h_i(x_1, x_2, \ldots, x_n) = b_i$ cannot be replaced by an equivalent constraint of the form $\sum_j a_{ij}{}^A(x_1, x_2, \ldots, x_n) \cdot x_j = b_i$; that is, whereas in the separable case the value of a function can be built up as a sum of products of the average rates of increase $a_{ij}{}^A$ of each variable and the magnitude x_j of the corresponding variable, this is not so in the nonseparable case. Wegner also adds that he would not recommend this method for the solution of practical problems, but that the approach has possibilities for further development. The method essentially reduces nonlinear programming to the solution of a sequence of sets of nonlinear simultaneous equations, where each set of equations differs from its predecessor by one variable, and the criteria for addition and deletion are nonlinear generalizations of those used in the simplex method. The method could be modified for nonseparable constraints by applying the nonlinear solution procedure to the constraints in their original form. The marginal substitution criterion for addition of a variable would remain the same, but a different procedure for deletion of a variable would be required.

gives the rate of increase of the objective function f.   Moreover, if

$$\sum_{i=1}^{m} a_{ik}{}^{M} c_i - c_k < 0$$

then x_k increases the objective function when introduced into the solution. It is further assumed that an initial basic feasible solution can be obtained from which $a_{ij}{}^{A}$ and $a_{ij}{}^{M}$ are computed.

Let $x^{(r)} = (x_1{}^{(r)}, \ldots, x_m{}^{(r)}, x_{m+1}{}^{(r)})$ be the rth approximation to the new basic feasible solution, and let $x_r \equiv (x_1)_r, \ldots, (x_{m+1})_r$ be the solution obtained by substituting $x^{(r)}$ into the original tableau and transforming to give a basic solution in m of the variables (including x_{m+1}). In general the following equations must be satisfied at the new basic solution:

$$x_i{}^{(r)} = (x_i)_r x^{(r)} \qquad i = 1, \ldots, m+1 \qquad (3\text{-}124)$$

The equation in the variable excluded from the basic feasible solution is an identity, and hence (3-124) is a system of m simultaneous equation in m unknowns.

We now detail the steps of the algorithm.

Step 1:   Let $(x_1{}^0, \ldots, x_m{}^0)$ be a basic feasible solution.   Compute

$$F_j = \sum_{i=1}^{m} (a_{ij}{}^{M}) c_i - c_j \qquad j = 1, \ldots, m$$

If all the F_j's are positive, we have already reached a local optimum. Otherwise choose x_k such that $|F_k|$ is maximum among $F_k < 0$.

Step 2:   Obtain a set of nonnegative values $x_1, \ldots, x_m, x_k$ with $x_j = 0$ for $j = 1, 2, \ldots, m$ which, when used to calculate $a_{ij}{}^{A}$ in the initial tableau, gives a solution identical to the assumed values of the variables when the initial tableau is solved for these variables.   This is achieved by the following procedure:

Let $x^{(0)} = (x_1{}^{(0)}, \ldots, x_m{}^{(0)})$ be the old basic feasible solution, and let $x_0 = [(x_1)_0, \ldots, (x_m)_0, (x_k)_0]$ be the new one formed when x_k is introduced into the tableau based on the old basic feasible solution.   One of the quantities $(x_p)_0$ will be zero.   Let this be dropped from x_0, and number the remaining variables consecutively; thus

$$x_0 = [(x_1)_0, \ldots, (x_m)_0]$$

Now consider the solution of the m equations

$$(x_i)_r - x_i{}^r = 0 \qquad i = 1, \ldots, m \qquad (3\text{-}125)$$

where the subscript and the superscript r refer to the rth iterate.

Let $\varphi_i\ (x_1{}^r,\ \ldots\ ,x_m{}^r) = x_i{}^r - (x_i)_r$. Then the approximate derivative is given as follows:

$$\left(\frac{\partial \varphi_i}{\partial x_j}\right)_{x^r} = \frac{1}{x_j{}^r}\ [\varphi_i(x_1{}^r,\ \ldots\ ,\ x_j{}^r + dx_j{}^r,\ \ldots\ ,\ x_m{}^r) - \varphi_i(x_1{}^r,\ \ldots\ ,\ x_m{}^r)]$$

The derivative $\partial\varphi_i/\partial x_j$ can be approximately calculated for all i by evaluating

$$\varphi_i(x_1{}^r,\ \ldots\ ,\ x_j{}^r + dx_j{}^r,\ \ldots\ ,\ x_m{}^r)$$

for all i in the initial tableau and substituting in the above. All the coefficients $(\partial\varphi_i/\partial x_j)_{x^r}$ can be calculated by carrying out a simplex-type transformation for the tableau obtained by adding $dx_j{}^r$ to each corresponding variable j in turn.

Using the Newton-Raphson method for m equations, we arrive at a new solution. If the new solution has no negative variables, it is a basic feasible solution. If a negative variable appears, a wrong variable x_k had been excluded at the beginning of the iteration procedure. We return to that step, and, instead of x_k, we choose the next variable for which $|F_k|$ is maximum among the remaining $F_k < 0$.

Step 3: Let F_k be the value obtained using the coefficients corresponding to the new values of the variables in that tableau which gives the solution for the previous set of variables. The previous solution is to be discarded if F_k is negative. Otherwise calculate the value of the objective function at the new basic solution. If f shows improvement over the previous value, discard the previous solution; otherwise, keep it.

The three steps are to be repeated unless either an optimum is reached or all negative F_j become positive when the corresponding variable is introduced into the solution. If the surface can be approximated by a quadratic surface in this region, then an interpolation procedure for finding an approximation to the optimum solution is as follows:

Let $x^{(0)}$ be the initial basic feasible solution, and let $x^{(1)}, x^{(2)},\ \ldots\ , x^{(r)}$ be the r basic feasible solutions obtained by introducing the r variables which make F_j change sign. Let $x_{0k}, k = 1,\ \ldots\ , r$, be the coordinate axes in the direction of $x^{(k)} - x^{(0)}$, and let the distance between $x^{(k)}$ and $x^{(0)}$ be a unit distance. Let F_{0k} be the value of F_k before x_k was introduced into the solution, and let $F_{hk}\ (>0)$ be the value of F_k after x_h has been introduced into the solution.

In the r-dimensional space defined above, F_{0k} is the rate of increase of the objective function in the direction x_0 at the basic feasible solution $x^{(0)}$.

Let $\partial f/\partial x_{0k} = F_k(x_{01},\ \ldots\ ,x_{0r})$ be the rate of increase of f at an arbitrary point. Then the conditions for an optimum are given by $F_k(x_{01},\ \ldots\ ,x_{0r}) = 0, k = 1,\ \ldots\ , r$.

Using the Newton-Raphson method, we obtain

$$\sum_{p=1}^{m} \left(\frac{\partial F_k}{\partial x_p}\right) dx_0 = -F_{0k} \qquad (3\text{-}126)$$

Since $(\partial F_k/\partial x_p)x^{(0)}$ can be approximated by $F_{pk} - F_{0k}$, (3-126) can be approximated by

$$\sum_{p=1}^{m} (F_{pk} - F_{0k}) dx_{0k} = -F_{0k} \qquad k = 1, \ldots, r$$

Since the solution region is convex in the neighborhood of a nonbasic optimum solution, the interpolated solution lies, in general, inside the feasible region.

Example (Wegner):   Maximize

$$f = .5x_1 + x_2$$

subject to
$$x_1, \, x_2 \geq 0$$
$$.15x_1 + .1x_2 + .025x_2{}^2 \leq 1$$
$$.25x_1 + .1x_2 \leq 1$$

The initial tableau is given by

	.5	1	0	0
1	.15	$.1 + .025x_2$	1	0
1	.25	.1	0	1
0	$-.5$	-1	0	0

$$a_{12}(x_2)x_2 = .1x_2 + .025x_2{}^2$$
$$a_{12}{}^A = .1 + .025x_2$$
$$a_{12}{}^M = .1 + .05x_2$$

Initially,
$$x_2 = 0 \qquad a_{12}{}^A = a_{12}{}^M = .1$$

Following step 2, we bring x_1 into the solution, obtaining the following tableau (the tableaux of average and marginal coefficients will be written as a single tableau for convenience):

	.5	1	0	0
4	0	.04, .04	.1	$-.6$
4	1	.4	0	4
2	0	$-.8$	0	2

Since x_2 is the only variable with negative F_j, it must now be introduced. The iterative procedure defined by the following formula is used:

$$x^{r+2} = x^r - \frac{(x_r - x^r)(x^{r+1} - x^r)}{(x_{r+1} - x^{r+1}) - (x_r - x^r)}$$

This gives the following set of iterates (initial approximation $x_2{}^0 = 0$, $x_2 = 1$), which indicate that convergence is rapid once the region of solution is located:

Iteration r	0	1	2	3	4	5	6
$x_2{}^r$	0	1	2.174	2.900	3.238	3.277	3.279
$a_{12}{}^r$	.1	.125	.1544	.1725	.1810	.1819	
$(x^2)_r$	10	6.4	4.237	3.566	3.306	3.281	

The tableau of average and marginal substitution coefficients corresponding to this basic feasible solution is:

	.5	1	0	0
3.279	0	1, 1	8.197, 4.902	$-4.918, -2.941$
2.689	1	0, 0	$-3.279, -1.961$	5.967, 5.176
4.624	0	0	3.922	$-.353$

F_4 is negative; hence the fourth variable should be reintroduced into the solution.

The value of x_2 corresponding to the new set of basic variables is 4.633, and $x_4 = .537$. When these values are used to recalculate the coefficients for the basic feasible solution represented by the previous tableau, the following tableau is obtained, indicating that at this level it is no longer profitable to introduce x_4:

	.5	1	0	0
2.567	0	1	6.418, 3.681	$-3.854, -2.208$
2.973	1	0	2.567, 1.472	5.590, 4.883
			4.417	.234

The above tableau indicates that the true optimum lies between these two basic feasible solutions, so that the interpolation formula should be used to find an approximation to the optimum.

Let $x_i{}^{(1)}$, $x_i{}^{(2)}$ represent the values of the variables at the enclosing basic feasible solutions, and let $x_i{}^{(3)}$ represent the interpolated value:

$$x_1{}^{(3)} = x_1{}^{(1)} - \frac{.353}{.587} x_2{}^{(1)} - x_2{}^{(2)} = 1.073$$

$$x_2{}^{(3)} = x_2{}^{(1)} - \frac{.353}{.587} x_2{}^{(1)} - x_2{}^{(2)} = 4.093$$

$$x_4{}^{(3)} = \frac{.353}{.587} x_4{}^{(2)} = .323$$

The value of x is increased by increasing one or more of the variables with a positive-value index until the boundary is reached. Hence the optimum is

$$x_1 = 1.153 \qquad x_2 = 4.093 \qquad x_4 = .323$$
$$f = 4.669 \text{ (approx)}$$

Houthakker's Capacity Method[40]

Houthakker considers the problem of maximizing a convex quadratic function $f(x) = a + x' + xAx'$ subject to

$$x \geq 0 \tag{3-127}$$
$$x_1 + x_2 + \cdots + x_n = \beta \leq \beta^* \tag{3-128}$$

and

$$b_{i1}x_1 + b_{i2}x_2 + \cdots + b_{in}x_n - b_i \leq 0 \qquad i = 1, \ldots, m \tag{3-129}$$

where $A = (a_{ij})$ is an $n \times n$ negative-definite matrix, and $a = (a_1, \ldots, a_n)$ is a constant. It is further assumed that the b_{ij} are all nonnegative, β^* and the b_i are all positive, and β is a variable, ranging from 0 to β^*.

The constraint (3-128) is sometimes part of the original problem; otherwise it is introduced artificially for this method. This does not change the basic nature of the problem in any way, since this constraint only determines the route by which the optimum is reached when β^* is sufficiently large. It will be seen that at the optimum this constraint can be ignored. This is shown explicitly in the example considered below.

In this method we start from the feasible point $x_1 = 0, \ldots, x_n = 0$ and then change the value of certain of the x_j at different stages so that f increases, until the maximum is reached, without ever leaving the feasible region. The x_j whose values are changed, like Beale's nonbasic variables, are different from stage to stage. Furthermore $\partial f / \partial x_j$ is required to be the same at any stage for all the x_j which are being increased. At the start the variables with the largest derivatives at the initial point are chosen as the effective set. At each stage we determine how the variables are changed according to whether:

1. Equality does not hold for any of the linear constraints.
2. Equality does hold for some of the linear constraints.

The set of constraints for which equality holds is called the *effective set* at any stage. Geometrically, the second circumstance means that the iterations are performed from one point on the surface of the polyhedron, determined by the constraints, to another such point. The variable capacity β whose range is the closed interval $[0, \beta^*]$ determines the different stages at values $\beta_0, \beta_1, \ldots, \beta_g$, called *critical values* of β. Associated with each critical value β_g there is an effective set Γ_g of variables consisting of those x_j which assume positive values for $\beta_\sigma < \beta < \beta_{\sigma+1}$, and an effective set of constraints Δ_σ to each of whose members, $\sum_j b_{ij}x_j - b_i = 0$, corresponds a parameter λ_i, whose role will be explained later. This method consists of finding successive critical values of β and their corresponding effective sets.

Making use of the second constraint and the relation $x \geq 0$, we see that at $\beta = 0$ all the variables are zero. We shall increase those x_j for which the $a_j = \partial f/\partial x_j\big|_{x=0}$ are largest. These x_j form the effective set corresponding to $\beta_0 = 0$. Let Γ_0 denote this set. Γ_0 may contain one or more x_j, but it is not empty. The effective set of constraints Δ_0 (i.e., those for which equality holds) is empty at this stage. We increase the $x_j \epsilon \Gamma_0$ until one of the following takes place:

1.   One or more of the constraints are violated.

2. $\partial f/\partial x_h$ for $x_h \notin \Gamma_0$ becomes equal to $\partial f/\partial x_j$ for $x_j \epsilon \Gamma_0$ when evaluated at the last effective set.

3. $\partial f/\partial x_j$ becomes equal to zero for all $x_j \epsilon \Gamma_0$.

The value of β determines which of the possibilities is to occur first. Let $\beta = \beta_1$ be the value corresponding to the new effective set. If case 1 occurs first, we introduce a variable λ_i for each of the constraints about to be violated and let Δ_1 be the set of all these λ_i. If case 2 occurs first, we introduce all those x_h into our effective set, which we denote by Γ_1. Case 3 indicates that we have reached the optimum.

Next we equate to λ the derivatives of $f + \sum_i \lambda_i \sum_j (b_{ij}x_j - b_i)$ with respect to all the x_j, equate to zero its derivatives with respect to λ_i, and solve for x_j and λ_i. The unknown λ is then determined in terms of β. Note that $\Sigma(b_{ij}x_j - b_i)$ is an equality constraint (i.e., an effective constraint).

There are five cases to consider. Each leads to a value of β. The smallest of these values is taken as the next critical value, according to which of the following occurs:

1.   The rate of increase of f with respect to all these variables becomes zero, and we conclude that we have reached the maximum, that is, $\lambda = 0$.

2.   One or more of the variables of Γ may become zero; they are dropped from the effective set. They are equated to zero one at a time in the second constraint giving a value of β. The smallest β is chosen, and its corresponding variable is dropped.

3.   One or more of the effective constraints may cease to be effective (i.e., its λ_i is zero) and is dropped from Δ_1. Since λ_i is expressed in terms of β, putting $\lambda_i = 0$ determines β.

4.   The derivative with respect to one or more of the variables $x_h(x_h \notin \Gamma_1)$ may become equal to the derivative with respect to the varia-

bles $x_j \epsilon \Gamma_1$. They are equated, and the value of β is determined. If there are several of these, the minimum β is taken. We add those x_h to the effective set of variables.

5. Some new constraints may become effective. This is determined by substituting the variables which are functions of β in each inequality, which is forced to be an equality, and then solving for β; the smallest β is chosen.

The iterations are repeated until the optimum is reached. The process converges in a finite number of steps.

Following is a formal outline of the steps of the algorithm.

Step 1 (*initial step*): $\beta_0 = 0$, and Γ_0 consists of all x_j for which a_j is largest.

Step 2: Express the x_j and λ_i contained in the current effective sets Γ_g and Δ_g, $g = 0, 1, \ldots,$ in terms of β by solving the following set of equations (λ denotes the common rate of increase of the variables contained in Γ_g subject to the effective constraints):

$$\lambda = a_j + \sum_i b_{ij}\lambda_i + \sum_k a_{jk}x_k \qquad (3\text{-}130)$$

$$b_i = \sum_j b_{ij}x_j$$

$$\beta = \sum_j x_j$$

where i runs over the components of Δ_g (if not empty; obviously at the first step Δ_g is empty), and j and k over the components of Γ_g (which is never empty). Solving the above set of equations, obtain $x_j = c_j\beta + d_j$, where $\Sigma c_j = 1$, $\Sigma d_j = 0$. It is easy to see that

$$\lambda_i = p_i\beta + q_i \qquad (3\text{-}131)$$

where p_i and q_i are constants.

Step 3a: Compute $\beta_{g0} = -n/m$ and

$$\lambda = m\beta + n \qquad (3\text{-}132)$$

where
$$m = \sum_k a_{jk}c_k + \sum_i b_{ij}p_i \qquad (3\text{-}133)$$

$$n = \sum_k a_{jk}d_j + \sum_i b_{ij}q_i + a_j \qquad (3\text{-}134)$$

Step 3b: For any x_j ($\epsilon \Gamma_g$) for which $c_j < 0$, compute $\beta_{gj} = -d_j/c_j$ (at this value x_j becomes equal to zero), where the subscript gj indicates the gth step and the jth variable.

Step 3c: Compute $\beta_{gj} = -q_i/p_i$ for all λ_i contained in Δ_g for which $p_i < 0$ (if any).

Step 3d: For every x_h ($x_h \notin \Gamma_g$) compute β_{gh} from the equation

$$a_h + \sum_j a_{hj}x_j + \sum_i b_{ih} = \lambda$$

Thus

$$\beta_{gh} = \frac{n - \sum_j a_{hj}d_j - \sum_i b_{ih}q_i - a_h}{\sum_j a_{hj}c_j + \sum_i b_{ih}p_i - m}$$

Step 3e: For every constraint not included in Δ_g, compute β_{gh} from the equation $\sum_j b_{ij}x_j = b_i$, where j runs over the elements of Γ_g.   Hence

$$\beta_{gi} = \frac{b_i - \sum_j b_{ij}d_j}{\sum_j b_j c_j}$$

Step 4:   Find the smallest of the values β_{g0}, β_{gi}, β_{gh} computed in step 3, and ignore those less than or equal to β_g.   If the smallest value is less than β^*, there are five cases to be considered.

(*a*)   If β_{g0} from step 3a is smallest, the problem is solved by putting $\beta = \beta_{g0}$ in the last derived set of equations in step 3.   The capacity constraint $\left(\sum_j x_j \leq \beta^*\right)$ is not effective at the optimum.

(*b*)   If β_{gj} from step 3b is smallest, put $\beta_{gj} = \beta_{g+1}$.   Drop the corresponding x_j from the set Γ_g to form the new effective set $\Gamma_{(g+1)}$ with $\Delta_{(g+1)} = \Delta_g$.   Repeat step 2.

(*c*)   If β_{gi} from step 3c is smallest, put $\beta_{gi} = \beta_{g+1}$.   Drop the corresponding constraint and its parameter λ_i from the set Δ_g to form the new effective set $\Delta_{(g+1)}$ with $\Gamma_{(g+1)} = \Gamma_g$.   Repeat step 2.

(*d*)   If β_{gh} from step 3d is smallest, put $\beta_{gh} = \beta_{g+1}$.   Add the corresponding variable x_h to the set Γ_g to form the new effective set $\Gamma_{(g+1)}$ with $\Delta_{(g+1)} = \Delta_g$.   Repeat step 2.

(*e*)   If they all exceed β^* or are all less than β_g, put $\beta = \beta^*$, and obtain the corresponding values of the variables.   An optimum has been reached.

Example: Maximize

$$f = 6 + 12x_1 + 10x_2 + 4x_3 - 4x_1{}^2 - 2x_1x_2 - 8x_1x_3 - 3x_2{}^2 - 6x_3{}^2$$

subject to the constraint $x_1 + x_2 + x_3 = 4$ and $x_i \geq 0$, $i = 1, 2, 3$.

Since x_1 is the variable with the largest coefficient, we increase x_1 according to the derivative test previously described.   Now

$$\frac{\partial f}{\partial x_1} = 12 - 2x_2 - 8x_3 - 8x_1$$

Hence
$$\frac{\partial f}{\partial x_1} = 0 = 12 - 8\beta \qquad \text{or} \qquad \beta = \frac{3}{2}$$

Further increase of x_1 after this level would decrease f. If within this range no other variable is increased, the maximum of f would be reached without the constraint becoming effective.

However, other variables can be introduced. To see which, we compute $\partial f/\partial x_j$, and introduce x_j once $\partial f/\partial x_j$ $(j \neq 1)$ becomes equal to $\partial f/\partial x_1$. At the point of introduction of x_j $(j \neq 1)$, x_j is itself zero, and $x_1 = \beta$. We find

$$\frac{\partial f}{\partial x_1} = 12 - 8\beta = \frac{\partial f}{\partial x_2} = 10 - 2\beta \qquad \beta_{0,2} = \tfrac{1}{3}$$

$$\frac{\partial f}{\partial x_1} = 12 - 8\beta = \frac{\partial f}{\partial x_3} = 4 - 8\beta$$

In this case they can never be equal. Hence the new effective set is $(x_1, x_2) = \Gamma_1$.

We proceed, using $\beta_1 = \tfrac{1}{3}$ and $\Gamma_1 = (x_1, x_2)$. The variables x_1, x_2 can be expressed as functions of β by solving the following set of equations:

$$\frac{\partial f}{\partial x_1} = 12 - 2x_2 - 8x_1 = \lambda$$

$$\frac{\partial f}{\partial x_2} = 10 - 2x_1 - 6x_2 = \lambda$$

$$x_1 + x_2 = \beta$$

Solving the above set, we obtain

$$x_1 = \tfrac{1}{5} + \tfrac{2}{5}\beta$$
$$x_2 = -\tfrac{1}{5} + \tfrac{3}{5}\beta$$
$$\lambda = {}^{54}\!\!/_5 - {}^{22}\!\!/_5\beta$$

Let $\lambda = 0$; then $\beta_{1,0} = {}^{54}\!\!/_{22} = {}^{27}\!\!/_{11}$.

Since neither x_1 nor x_2 is a decreasing function of β, step $3b$ is unnecessary. If either x_1 or x_2 were a decreasing function of β, we should put $x_1 = 0$ or $x_2 = 0$ and obtain $\beta_{1,1}$ and $\beta_{1,2}$. The variable left out of Γ_1 is x_3; hence we calculate $\partial f/\partial x_3$, set it equal to λ, and calculate $\beta_{1,3}$. We obtain

$$\frac{\partial f}{\partial x_3} = 4 - 8x_1 - 12x_3 = {}^{54}\!\!/_5 - {}^{22}\!\!/_5\beta$$

or
$$4 - 8(\tfrac{1}{5} + \tfrac{2}{5}\beta) = {}^{54}\!\!/_5 - {}^{22}\!\!/_5\beta$$

(since $x_3 = 0$ at that stage). We solve for β, obtaining $\beta = 7$. Hence $\beta_{1,3} = 7$.

Obviously $\beta_{1,0}$ is smaller. Hence the optimum would be reached when $\lambda = 0$ or $\beta = {}^{25}\!\!/_{11}$. If $\beta_{1,3}$ had been smaller, we would have introduced x_3 in the effective set and proceeded as usual. The optimum values are obtained as follows:

$$x_1 = \tfrac{1}{5} + \tfrac{2}{5} \, {}^{27}\!/_{11}$$
$$x_2 = -\tfrac{1}{5} + \tfrac{3}{5} \, {}^{27}\!/_{11}$$

or $x_1 = {}^{13}\!/_{11}$ $x_2 = {}^{14}\!/_{11}$ $x_3 = 0$

We obtain the optimum value of f by substituting in f.

If β^* had been 2 instead of 4, we would have put $\beta = 2$ at this stage and would still have arrived at optimum values for x_1 and x_2. This exemplifies the fact that the capacity constraint does not alter the nature of the problem, even if it is artificially introduced.

Remark: Boot[10a] has shown that the Houthakker capacity procedure works with negative as well as positive coefficients in the matrix of side conditions. In either case, however, the procedure as given by Houthakker may fail when, say at one and the same value of β, both Γ and Δ should change, or Γ should change twice or Δ should change twice. These difficulties can be overcome by a minor change in the method.

3-7 OPTIMIZATION PROBLEMS WITH INFINITELY MANY CONSTRAINTS

Consider the following problem: Minimize $f(x)$, $x = (x_1, \ldots, x_n)$, in a region R defined by the infinite system of inequalities $G(x,\alpha) \geq 0$, where α is a parametric vector $\alpha = (\alpha_1, \ldots, \alpha_m)$ belonging to a set S of m-dimensional space. Ordinary problems with a finite number of constraints may be put in this form. For example, $g_1(x) \geq 0$, $g_2(x) \geq 0$ can be written, using a single parameter, as

$$G(x,\alpha) \equiv \alpha g_1(x) + (1 - \alpha)g_2(x) \geq 0$$

where the first inequality is obtained for $\alpha = 1$, and the second for $\alpha = 0$. It may also be written $(2 - \alpha)g_1(x) + (\alpha - 1)g_2(x) \geq 0$.

Assume that $f(x)$ and $G(x,\alpha)$ have continuous derivatives with respect to x_j. The following two theorems are due to Fritz John.[42a]

Theorem 3-15 If $\bar{x}$ satisfies $G(x,\alpha) \geq 0$ for all admissible values of α and yields a minimum $f(x)$ on R, then there exists a finite set of points

$$\alpha^{(1)}, \ldots, \alpha^{(p)} \qquad 1 \leq p \leq n$$

and numbers $\lambda_0 \geq 0$ $\lambda_i > 0$ $i = 1, \ldots, p$ (3-135)

such that $G(\bar{x}, \alpha^{(i)}) = 0$ $i = 1, \ldots, p$ (3-136)

and the function

$$F(x,\lambda) = \lambda_0 f(x) - \sum_{i=1}^{p} \lambda_i G(x,\alpha^{(i)}) \qquad (3\text{-}137)$$

s stationary at $\bar{x}$, that is,

$$\frac{\partial F}{\partial x_j}\bigg|_{\bar{x}} = 0 \qquad j = 1, \ldots, n$$

Proof: To sketch the proof of the theorem, let S' be the set of α for which $G(\bar{x},\alpha) = 0$. The first step is to show (by contradiction) that

$$df = \sum_{j=1}^{n} \frac{\partial f}{\partial x_j}\bigg|_{\bar{x}} dx_j < 0$$

$$dG = \sum_{j=1}^{n} \frac{\partial G}{\partial x_j}\bigg|_{\bar{x}} dx_j > 0 \qquad \text{for } \alpha \text{ in } S'$$

has no solution $(dx_1, \ldots, dx_n)$. This implies that the set T consisting of

$$q = \left(-\frac{\partial f(\bar{x})}{\partial x_1}, \ldots, -\frac{\partial f(\bar{x})}{\partial x_n}\right)$$

and

$$\left(\frac{\partial G(\bar{x},\alpha)}{\partial x_1}, \ldots, \frac{\partial G(\bar{x},\alpha)}{\partial x_n}\right)$$

does not lie in an open half-space with a bounding hyperplane through the origin. Thus the origin is a point of the convex hull of T (that is, the origin is in the intersection of all convex sets containing T). Because of the assumptions, T is closed and bounded. Therefore it is the convex hull of its extreme points. These are points that are not interior to any segment in the set. Since any point of the convex hull of T belongs to a simplex with vertices in T, and thus the origin also does, it follows that 0 is a convex combination of at most $n + 1$ points of T. Relations (3-135) to (3-137) express this fact analytically.

Theorem 3-16 (*sufficient condition for a minimum*): If $\bar{x}$ is a stationary point of $F(x,\lambda)$ with $\lambda_0 \geq 0$, $\lambda_1, \ldots, \lambda_p > 0$, and if the matrix

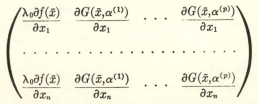

has rank n at $\bar{x}$, then $\bar{x}$ is a relative minimum of $f(x)$ subject to the finite set of inequalities $G(x,\alpha^{(i)}) \geq 0$, $i = 1, \ldots, p$, and consequently also a relative minimum on $G(x,\alpha) \geq 0$.

Proof: This theorem is proved by contradiction. If $\bar{x}$ is not a relative minimum in R, then there is a sequence of positive numbers a_k

and a set of points $dx^{(k)} = (dx_1^{(k)}, \ldots, dx_n^{(k)})$ such that

$$\lim_{k \to \infty} a_k = 0 \qquad \sum_{j=1}^{n} (dx_j^{(k)})^2 = 1$$

and
$$f(\bar{x} + a_k dx^{(k)}) < f(\bar{x})$$
$$G(\bar{x} + a_k dx^{(k)}, \alpha^{(i)}) \geq 0 \qquad i = 1, \ldots, p$$

Then for various θ's in $0 \leq \theta \leq 1$ we have

$$df(\bar{x} + \theta a_k \, dx^{(k)}) = \sum_{j=1}^{n} \frac{\partial f}{\partial x_j} (\bar{x} + \theta a_k \, dx^{(k)}) \, dx_j \leq 0$$

and similarly, $dG(\bar{x} + \theta a_k \, dx^{(k)}, \alpha^{(i)}) \geq 0 \qquad i = 1, \ldots, p$

A suitably chosen subsequence of $dx^{(k)}$ converges to a vector $dx \neq 0$ for which $df(\bar{x}) \leq 0$ and $dG(\bar{x}, \alpha^{(i)}) \geq 0$ for $i = 1, \ldots, p$. Because $\bar{x}$ is a stationary point,

$$0 = dF(\bar{x}, \lambda) = \lambda_0 \, df(\bar{x}) - \sum_{i=1}^{p} \lambda_i \, dG(\bar{x}, \alpha^{(i)})$$

Since $\qquad \lambda_0 \geq 0 \qquad\qquad \lambda_i > 0 \qquad i = 1, \ldots, p$
we have $\quad \lambda_0 df(\bar{x}) = 0 \qquad dG(\bar{x}, \alpha^{(i)}) = 0 \qquad i = 1, \ldots, p$

Consider the last system of equations. Since the rank of the matrix of partial derivatives is n, a solution $dx \neq 0$ does not exist. This contradicts the fact that our sequence converges to $dx \neq 0$. Thus $\bar{x}$ is a relative minimum.

Theorem 3-16 shows that with appropriate restrictions a relative minimum of f subject to $G(x, \alpha) \geq 0$ is also a relative minimum subject to $G(x, \alpha^{(i)}) > 0$, $i = 1, \ldots, p$. John considers $f(x) = -x^2$ and

$$G(x, \alpha) \equiv \alpha^2 - \alpha x^2 \geq 0 \qquad 0 \leq \alpha \leq 1$$

The constraints are satisfied only for $x = 0$, which is the relative minimum of f. On the other hand, if the constraints are $G(x, \alpha^{(i)}) \geq 0$, $i = 1, \ldots, p$, with $0 \leq \alpha^{(i)} \leq 1$, then points in the neighborhood of $x = 0$ are admissible, and $f(x)$ has no relative minimum in this set at $x = 0$.

Theorems 3-15 and 3-16 are used to show that, corresponding to the ellipsoid of smallest volume containing a set in n dimensions, there is a concentric ellipsoid which is homothetic to it, at the ratio $1/n$, contained in the convex hull of the set. Since the boundary of the set may be a convex surface, one concludes that a closed convex surface lies between two concentric homothetic ellipsoids of ratio $1/n$. Two figures are homothetic if there exists a point such that the length of all rays drawn from the point to the two figures have constant proportion.

REFERENCES

1. Ablow, C. M., and G. Brigham: An Analog Solution of Programming Problems, *Operations Res.*, vol. 3, no. 4, pp. 388–394, November, 1955.
2. Arrow, K. J.: A Gradient Method for Approximating Saddle Points and Constrained Maxima, The Rand Corporation, Santa Monica, Calif., 1951.
3. ——— and L. Hurwicz: Reduction of Contrained Maxima to Saddle Point Problems, in J. Neyman (ed.), "Proceedings of the Third Berkeley Symposium on Mathematical Statistics and Probability," vol. 5, pp. 1–20, University of California Press, Berkeley, Calif., 1956.
4. ———, ———, and H. Uzawa: "Studies in Linear and Nonlinear Programming," Stanford Mathematical Studies in the Social Sciences II, Stanford University Press, Stanford, Calif., 1958.
5. ———, S. Karlin, and H. Scarf: "Studies in the Mathematical Theory of Inventory and Production," Stanford University Press, Stanford, Calif., 1958.
6. Barankin, E. W., and R. Dorfman: On Quadratic Programming, in Berkeley, Series in Statistics, vol. 2, pp. 285–317, University of California Press, Berkeley, Calif., 1958.
7. Beale, E. M. L.: On an Iterative Method for Finding a Local Minimum of a Function of More than One Variable, *S.T.R.G. Rept.* no. 24, 1958.
8. ———: On Optimizing a Convex Function Subject to Linear Inequalities, *J. Roy. Statist. Soc.*, vol. 17, pp. 173–184, 1955.
9. ———: On Quadratic Programming, *Naval Research Logistics Quart.*, vol. 6, pp. 227–243, 1959.
10. Boot, J. C. G.: Notes on Quadratic Programming, The Kuhn Tucker Theorem and Theil-van DePanne. Conditions, Degeneracy and Equality Constraints, *Management Sci.*, vol. 7, p. 85, 1961.
10a. ———: Binding Constraint Procedures of Quadratic Programming, *Netherland School of Economics Rept.* 6207, Rotterdam, 1963.
11. Brown, R. R.: Gradient Methods for the Computer Solution of System Optimization Problems, *Mass. Inst. Technol., Dept. of Electrical Eng., WADC Tech. Note* 7-159, September, 1957.
12. Carroll, C. W.: The Created Response Surface Technique for Optimizing Nonlinear Restrained Systems, *Operations Res.*, vol. 9, no. 2, pp. 169–184, March–April, 1961.
13. Charnes, A., and W. W. Cooper: "Management Models and Industrial Applications of Linear Programming," vols. I and II, John Wiley & Sons, Inc., New York, 1961.
14. ——— and C. E. Lemke: Minimization of Non-linear Separable Convex Functionals, *Naval Research Logistics Quart.*, vol. 1, pp. 301–302, 1954.
15. Cheney, W., and A. A. Goldstein: Proximity Maps for Convex Sets, *Symposium on Mathematical Programming*, Rand Corporation, March, 1959.
16. Dantzig, G. B.: General Convex Objective Forms, Rand Corporation Paper -1664, April, 1959.
17. Debreu, G. E.: Definite and Semidefinite Quadratic Forms, *Econometrica*, vol. 0, pp. 295–300, 1952.
18. Deland, E. C.: Continuous Programming Methods on an Analog Computer, Rand Corporation Paper P-1815, September, 1959.
19. Dennis, J. R.: "Mathematical Programming and Electrical Net-works," *M.I.T. Internal Tech. Rept.* 10, 1958, and John Wiley & Sons, Inc., New York, 1959.
20. Dorn, W. S.: A Duality Theorem for Convex Programs, *IBM J. Research and Development*, vol. 4, no. 4, October, 1960.

21. ————: Duality in Quadratic Programming, *New York University Rep* 08676 (Physics), 1958.

21a. ————: Non-linear Programming—A Survey, *Management Sci. J.*, vo 9, no. 2, pp. 171–208, January, 1963.

21b. ————: On Lagrange Multipliers and Inequalities, *Operations Res.*, vol. 9 no. 1, January–February, 1961.

22. Dornheim, F. R.: Optimization Subject to Nonlinear Constraints Using th Simplex Method and Its Application to Gasoline Blending, Sinclair Research Labora tories, Inc., Harvey, Ill., presented at Optimization Techniques Symposium, Nev York University, May 18, 1960.

22a. Eggleston, H. G.: "Convexity," Cambridge Tract 47, Cambridge, Univer sity Press, Cambridge, 1958.

23. Farkas, J.: Uber die Theorie der Einfachen Ungleichungen, *Reine Angeu Math.*, vol. 124, pp. 1–24, 1902.

24. Fenchel, W.: "Convex Cones, Sets and Functions," Princeton Universit; Press, Princeton, N.J., 1953.

25. Fiacco, A. V.: Comments on the paper of C. W. Carroll, *Operations Res.*, vo 9, no. 2, pp. 184–185, March–April, 1961.

26. ———— and G. P. McCormick: A Method for Nonlinear Programming: A Extension of Carroll's, in preparation.

27. ————, N. M. Smith, and D. Blackwell: A More General Method fo Nonlinear Programming, Seventeenth National Meeting of the Operations Researc Society of America, New York, May 20, 1960.

28. Frank, E.: On the Calculation of the Roots of Equations, *J. Math. and Phys* vol. 34, pp. 187–197, 1955.

29. Frank, M., and P. Wolfe: An Algorithm for Quadratic Programming, *Nava Research Logistics Quart.*, vol. 3, nos. 1 and 2, pp. 95–110, March–June, 1956.

30. Frisch, Ragnar: The Multiplex Method for Linear Programming, *Univer setetets Socialokonomiske Institut (Oslo) Memorandum*, October, 1955.

31. Gale, D.: Convex Polyhedral Cones and Linear Inequalities, in T. C. Koop man (ed.), "Activity Analysis of Production and Allocation," John Wiley & Sons Inc., New York, 1951.

32. Gass, S. I.: "Linear Programming," McGraw-Hill Book Company, Inc New York, 1958.

33. Gerstenhaber, M.: Theory of Convex Polyhedral Cones, in T. C. Koopma (ed.), "Activity Analysis of Production and Allocation," John Wiley & Sons, Inc New York, 1951.

34. Gol'shteyn, E. G.: An Infinite-dimensional Analog of the Problem of Linea Programming and Its Applications to Certain Questions in the Theory of Approxi mations, U.S. Joint Publications Research Service, 11309, December, 1961.

35. Green, J. W.: Recent Applications of Convex Functions, *Am. Math. Monthly* 1953.

36. Griffith, R. E., and R. A. Stewart: A Non-linear Programming Techniqu for the Optimization of Continuous Processing Systems, *Management Science*, vol. 7 p. 379, 1961.

37. Hadley, George: "Linear Programming," Addison-Wesley Publishing Com pany, Inc., Reading, Mass., 1962.

37a. Hanson, M. A.: A Duality Theorem in Nonlinear Programming, *Austral. J Statist.*, vol. 3, pp. 64–72, 1961.

38. Hartley, H. O.: Nonlinear Programming by the Simplex Method, *Econo metrica*, vol. 29, pp. 223–237, 1961.

39. Hildreth, C.: A Quadratic Programming Procedure, *Naval Research Logistic Quart.*, vol. 4, pp. 79–85, 1957.

40. Houthakker, H. A.: The Capacity Method of Quadratic Programming, *Econometrica*, vol. 28, pp. 62–87, 1960.

41. Huard, P.: Dual Programs, *IBM J.*, vol. 6, no. 1, pp. 137–139, 1962.

42. Hurwicz, L., and H. Uzawa: A Note on the Lagrangian Saddle-points, *Univ. of Minn. Tech. Rept.* 1, September, 1958.

42a. John, F.: Extremum Problems with Inequalities as Subsidiary Conditions, in "Studies and Essays," Courant anniversary volume, pp. 187–204, Interscience Publishers, Inc., New York, 1948.

43. Karlin, S.: "Mathematical Methods and Theory in Games, Programming and Economics," vol. I, Addison-Wesley Publishing Company, Inc., Reading, Mass., 1959.

44. Kelley, J. E., Jr.: The Cutting Plane Method for Solving Convex Programs, *J. Soc. Ind. Appl. Math.*, vol. 8, pp. 703–712, 1960.

45. Kose, T.: Solutions of Saddle Value Problems by Differential Equations, *Econometrica*, vol. 24, no. 1, pp. 59–70, 1956.

46. Kuhn, H. W., and A. W. Tucker: Non-linear Programming, in J. Neyman (ed.), "Proceedings of the Second Berkeley Symposium on Mathematical Statistics and Probability, University of California Press, Berkeley, Calif., 1951.

47. Mangasarian, O. I.: Duality in Nonlinear Programming, *Quart. Appl. Math.*, vol. 20, no. 3, October, 1962.

48. Mann, H. B.: Quadratic Forms with Linear Constraints, *Am. Math. Monthly*, vol. 50, pp. 430–433, 1943.

49. Manne, A. S.: Concave Programming for Gasoline Blends, Rand Corporation Paper P-383, April, 1953.

50. Markowitz, H. M.: The Optimization of a Quadratic Function Subject to Linear Constraints, *Naval Research Logistics Quart.*, vol. 3, pp. 111–133, 1956.

51. ———: Portfolio Selection, *J. Finance*, vol. 7, p. 77, 1952.

52. Metzger, R. W.: Elementary Mathematical Programming, John Wiley & Sons, Inc., New York, 1958.

53. Miller, C. E.: The Simplex Method for Local Separable Programming, Standard Oil Company of California *Report of the Electronic Computer Center*, August, 1960.

54. Phipps, C. G.: Maxima and Minima under Restraint, *Am. Math. Monthly*, vol. 59, pp. 230–235, 1952.

55. Pietrzykowski, Tomasz: On an Iteration Method for Maximizing a Concave Function on a Convex Set, *Prace ZAM*, ser. A, no. 13, 1961.

56. Pyne, I. B.: Linear Programming on an Electronic Analogue Computer, *AIEE Trans. Annual*, part I, pp. 139–143, 1956.

57. Riley, V. and S. I. Gass: "Linear Programming and Associated Techniques, A Comprehensive Bibliography on Linear, Non-linear and Dynamic Programming," Operations Research Office, Johns Hopkins University, 1958.

58. Rosen, J. B.: Stability of Differential Equations and Equivalent Non-linear Programming Problem, *Notices Am. Math. Soc.*, vol. 7, p. 996, 1960.

59. ———: The Gradient Projection Method, Part I: Linear Constraints, *J. Soc. Ind. Appl. Math.*, vol. 8, no. 1, pp. 181–217, March, 1960.

60. ———: The Gradient Projection Method for Nonlinear Programming, Part II: Non-linear Constraints, p. 954, Shell Development Co., Emeryville, Calif., 1961.

61. Saaty, T. L.: "Mathematical Methods of Operations Research," McGraw-Hill Book Company, Inc., New York, 1959.

62. Shapiro, Marvin, and G. B. Dantzig: Solving the Chemical Equilibrium Problem Using the Decomposition Principle, Rand Corporation Paper P-2056, August, 1960.

63. Simonnard, M.: "Programmation Lineaire," Dunod, Paris, 1962.

64. Slater, M.: Lagrange Multipliers Revisited, A Contribution to Non-linear Programming, *Cowles Commission Discussion Paper Math.* 403, 1950.

65. Stone, M. H.: "Convexity," The University of Chicago Press, Chicago, 1946.

66. Theil, H., and C. Van DePanne: Quadratic Programming as an Extension of Conventional Quadratic Maximization, *Management Sci.*, vol. 7, no. 1, pp. 1–20, 1960.

67. Vajda, S.: "Mathematical Programming," Addison-Wesley Publishing Company, Inc., Reading, Mass., 1961.

68. Warga, J.: Convex Minimization Problems I, Revision I, Avco Manufacturing Corp., Wilmington, Mass., *RAD Tech. Memo* TM-59-20, February, 1959.

69. ———: Convex Minimization Problems II, The Transportation Problem I, Avco Manufacturing Corp., Wilmington, Mass., *RAD Tech. Memo* TM-59-21, February, 1959.

70. ———: Convex Minimization Problems III, Linear Programming, Revision I, Avco Manufacturing Corp., Wilmington, Mass., *RAD Tech. Memo* TM-59-22, February, 1959.

71. Wegner, P.: Some Non-linear Programming Solution Procedures, unpublished M.A. thesis, Pennsylvania State University, 1959.

72. ———: A Non-linear Extension of the Simplex Method, *Management Sci.*, vol. 7, pp. 43−55, 1960.

73. Weyl, H.: Elementare Theorie der Konvexen Polyeder, *Comm. Math. Helv.*, vol. 7, pp. 290–306, 1934–1935 [English translation: The Elementary Theory of Convex Polyhedra, in H. W. Kuhn and A. W. Tucker (eds.), "Contributions to the Theory of Games," Annals of Mathematics Studies No. 24, Princeton University Press, Princeton, N.J., 1950].

73a. White, W. B., S. M. Johnson, and G. B. Dantzig: Chemical Equilibrium in Complex Mixtures, *J. Chem. Phys.*, vol. 28, pp. 751–755, May, 1958.

74. Witzgall, C.: Gradient-projection Methods for Linear Programming, *Princeton University and I.B.M. Rept.* 2, August, 1960.

75. Wolfe, P.: A Duality Theory for Non-linear Programming, *Quart. Appl. Math.*, vol. 19, no. 3, October, 1961.

76. ———: Accelerating the Cutting Plane Method for Nonlinear Programming, *J. Soc. Ind. Appl. Math.*, vol. 9, no. 3, pp. 481–488, September, 1961.

77. ——— (ed.): The Rand Symposium on Mathematical Programming: Linear Programming and Recent Extensions, Rand Corporation, Santa Monica, Calif., March 16–20, 1959.

78. ——— (ed.): Recent Developments in Non-linear Programming, Part I, Rand Corporation Paper 2063, 1961.

79. ———: The Generalized Simplex Method, Rand Corporation Paper P-1818, May, 1959.

80. ———: The Simplex Method for Quadratic Programming, Notes on Linear Programming and Extensions, Part II, Rand Corporation Paper RM 2388, June 5, 1959.

81. ———: The Simplex Method for Quadratic Programming, *Econometrica*, vol. 27, no. 3, pp. 382–398, 1959.

82. ———: Some Simplex-like Nonlinear Programming Procedures, *Operations Res.*, vol. 10, no. 4, pp. 438–447, July–August, 1962.

83. ———: Recent Developments in Nonlinear Programming, Rand Corporation Report R-401-PR, May, 1962.

84. Zoutendijk, G.: Maximizing a Function in a Convex Region, *J. Roy. Statist. Soc.*, vol. 21, pp. 338–355, 1949.

85. ———: "Methods of Feasible Directions," Elsevier Publishing Company, Amsterdam, 1960.

FOUR

NONLINEAR ORDINARY DIFFERENTIAL EQUATIONS

4-1 INTRODUCTION

A differential equation provides a nucleus about which a mathematical analyst pursues his investigations into the existence, characterization, and construction of solutions and a bridge of understanding which enables the scientist to represent his natural phenomena and test and predict their change and development. For a large class of scientific problems, it is the oscillatory nature of the phenomenon under study that is of the greatest interest and must be carefully examined. Consequently the principal ideas stressed in this chapter are those connected with periodicity and stability, ideas of paramount importance in this day of thriving technology. We shall be concerned with the behavior of solutions in the small, in particular near singularities, as well as in the large, in pursuit of some periodic solutions whose existence is confirmed by the presence of closed curves known as limit cycles. The stability of a solution is essential to many scientific and engineering applications. It is often desirable that the system settles into a steady state. Here one is concerned with the behavior of the solution after a long lapse of time. Lyapunov's theory provides criteria for stability without the actual construction of a solution. However, as we shall see in the case of linear (and some nonlinear) systems, it is possible to examine stability through knowledge of the general nature of the solution.

After illustrating how nonlinear equations arise in practice, we shall give a general existence theorem. All that this theorem ensures is the existence of a solution. On the other hand if we ask "When does a differential equation have a periodic solution?" we cannot answer with such generality. Existence theorems for periodic solutions of particular classes of differential equations have been given. Some of these existence theorems are briefly mentioned. There are several methods for obtaining approximate periodic solutions of nonlinear equations; a few of them will be illustrated here. Most of these methods hinge on the idea of a so-called *bifurcation* equation, which is derived from a given differential equation with a parameter. Periodic solutions of the equation for small values of the parameter are obtained by successive iterations initially dependent on the periodic solution of the linear equation resulting from equating the parameter to zero. The solution of the linear equation bifurcates into a family of solutions for the nonlinear case.

After the existence theorem, we introduce concepts related, for the most part, to linear equations with constant coefficients. Illustrations of these concepts, most of which revolve about periodic solutions and stability, are included. We then contrast this with the behavior, construction, and stability of solutions of nonlinear equations. We point out that, unlike the linear case, the principle of superposing particular solutions to obtain the general solution does not apply. Here again we seek periodic solutions.

In a short section we give a brief description of the local behavior of solutions of second-order equations near singularities. This is followed by a discussion of limit cycles. Then follows a study of periodic solutions of linear systems with periodic coefficients, with remarks on the same idea for simple nonlinear systems.

The analysis of stability through the method of Lyapunov occupies the next section. The fact that this theory is essential for application is also brought out in Chap. 5. In the last section of this chapter we turn to the question of the construction of exact and approximate solutions. However, by then at least two methods of constructing solutions will have been illustrated: a successive-approximation method due to Picard given in the existence section, and the perturbation method (which is related to contraction mappings) of Poincaré, Lindstedt, and others.

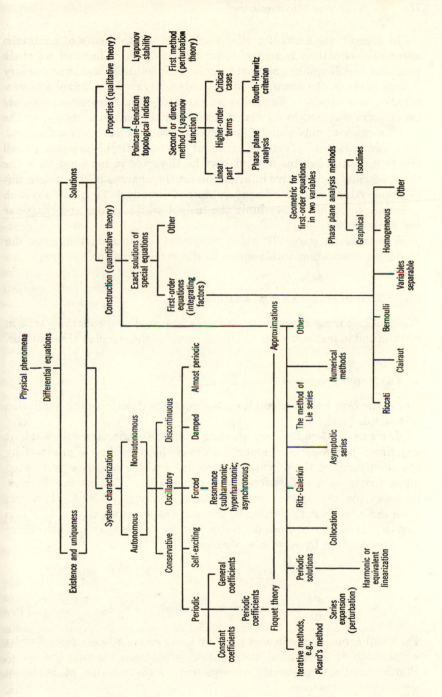

177

In general, the search for closed solutions in the theory of nonlinear differential equations is an ambitious undertaking.   Therefore we shall present a number of approximate methods of solution, including perturbation methods, the linearization method of Krylov-Bogoliubov and van der Pol, and others.   We shall also exemplify the rare cases in which closed-form solutions exist, e.g., certain Ricatti equations.

The study of differential equations is closely tied to application, and is in turn enriched by knowledge about the natural phenomenon giving rise to it.   Although the analyst may be thwarted in his quest for solutions, adequate qualitative information for the analysis of the underlying physical systems can sometimes be obtained directly.   To ignore physical concepts would ultimately limit the insight which guides and sharpens the analytic tool.

The table on page 177 which concludes this introduction gives the reader an orientation with respect to the subject matter.

4-2   SOME NONLINEAR EQUATIONS

We shall give three examples to illustrate how nonlinear equations arise in physics and in mathematical disciplines such as the calculus of variations.

a   *The Simple Pendulum*

One of the best-known examples is the equation describing the oscillation of a simple pendulum.

Let s be the length of arc starting from the origin, and θ the angle at any time; the acceleration along the arc of the pendulum is produced by the component of the downward force (due to the weight w) projected along the tangent.   This force is $w \sin \theta$.   Note that as the distance increases (on the upward motion), the velocity is positive.   However since the velocity itself decreases, the acceleration is negative.   It is also negative when the distance decreases (on the downward motion).   Since the mass is given by $m \equiv w/g$, we have

$$\frac{w}{g} \frac{d^2 s}{dt^2} = -w \sin \theta$$

or $\qquad \dfrac{d^2\theta}{dt^2} + \dfrac{g}{L} \sin \theta = 0 \qquad$ since $\qquad s = L\theta \qquad$ (4-1)

For small θ, one approximates $\sin \theta$ by θ and solves a linear second-order equation in θ.   In general the solution is obtained through the use of elliptic functions frequently encountered in the solution of nonlinear second-order equations.

If we integrate (4-1) once, we obtain

$$\frac{1}{2}\left(\frac{d\theta}{dt}\right)^2 - \frac{g}{L}\cos\theta = -\frac{g\cos\mu}{L} \tag{4-2}$$

where the right-hand side is obtained by evaluating the constant of integration at μ, the angle of maximum displacement from the origin.   Note

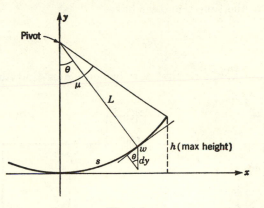

Fig. 4-1

that $d\theta/dt \big|_{\theta=\mu} = 0$.   To reduce the solutions to standard form, let

$$k = \sin\frac{\mu}{2} \qquad \cos\theta = 1 - 2k^2\sin^2\varphi$$

and

$$dt = \sqrt{\frac{L}{g}}\,\frac{d\varphi}{(1 - k^2\sin\varphi)^{\frac{1}{2}}} \tag{4-3}$$

Exercise 4-1   Verify this expression.   Show that

$$\cos\theta - \cos\mu = 2k^2\cos^2\varphi$$
$$\sin\theta = 2k\sin\varphi(1 - k^2\sin^2\varphi)^{\frac{1}{2}}$$

and hence compute $d\theta$ from the second expression in terms of $d\varphi$.

This expression can now be used to obtain the time required to traverse an angle θ (from which φ can be obtained); i.e.,

$$t = \sqrt{\frac{L}{g}}\int_0^\varphi \frac{d\varphi}{(1 - k^2\sin^2\varphi)^{\frac{1}{2}}} \tag{4-4}$$

b   *The Equation of van der Pol*

Consider a simple circuit (see Fig. 4-2; consider only the inner circuit, without the generator) with a coil inductance L (measured in henrys) and a resistance R.   We have, according to Ohm's law,

$$E_R = RI$$

Current variations induce counterdrop in potential E_L across the coil (whose own resistance is negligible; otherwise it would in turn contribute a term similar to RI) which is proportional to the rate of change of current with respect to the time t, and we have

$$E_L = L\frac{dI}{dt}$$

Now from Kirchhoff's first law, which states that the sum of the

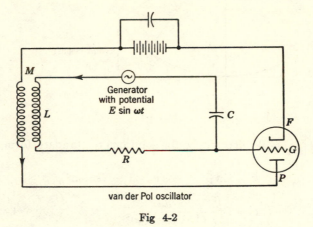

van der Pol oscillator

Fig 4-2

potential drops must be zero, we have

$$L\frac{dI}{dt} + RI = E$$

where E is the electromotive force supplied by the battery.

Exercise 4-2   Obtain $I = (E/R)(1 - e^{-Rt/L})$ for the solution of the above equation, where E is constant at $t = 0$.

If we insert, in series with the above circuit, a capacitor with capacitance C (measured in farads), then the electric charge on it is given by Q (measured in coulombs), and if E_C is the potential drop across the capacitor, we have

$$Q = CE_C$$

Because the current is by definition the rate of flow of electric charge, we have

$$I = \frac{dQ}{dt}$$

Thus
$$E_C = \frac{1}{C} \int I \, dt$$

and, by Kirchhoff's first law,

$$L \frac{dI}{dt} + RI + \frac{1}{C} \int I \, dt = E \tag{4-5}$$

When there is mutual inductance between two circuits, such as when a transformer is used, then (from the definition of inductance and Kirchhoff's law) the potential drop generated by mutual inductance is subtracted from the total if the currents of the two circuits are in the same direction through the coils.   Otherwise they are added.

Thus if a circuit has no battery, but its current is generated through mutual inductance M, E in the last equation is replaced by $M \, dI_a/dt$, where I_a is the current in the neighboring mutual-inductance circuit.   If, in addition, the circuit has its own electromotive force, such as might be produced by a generator, then this force is usually described by a periodic function $E \sin \omega t$, which must also be subtracted.   If these ideas are applied to the inner circuit of the figure marked with R, L, C and the generator, we obtain

$$L \frac{dI}{dt} + RI + \frac{1}{C} \int_0^t I \, dt - M \frac{dI_a}{dt} = E \sin \omega t \tag{4-6}$$

Experimentally it is found that

$$I_a = \sigma \left(E_C - \frac{E_C^3}{3E_G^2} \right)$$

where σ is the transconductance of the tube, and E_G is the grid potential corresponding to anode saturation.[57,87]

If we can ignore the right side (i.e., if the generator does not operate), substitute

$$\frac{E_C}{E_G} \equiv u$$

$$\alpha = \frac{1}{LC} \qquad \beta = \frac{\sigma M}{LC} - \frac{R}{L}$$

$$\gamma = \frac{\sigma M}{3LC}$$

simplify, and then put

$$u = y \sqrt{\frac{\beta}{3\gamma}}$$

Simplify further, and finally replace t with $t/\sqrt{\alpha}$ and $\beta/\sqrt{\alpha}$ with ϵ (which is usually a small positive parameter). We then have van der Pol's equation,

$$\frac{d^2y}{dt^2} - \epsilon(1 - y^2) \frac{dy}{dt} + y = 0 \tag{4-7}$$

c Variational Equations

Nonlinear equations occasionally occur indirectly, for example in the solution of Euler's equation arising from a problem in the calculus of variations.

Assume that $\bar{y}(x)$ yields a relative minimum to the integral

$$I = \int_{x_1}^{x_2} f[x,y(x),y'(x)] \, dx \tag{4-8}$$

and let $\eta(x)$ be an arbitrary curve that is continuous and twice differentiable for $x_1 \leq x \leq x_2$, with $\eta(x_1) = \eta(x_2) = 0$.

Consider the curves

$$y(x,\epsilon) = \bar{y}(x) + \epsilon\eta(x) \tag{4-9}$$

where ϵ is a small parameter. The integral has a minimum value at $\epsilon = 0$, that is, $dI/d\epsilon = 0$ when $\epsilon = 0$, which is a necessary condition for a minimum. Now

$$\frac{dI}{d\epsilon} = \int_{x_1}^{x_2} \left(\frac{\partial f}{\partial y} \frac{dy}{d\epsilon} + \frac{\partial f}{\partial y'} \frac{dy'}{d\epsilon} \right) dx \tag{4-10}$$

(Note that x does not depend on ϵ.)

If we integrate the second term by parts and use

$$\frac{dy}{d\epsilon} = \eta(x) \qquad \frac{dy'}{d\epsilon} = \eta'(x)$$

we have $\displaystyle \int_{x_1}^{x_2} \frac{\partial f}{\partial y'} \eta'(x) \, dx = \frac{\partial f}{\partial y'} \eta(x) \Big|_{x_1}^{x_2} - \int_{x_1}^{x_2} \frac{d}{dx} \frac{\partial f}{\partial y'} \eta(x) \, dx$

where y still has the form $y(x,\epsilon)$.

Now if we apply the condition $\eta(x_1) = \eta(x_2) = 0$, substitute in $dI/d\epsilon$, and put $\epsilon = 0$, we obtain

$$\int_{x_1}^{x_2} \left[\frac{\partial}{\partial y} f(x,\bar{y},\bar{y}') - \frac{d}{dx} \frac{\partial}{\partial y'} f(x,\bar{y},\bar{y}') \right] \eta(x) = 0 \tag{4-11}$$

Since $\eta(x)$ is arbitrary, its coefficient must vanish (see remark); this gives Euler's equation:

$$\frac{d}{dx}\frac{\partial f}{\partial y'} - \frac{\partial f}{\partial y} = 0 \tag{4-12}$$

As can be seen from the exercises, special application of this equation can be nonlinear.

Remark: If $g(x)$ is continuous, and if for every twice-differentiable function $\eta(x)$ such that

$$\eta(x_1) = \eta(x_2) = 0$$

we have

$$\int_{x_1}^{x_2} g(x)\eta(x)\,dx = 0$$

then $g(x) = 0$. For if there is one value $x_1 \le x' \le x_2$ at which $g(x') \ne 0$ then suppose $g(x') > 0$; because of continuity there is an interval $x'_1 \le x \le x'_2$ in which $g(x) > 0$.

Define

$$\eta(x) = \begin{cases} (x'_1 - x)^4(x'_2 - x)^4 & x'_1 \le x \le x'_2 \\ 0 & \text{otherwise} \end{cases}$$

Then $\quad 0 = \int_{x_1}^{x_2} g(x)\eta(x) = \int_{x_1}^{x_2} (x'_1 - x)^4(x'_2 - x)^4 g(x)\,dx > 0$

which is a contradiction.

Exercise 4-3 Find the shortest curve between two given points in the plane; i.e., minimize

$$\int_{x_1}^{x_2} \sqrt{1 + y'^2}\,dx$$

and obtain the equation

$$\frac{y'}{\sqrt{1 + y'^2}} = \text{const}$$

Exercise 4-4 By rotating about the x axis the curve $y(x)$ which passes through $x = 0$ and $x = x_1$, obtain the minimum surface of revolution. To do this, first obtain

$$\sqrt{1 + y'^2} - \frac{d}{dx}\frac{yy'}{\sqrt{1 + y'^2}} = 0$$

4-3   EXISTENCE AND UNIQUENESS FOR FIRST - ORDER SYSTEMS

We shall be examining mostly first-order systems, because the differential equation of order n

$$\frac{d^n y}{dx^n} = f\left(x, y, \frac{dy}{dx}, \cdots, \frac{d^{n-1}y}{dx^{n-1}}\right) \tag{4-13}$$

is equivalent to the system of first-order equations

$$\frac{dy}{dx} = y_1$$

$$\frac{dy_1}{dx} = y_2$$

$$\cdots\cdots$$

$$\frac{dy_{n-2}}{dx} = y_{n-1}$$

$$\frac{dy_{n-1}}{dx} = f(x,y,y_1, \ldots ,y_{n-1})$$

(4-14)

Equivalence means that any solution of the system defines a solution of the equation, and conversely. In fact any system of differential equations of order greater than 1 may be reduced to a first-order system by first solving explicitly for the highest-order derivative. This process may lead to certain difficulties, e.g., extracting roots.

If, in the equations of a first-order system, the independent variable t appears explicitly on the right side, then we write

$$\dot{x}_i = f_i(x_i, \ldots ,x_n, t) \qquad i = 1, \ldots , n$$

which is called a *nonautonomous* system. Without the explicit appearance of t we have

$$\dot{x}_i = f_i(x_i, \ldots ,x_n) \qquad i = 1, \ldots , n$$

which is known as an *autonomous* system. Note that a nonautonomous system can be reduced to a special autonomous system by replacing t on the right by the dependent variable x_{n+1} and adjoining to the resulting system the equation $dx_{n+1}/dt = 1$. However, we shall always distinguish between the two systems, as it is sometimes easier to investigate properties of given nonautonomous systems.

The systems investigated in the last four exercises are autonomous. The nonautonomous case is illustrated by Mathieu's equation,

$$\ddot{y} + (a + q \cos 2t)y = 0$$

where a and q are real constants, and by Hill's equation,

$$\ddot{y} + a_1(t)\dot{y} + a_2(t)y = 0$$

with $a_1(t)$ and $a_2(t)$ periodic with the same period T, that is,

$$a_i(t) = a_i(t + T) \qquad i = 1, 2$$

Exercise 4-5 Verify the property of an autonomous system that when its solution is subjected to a translation in time, the result is again a solution.

Exercise 4-6 Reduce the equations of Mathieu and Hill to first-order systems.

Exercise 4-7 By consulting the references, describe a physical problem which gives rise to a nonautonomous system.

We now investigate the question of the existence of a solution and its uniqueness for the first-order system

$$\dot{x}_i = f_i(x_1, \ldots, x_n, t) \qquad i = 1, \ldots, n$$

with the initial condition $x_1^0, \ldots, x_n^0$ at t_0. In Chap. 1, an existence theorem was given using functional analysis. Because of its importance, the somewhat different classical proof will be presented here.

A domain D is a set of points such that if $(t_1, x_1^1, \ldots, x_n^1)$ is in D, then for small $\epsilon > 0$ all points P which satisfy $|t - t_1| < \epsilon$, $|x_i - x_i^1| < \epsilon$, $i = 1, \ldots, n$, are also in D, and any two points of D can be connected by a continuous path lying in D. We shall use the function $f(t,x)$, where x denotes $(x_1, \ldots, x_n)$. An integral of the above system is a function $F(t,x)$ which is constant along each solution curve $x_1(t), \ldots, x_n(t)$, but not necessarily identically constant.

Theorem 4-1[20,34] Let the functions f_i be continuous in a domain D (a rectangle) defined by

$$|t - t_0| < a \qquad |x_i - x_i^0| < a_i \qquad i = 1, \ldots, n$$

and let the following Lipschitz condition be satisfied for any two points $\bar{x}$ and $\bar{\bar{x}}$ (with the same value of t) in D:

$$|f(t,\bar{x}) - f(t,\bar{\bar{x}})| < \sum_{j=1}^{n} A_j |\bar{x}_j - \bar{\bar{x}}_j|$$

Then in a suitable interval $|t - t_0| < \bar{a} < a$, the system has a unique solution $x_i(t)$ which satisfies $x_i(t_0) = x_i^0$, $i = 1, \ldots, n$.

Proof: Note first that (when f is differentiable) the Lipschitz condition may be replaced by the requirement that $\partial f_i / \partial x_j$, $j = 1, \ldots, n$, be continuous in D for $i = 1, \ldots, n$, from which the Lipschitz condition can be obtained by using the mean-value theorem; i.e.,

$$f_i(t,\bar{x}) - f_i(t,\bar{\bar{x}}) = (\bar{x}_1 - \bar{\bar{x}}_1) \frac{\partial f_i(t,\hat{x})}{\partial x_1} + \cdots + (\bar{x}_n - \bar{\bar{x}}_n) \frac{\partial f_i(t,\hat{x})}{\partial x_n}$$

where $\hat{x}$ lies on a line segment between the two points $\bar{x}$ and $\bar{\bar{x}}$. Because of continuity, $|\partial f / \partial x_i| \leq A$, and we have a Lipschitz condition if we put $A_j = A$.

Again because of the continuity of f, a constant M can be found such that $|f_i| \leq M$, $i = 1, \ldots, n$.

We choose $\bar{a}$ to satisfy

$$0 < \bar{a} \le a \qquad \bar{a} \le \frac{a_i}{M} \qquad i = 1, \ldots, n$$

and confine the proof to the rectangle

$$|t - t_0| < \bar{a} \qquad |x_i - x_i{}^0| < a_i \qquad i = 1, \ldots, n$$

The solution of the system will be obtained as the limit of the sequence of iterations given by Picard's method:

$$x_i{}^k(t) = x_i{}^0 + \int_0^t f_i[s, x^{k-1}(s)] \, ds \qquad i = 1, \ldots, n$$

We must first show inductively that $x_i{}^k(t)$, the ith component of $x^k(t)$, lies in the above rectangle, a property which is true for $k = 0$. We assume the truth of it for k and show that it also holds for $k + 1$. Using Picard's formula, we have

$$|x_i{}^{k+1}(t) - x_i{}^0| \le M|t - t_0| < M\bar{a} \le a_i$$

which is in the rectangle.

Our task now is to show that the sequence $x_i{}^k(t)$, which, from its definition, is continuous, converges uniformly to a limit $x_i(t)$ [which implies that $x_i(t)$ is continuous] in the rectangle. We have, for $i = 1, \ldots, n$.

$$|x_i{}^1 - x_i{}^0| \le \left| \int_{t_0}^t f_i(s, x^0) \, ds \right| \le M|t - t_0|$$

$$|x_i{}^2 - x_i{}^1| \le \left| \int_{t_0}^t [f_i(s, x^1) - f_i(s, x^0)] \, ds \right|$$

$$< \int_{t_0}^t \sum_{j=1}^n A_j |x_j{}^1(s) - x_j{}^0| \, ds$$

$$\le \int_{t_0}^t \sum_{j=1}^n A_j M |s - s_0| \, ds = M \frac{(t - t_0)^2}{2} \sum_{j=1}^n A_j$$

and, inductively,

$$|x_i{}^k - x_i{}^{k-1}| < M \frac{|t - t_0|^k}{k!} \left(\sum_{j=1}^n A_j \right)^{k-1} \le \frac{M \bar{a}^k}{k!} \left(\sum_{j=1}^n A_j \right)^{k-1}$$

This establishes the uniform convergence of the series

$$x_i(t) = x_i{}^0 + (x_i{}^1 - x_i{}^0) + (x_i{}^2 - x_i{}^1) + \cdots \qquad i = 1, \ldots, n$$

and hence the continuity of $x_i(t)$. Thus

$$x_i(t) = x_i{}^0 + \int_{t_0}^t f_i[s, x(s)] \, ds$$

from which $x_i(t_0) = x_i{}^0$ and

$$\frac{dx_i}{dt} = f_i(t, x_1, \ldots, x_n) \qquad i = 1, \ldots, n$$

To prove the uniqueness of $x_i(t)$, assume that $z_i(t)$ is another solution for which $z_i(t_0) = x_i^0$, $i = 1, \ldots, n$. Choose $\bar{a}$ such that both solutions lie in the rectangular region defined above.

From the relation

$$z_i(t) = x_i^0 + \int_{t_0}^t f_i[s, z(s)] \, ds$$

and because of Lipschitz condition,

$$|x_i(t) - z_i(t)| \leq \int_{t_0}^t \sum_{j=1}^n A_j |x_j(s) - z_j(s)| \, |ds| \leq B|t - t_0| \sum_{j=1}^n A_j$$

where
$$B = \max_i |x_i(t) - z_i(t)|$$

in the rectangle.

Thus, because of the relation between the quantity on the left and that on the right, we have

$$|x_i(t) - z_i(t)| \leq \int_{t_0}^t \Big(\sum_{j=1}^n A_j \Big)^2 B|t - t_0| \, |ds|$$

$$\leq B \Big(\sum_{j=1}^n A_j \Big)^2 \frac{|t - t_0|^2}{2}$$

and, inductively,

$$|x_i(t) - z_i(t)| \leq B \Big(\sum_{j=1}^n A_j \Big)^k \frac{|t - t_0|^k}{k!} \qquad i = 1, \ldots, n$$

As $n \to \infty$, the right side tends to zero, and hence $x_i(t) = z_i(t), i = 1, \ldots, n$, and the solution through the initial point $x_i^0, i = 1, \ldots, n$, is unique.

In a paper by Brauer and Sternberg,[5a] an example (due to Müller) is given in which, for the system $\dot{x} = f(x,t)$, $x(0) = 0$, the usual iterations

$$x_{n+1} = \int_0^t f(x_n, t) \, dt$$

do not converge, even though there is a unique solution.

Exercise 4-8 Show that the equation

$$a_n \frac{d^n y}{dt^n} + a_{n-1} \frac{d^{n-1} y}{dt^{n-1}} + \cdots + a_0 y = 0$$

can be reduced to the system

$$\dot{x}_1 = x_2$$
$$\cdots \cdots$$
$$\dot{x}_{n-1} = x_n$$
$$\dot{x}_n = -\frac{1}{a_n} (a_{n-1} x_n + \cdots + a_0 x_1)$$

Exercise 4-9   Kepler's laws of planetary motion are given by

$$m \left[\frac{d^2r}{dt^2} - r \left(\frac{d\theta}{dt} \right)^2 \right] + \frac{\gamma m M}{r^2} = 0$$

$$m \left(2 \frac{dr}{dt} \frac{d\theta}{dt} + r \frac{d^2\theta}{dt^2} \right) = 0$$

where m = mass of a planet
  M = mass of earth
  γ = gravitational constant
  r, θ = polar coordinates of the planet with the earth at the origin
Reduce these equations to a first-order system.

Exercise 4-10   Verify that van der Pol's equation may be reduced to either of the two first-order systems

$$\dot{y} = x \qquad \dot{x} = \epsilon(1 - y^2)x - y$$

$$y = -\dot{x} \qquad \dot{y} = \epsilon \left(y - \frac{y^3}{3} \right) + x$$

Exercise 4-11   Reduce the equations of an airplane with autopilot,

$$\ddot{y} + M\dot{y} + Nz = K$$
$$z = \phi(\sigma)$$
$$\sigma = ay + E\dot{y} + s^2\ddot{y} - \frac{z}{e}$$

to a first-order system; i.e., let $x_1 = y$, $x_2 = \dot{y}$, and $x_3 = z$.

Picard's method may be used for calculating a solution numerically. For example, to solve the equation

$$\dot{x} = xt - x$$

with $x(0) = 3$, we write

$$x = 3 + \int_0^t (xt - x) \, dt \tag{4-15}$$

and putting $x = 3$ in the integrand, we obtain

$$x^{(1)} = 3 + \int_0^t (3t - 3) \, dt = 3 + \tfrac{3}{2}t^2 - 3t$$

Putting $x = 3 + \tfrac{3}{2}t^2 - 3t$ in the integrand of (4-15) and integrating again gives $x^{(2)}$, etc.

Exercise 4-12   Carry out the computation to $x^{(4)}$.

4-4   LINEAR EQUATIONS—OSCILLATORY MOTION, STABILITY

A function $x(t)$ is periodic with period T if it satisfies the condition $x(t + T) = x(t)$.   If a particle moves on a line in such a way that its motion is described by

$$x = A \sin (\omega t + \varphi) + a \qquad (4\text{-}16)$$

where x is its distance from a fixed point on the line, and t is the time, the motion is called *simple harmonic*. The function $\sin (\omega t + \varphi)$ oscillates about the t axis, its amplitude is A (it is the maximum of $x - a$), and clearly it is periodic. The period T of the motion is the time required for the angle $\omega t + \varphi$ to change by 2π. Thus $T = 2\pi/\omega$. The frequency is the number of times the angle changes by 2π per unit time, and hence it is given by $1/T$. A motion described by $x = A e^{-\alpha t} \sin (\omega t + \varphi)$, with $\alpha > 0$, is a damped oscillatory motion. It oscillates because of the sine function and is damped by the exponential factor gradually tending to zero as t increases.

Consider an equation of the form

$$\ddot{x} + a\dot{x} + bx = f(t) \qquad \dot{x} = \frac{dx}{dt}$$

Three types of motion are represented by this equation: free motion (associated with bx, which gives the restoring force), damped motion (associated with $a\dot{x}$, which gives the damping force) and forced motion [associated with $f(t)$]. These facts are easily verified by considering representative physical systems and writing the relevant differential equations. The left side is called the equation of motion of free oscillations, in which the natural frequencies of the undamped oscillation is obtained by putting $a = 0$. For example, a coil spring with one end fixed and a weight attached to the other end would oscillate freely once it was started from equilibrium. However, the presence of air acts as a damping force which gradually diminishes the amplitude of oscillation. If the support of the spring were to move up and down according to some law, then the motion of the spring would be forced. If $f(t) = 0$, the oscillations described by the equation are said to be *free*. The general equation of free oscillations is

$$\ddot{x} + \omega^2 x = g(x, \dot{x})$$

It includes both nonlinear damping and restoring forces. In many cases in which nonlinearity occurs, the deviations from linearity are small, and the motion has approximately linear character. There are other cases in which, for example, damping occurs in the small-deflection range and causes self-excited periodic oscillations with a different period from the undamped harmonic motion. In a self-excited oscillation, as $t \to \infty$ every solution tends to a periodic solution known as a *limit cycle*. If a periodic forcing function $f(t)$ is applied to a spring with nonlinear free-oscillation term, the response of the system, referred to as *resonance* to this force, is different from that in the linear case. It is known that for a linear system the response to a periodic force with period T (that is, the solution of the forced-motion equation) consists of a periodic factor with the same period T and of another periodic factor corresponding to the equation of free oscillations with period equal to the natural period of the system. When the frequencies of the two factors are the same, we have resonance. This resonance occurs when large vibrations are produced by small forces, such as the strong ringing of an ordinary water glass, sometimes to the breaking point, in response to musical sounds. Here resonance occurs when the natural period of vibration of the glass and the period of the musical sounds are equal. In the nonlinear case a periodic forcing function can excite higher or *ultra-harmonics* and *sub-harmonic* resonances, i.e., oscillations with frequencies that are, respectively, multiples of the frequency of the forcing function and fractions of it.

The general solution of a linear nonhomogeneous equation is obtained by adding to the general solution of the homogeneous part a particular solution of the nonhomogeneous equation, usually obtained by the method of variation of parameters. If the coefficients of the equation are variable, there is no adequate theory for deriving the solution, as there is when the coefficients are constants. However, if enough linearly independent particular solutions are obtained by some method, then the general solution is also obtained by the principle of superposition which will be discussed soon. Ordinarily, little can be said about the general solution of a nonlinear equation. Of course there are a few nonlinear equations, such as Riccati's equation, whose general solution is known.

Let us briefly illustrate the solution of the linear equation with a forcing term:

$$\ddot{x} + 2a\dot{x} + x = A \cos \omega t$$

As the reader will recall and as we shall see later, to obtain the solution x_h of the homogeneous equation describing a system with damped vibrations (e.g., a vibrating spring with one end fixed and a weight attached to the other end),

$$\ddot{x} + 2a\dot{x} + x = 0 \qquad -1 < a < 1$$

one obtains the two roots of

$$\lambda^2 + 2a\lambda + 1 = 0$$

which are

$$\lambda_1 = -a + (a^2 - 1)^{\frac{1}{2}} \qquad \lambda_2 \approx -a - (a^2 - 1)^{\frac{1}{2}}$$

Then $\quad x_h = c_1 e^{\lambda_1 t} + c_2 e^{\lambda_2 t} = A_1 e^{-at} \cos (\omega_0 t + \alpha) \qquad \omega_0^2 = 1 - a^2$

which describes a free system with damped oscillations. The factor e^{-at}

is called the *damping factor*, and a is the damping *constant*. Note that if $a = 1$, then $x_h = (c_1 + c_2 t)e^{-t}$.

Exercise 4-13 Verify the last equation, and find the defining expression for the amplitude A and the phase angle α.

To obtain a particular solution x_p of the nonhomogeneous equation, we apply Lagrange's method of variation of parameters and hence assume a solution of the form $x_p = c_1 e^{\lambda_1 t} + c_2 e^{\lambda_2 t}$. We then replace c_1 and c_2 with parameters P and Q, and we have on differentiation

$$\dot{x}_p = \dot{P}e^{\lambda_1 t} + \lambda_1 P e^{\lambda_1 t} + \dot{Q}e^{\lambda_2 t} + \lambda_2 Q e^{\lambda_2 t}$$

If we set

$$\dot{P}e^{\lambda_1 t} + \dot{Q}e^{\lambda_2 t} = 0$$

and differentiate again, we obtain

$$\ddot{x}_p = \lambda_1{}^2 P e^{\lambda_1 t} + \lambda_2{}^2 Q e^{\lambda_2 t} + \lambda_1 \dot{P}e^{\lambda_1 t} + \lambda_2 \dot{Q}e^{\lambda_2 t}$$

Substitution in the equation yields

$$\lambda_1{}^2 P e^{\lambda_1 t} + \lambda_2{}^2 Q e^{\lambda_2 t} + \lambda_1 \dot{P}e^{\lambda_1 t} + \lambda_2 \dot{Q}e^{\lambda_2 t} + 2a(\lambda_1 P e^{\lambda_1 t} + \lambda_2 Q e^{\lambda_2 t}) \\ + P e^{\lambda_1 t} + Q e^{\lambda_2 t} = A \cos \omega t$$

Since each of $e^{\lambda_1 t}$ and $e^{\lambda_2 t}$ is a particular solution of the homogeneous equation when the forcing term is zero, the last equation becomes

$$\lambda_1 \dot{P}e^{\lambda_1 t} + \lambda_2 \dot{Q}e^{\lambda_2 t} = A \cos \omega t$$

which, when solved simultaneously with

$$\dot{P}e^{\lambda_1 t} + \dot{Q}e^{\lambda_2 t} = 0$$

gives

$$\dot{P} = \frac{A e^{-\lambda_1 t} \cos \omega t}{\lambda_1 - \lambda_2}$$

$$\dot{Q} = -\frac{A e^{-\lambda_2 t} \cos \omega t}{\lambda_1 - \lambda_2}$$

$$P = A e^{-\lambda_1 t} \frac{-\lambda_1 \cos \omega t + \omega \sin \omega t}{(\lambda_1{}^2 + \omega^2)(\lambda_1 - \lambda_2)} + a_1$$

$$Q = -A e^{-\lambda_2 t} \frac{-\lambda_2 \cos \omega t + \omega \sin \omega t}{(\lambda_2{}^2 + \omega^2)(\lambda_1 - \lambda_2)} + a_2$$

Thus the solution is obtained by the superposition of the free and forced oscillations:

$$x = x_h + x_p = A_1 e^{-at} \cos(\omega_0 t + \alpha) \\ + A \frac{(1 - \omega^2)\cos \omega t + 2a\omega \sin \omega t}{(1 - \omega^2)^2 + 4a^2\omega^2} \quad (4\text{-}17)$$

The amplitude of the forced oscillation (obtained in the same way as A_1

in the homogeneous case) is equal to the square root of the sum of the squares of the coefficients. It is given by

$$\frac{A}{[(1 - \omega^2)^2 + 4a^2\omega^2]^{1/2}}$$

There are several interesting observations that we can make about the foregoing solution. For example, if $a > 0$, as $t \to \infty$ the free oscillations are completely damped and are therefore "transient," while the forced oscillations build up to a steady state and become the dominant component of the solution. Thus the asymptotic stability (i.e., the limit as $t \to \infty$) of the solution is determined by examining this component, which in this case is periodic and is stable. In general a small damping constant is accompanied by slow arrival at the steady state, and large damping by fast decay. In addition a small damping constant is accompanied by large amplitude of the forced oscillation at resonance, and the smaller the constant the greater is the increase in the amplitude as the forced-oscillation frequency moves away from the resonance frequency.

To analyze the behavior of the solution as $t \to \infty$, it is perhaps advantageous to assume first that the damping force is zero, that is, $a = 0$. In that case the free oscillations are described by $A \cos (\bar{\omega}_0 t + \alpha)$, where $\bar{\omega}_0 = 1$ and there is no decay of the free oscillations as $t \to \infty$. The amplitude of the forced oscillations has its greatest value at $\omega = 1$, which is the same value as $\bar{\omega}_0$, and it is at this common value of the frequency that resonance occurs. When $a \neq 0$, the damped frequency ω_0 is always less than the frequency of the undamped oscillations $\bar{\omega}_0$. In general, damping is measured by means of the logarithmic decrement $\delta = aT$, where T is the period (assume that $\alpha = 0$, that is, that the origin is chosen so that both damped and undamped oscillations give $x = A_2$ at $t = 0$), and if Q is defined as π times the ratio of the maximum energy stored to the energy dissipated per cycle, then $Q = \pi/\delta$ and

$$\left(\frac{\omega_0}{\bar{\omega}_0}\right)^2 = 1 - \frac{1}{2Q^2}$$

Thus the decrease in ω_0, the frequency of the damped oscillation, increases with increased damping.

The amplitude of the forced oscillation when $a \neq 0$ is maximum when the forcing frequency ω is at the damped oscillation frequency. Resonance of an oscillatory system can be both desirable and undesirable. For example, driving a system at its resonance frequency is associated with a maximum dissipation of power because of the behavior of the amplitude under this condition.

Knowledge of the resonance frequency enables the introduction of appropriate tuning measures to obtain the maximum response. In other cases, involving faithful reproduction of musical sounds, for example, it is desirable to have the frequencies of reproduction away from the resonance frequencies. Of course large damping would also prevent resonance, and there is a choice of taking these measures.

Again, the well-known practice of an army falling out of step in crossing a bridge is a precaution taken to avoid a large response from the vibrating bridge at a possible resonance frequency.

The foregoing procedure for solving a linear forced equation is classical, and we have given it for variety and because it is elementary in not requiring matrix theory. The same result could have been obtained using the matrix form of the first-order system and the so-called *variation of parameters integral equation* encountered later in this section.

General Homogeneous Linear Systems with Constant Coefficients

Consider the homogeneous linear system

$$\dot{x}_i = \sum_{j=1}^{n} a_{ij}x_j \qquad i = 1, \ldots, n \tag{4-18}$$
$$x_i(t_0) = x_i^0$$

where the coefficients a_{ij} are constants. In vector notation we have

$$\dot{x} = Ax \qquad x(t_0) = x^0$$

Formally its solution may be written in the form

$$x = e^{At}C \qquad \text{where} \qquad C = x^0 \tag{4-19}$$

Our main task is to determine e^{At}.

It is well known that if the characteristic roots $\lambda_1, \ldots, \lambda_n$ of A are all distinct, then the solution can be written in the form

$$x_i = \sum_{j=1}^{n} c_{ij}e^{\lambda_j t} \tag{4-20}$$

In the case of multiple roots, the exponential corresponding to a root of multiplicity m is multiplied by a polynomial in t whose power does not exceed $m - 1$. The nature of the solution corresponding to multiple roots can be approached using the Jordan canonical form which is obtained by applying a similarity transformation to A. This procedure can be found in Ref. 67.

Alternatively, e^{At} can be obtained as a special case of the following spectral form from the calculus of bounded operators applied to an entire function $f(A)$ of a bounded operator A or from Sylvester's well-known theorem in this field:[21,18]

$$f(A) = \sum_{i=1}^{k} \sum_{m=0}^{m_i-1} \frac{(A - \lambda_i I)^m}{m!} f^{(m)}(\lambda_i) Z(\lambda_i) \tag{4-21}$$

Here k is the number of distinct characteristic roots of the matrix A, m_i is the multiplicity of the ith root λ_i of the characteristic equation, $|\lambda I - A| = 0$, $f^{(m)}(\lambda_i)$ is the mth-order formal derivative of f evaluated at λ_i, and the $Z(\lambda_i)$ are complete orthogonal idempotent matrices of the matrix A; that is, they have the properties

$$\sum_{i=1}^{k} Z(\lambda_i) = I \qquad Z(\lambda_i) Z(\lambda_j) = 0 \qquad i \neq j \qquad Z^2(\lambda_i) = Z(\lambda_i)$$

where I and 0 are the identity and null matrices, respectively.

When the characteristic roots are all distinct, one has, for an nth-order matrix A,

$$f(A) = \sum_{i=1}^{n} f(\lambda_i) Z(\lambda_i) \tag{4-22}$$

where

$$Z(\lambda_i) = \frac{\displaystyle\prod_{j \neq i} (\lambda_j I - A)}{\displaystyle\prod_{j \neq i} (\lambda_j - \lambda_i)} \tag{4-23}$$

To illustrate how (4-22) and (4-23) are obtained when f is a polynomial in A, note from the nth-degree polynomial $|\lambda I - A| = 0$ that A^n can be expressed in terms of lower powers of A and hence that f can always be reduced to a polynomial of degree not exceeding $n - 1$. If we write

$$f(A) = \sum_{i=1}^{n} \alpha_i \prod_{\substack{j=1 \\ j \neq i}}^{n} (A - \lambda_i I)$$

and multiply on the right successively by e_k, $k = 1, \ldots, n$ the characteristic vector of λ_k, and use the fact that $A e_k = \lambda_k e_k$ and hence that $f(A) e_k = f(\lambda_k) e_k$, we have

$$\alpha_k = \frac{f(\lambda_k)}{\displaystyle\prod_{j \neq i} (\lambda_k - \lambda_j)}$$

which gives the desired result.

If $f(A) = e^{At}$ and the characteristic roots of A are distinct, we have the spectral resolution of $f(A)$ given by

$$e^{At} = \sum_{i=1}^{n} e^{\lambda_i t} Z(\lambda_i) \tag{4-24}$$

The case of multiple characteristic roots is derived from the confluent form of Sylvester's theorem.   If we write, for brevity,

$$f(A) = \sum_{i=1}^{k} T(\lambda_i)$$

where k is the number of distinct roots, then

$$T(\lambda_i) = f(\lambda_i)Z_{m_i-1}(\lambda_i) + f'(\lambda_i)Z_{m_i-2}(\lambda_i) + \frac{f''(\lambda_i)}{2!}Z_{m_i-3}(\lambda_i) + \cdots \quad (4\text{-}25)$$

Here m_i refers to the multiplicity of the root λ_i, and

$$Z_{m_i}(\lambda_i) = \frac{1}{m_i!}\frac{d^{m_i}}{d\lambda^{m_i}}\frac{F(\lambda)}{\Delta_{m_i}(\lambda)}\Bigg|_{\lambda=\lambda_i} \quad (4\text{-}26)$$

where $\quad F^{(m)}(\lambda_i) = m!(-1)^{n-m-1}(\lambda_i I - A)^{m_i-m-1}\prod_{j\neq i}(\lambda_j I - A) \quad (4\text{-}27)$

gives the mth-order derivative of F, and

$$\Delta_{m_i}(\lambda) = \prod_{j\neq i}(\lambda - \lambda_j) \quad (4\text{-}28)$$

Note, for example, that

$$Z_1(\lambda) = \frac{d}{d\lambda}\frac{F(\lambda)}{\Delta(\lambda)} = \frac{\Delta(\lambda)F'(\lambda) - F(\lambda)\Delta'(\lambda)}{[\Delta(\lambda)]^2}$$

Exercise 4-14   Consider the system

$$\dot{x} = x + y \qquad \dot{y} = x - y$$

Show that $\lambda_1 = \sqrt{2}$, $\lambda_2 = -\sqrt{2}$; then justify the expression

$$e^{At} = \frac{e^{\lambda_1 t}}{\lambda_1 - \lambda_2}(A - \lambda_2 I) + \frac{e^{\lambda_2 t}}{\lambda_2 - \lambda_1}(A - \lambda_1 I)$$

and use it to obtain the solution.

Exercise 4-15   Using the above procedure, compute e^{At}, where

$$A = \begin{pmatrix} 0 & 1 & 0 \\ 0 & -2 & 0 \\ 0 & 1 & 0 \end{pmatrix}$$

We shall soon examine the question of the stability of solutions.   One form of stability, known as *asymptotic stability*, requires that

$$\lim_{t\to\infty} x_i = 0 \qquad i = 1, \ldots, n$$

i.e., that every solution tends to the zero solution as $t \to \infty$, and hence that every solution is asymptotically stable. Of course a nonzero solution may be the steady-state solution, as we have seen in the spring example.

The fact that the solution of a linear system is a sum of exponentials (some of which are multiplied by powers of t in the multiple-root case) yields asymptotic stability if the λ_i which are complex numbers have negative real parts. If any real part is positive, then the solution is unstable, since it becomes unbounded as $t \to \infty$. Critical cases occur if some λ_i are zero or are pure imaginary while the remaining ones have negative real parts. In conformity with existing abuse of language, we shall sometimes refer to a solution as being stable when in fact it is asymptotically stable. Different forms of stability will be introduced later, but the brief presentation given above will serve our immediate needs.

It is interesting to note that, for linear systems, asymptotic stability is always in the large (i.e., holds for any choice of initial conditions), whereas in nonlinear systems the extent of asymptotic stability depends on nonlinearities. It is here that Lyapunov's theory helps determine the size of the region of asymptotic stability. We shall illustrate this in Sec. 4-11.

From a computational standpoint, it is not easy to find the roots of an nth-degree polynomial.

The following well-known theorem provides a useful test for the asymptotic stability of a linear system.

Theorem (**Routh-Hurwitz**)[25,26] A necessary and sufficient condition that all the roots of the equation

$$|\lambda I - A| = a_0 \lambda^n + a_1 \lambda^{n-1} + \cdots + a_n = 0$$

with real coefficients, have negative real parts is that the following determinants all be positive:

$$D_1 = a_1 \qquad D_2 = \begin{vmatrix} a_1 & a_0 \\ a_3 & a_2 \end{vmatrix} \qquad D_3 = \begin{vmatrix} a_1 & a_0 & 0 \\ a_3 & a_2 & a_1 \\ a_5 & a_4 & a_3 \end{vmatrix}$$

$$D_n = \begin{vmatrix} a_1 & a_0 & 0 & 0 & \cdots & 0 \\ a_3 & a_2 & a_1 & a_0 & \cdots & 0 \\ \multicolumn{6}{c}{\dotfill} \\ a_{2n-1} & a_{2n-2} & a_{2n-3} & a_{2n-4} & \cdots & a_n \end{vmatrix} \equiv a_n D_{n-1}$$

that is, that $D_1, D_2, \ldots, D_{n-1}$ and a_0 and a_n are all greater than zero. Note that $a_p = 0$ for $p > n$, but this is the general form.

Exercise 4-16 Write out and develop these conditions for equations of the third and fourth degrees.

The idea of a fundamental system of solutions plays an essential role in the solution of the general linear equation, which we discuss below. A homogeneous linear equation whose coefficients may be arbitrary functions of time t, of order n (or, equivalently, a system of n first-order homogeneous equations), has n linearly independent solutions. A set of functions $w_i(t)$, $i = 1, \ldots, n$, is linearly dependent if there exist numbers α_i not all zero such that

$$\sum_{i=1}^{n} \alpha_i w_i(t) = 0 \qquad (4\text{-}29)$$

Otherwise the $w_i(t)$ are said to be linearly independent. Any set of n independent solutions is called a *fundamental* system of solutions.

The wronskian of $w_i(t)$, $i = 1, \ldots, n$, is the determinant

$$W(t) = \begin{vmatrix} w_1(t) & \cdots & w_n(t) \\ \dfrac{dw_1}{dt} & \cdots & \dfrac{dw_n}{dt} \\ \dfrac{d^{n-1}w_1}{dt^{n-1}} & \cdots & \dfrac{d^{n-1}w_n}{dt^{n-1}} \end{vmatrix} \qquad (4\text{-}30)$$

Note that this determinant is obtained by successive differentiation of the equation describing dependence. This yields a set of n linear homogenous equations in α_i. Thus the functions are linearly dependent if and only if the wronskian is zero (for all values of t), provided the functions are solutions of a nonsingular homogeneous linear equation.

If the $w_i(t)$ are linearly independent solutions, then every other solution $w(t)$ is a linear combination of them; i.e., there exist numbers β_i such that

$$w(t) = \sum_{i=1}^{n} \beta_i w_i(t)$$

Again by successive differentiations we obtain a set of equations. If we fix a value of t at t_0 and seek a solution for β_i, $i = 1, \ldots, n$, then since the wronskian is not zero, we are able to obtain a unique solution for β_i. Because of the existence of unique solutions to the general systems of differential equations under consideration, the solution corresponding to

$$\sum_{i=1}^{n} \beta_i w_i(t)$$

for arbitrary t must coincide with $w(t)$, since both have the same initial value at t_0.

Exercise 4-17 Show by the method of transforming a single nth-order equation to a system of first-order equations with the fundamental system of solutions $x_{ki}(t)$,

$i, k = 1, \ldots, n$, that we have, for the wronskian,

$$W(t) = \begin{vmatrix} x_{11}(t) & \cdots & x_{1n}(t) \\ \cdots\cdots\cdots\cdots\cdots \\ x_{n1}(t) & \cdots & x_{nn}(t) \end{vmatrix}$$

Exercise 4-18   Consider the system

$$\frac{dx_1}{dt} = p_{11}(t)x_1 + p_{12}(t)x_2$$

$$\frac{dx_2}{dt} = p_{21}(t)x_1 + p_{22}(t)x_2$$

and the wronskian of this system,

$$W(t) = \begin{vmatrix} x_{11}(t) & x_{12}(t) \\ x_{21}(t) & x_{22}(t) \end{vmatrix}$$

Show that

$$\frac{dW}{dt} = [p_{11}(t) + p_{22}(t)]W$$

and obtain Abel's identity,

$$W(t) = W(t_0) \exp \int_{t_0}^{t} [p_{11}(t) + p_{22}(t)]\, dt \tag{4-31}$$

Generalize to a system of n equations, obtaining

$$W(t) = W(t_0) \exp \int_{t_0}^{t} \sum_{k=1}^{n} p_{kk}(t)\, dt \tag{4-32}$$

Exercise 4-18a   Show that $w_1(t) = e^{at}$, $w_2(t) = \sin t$ are linearly independent.

The solution of the most general linear nonhomogeneous system is obviously obtainable from these more general considerations. In the interest of economy we treat the following nonlinear system, which will arise later in the chapter.

Consider the system

$$\dot{x} = A(t)x + \epsilon f(x,t) \qquad x(0) \equiv x_0 \tag{4-33}$$

where the vector f is periodic with period T, $x = (x_1, \ldots, x_n)$, and the coefficients of $A(t)$ are also periodic with period T.

Let $X(t)$ be a fundamental system of solutions of the linear part

$$\dot{x} = A(t)x$$

satisfying $$X(0) = I$$

Using the method of variations of parameters, put

$$x = Xy$$

Then $$\dot{x} = X\dot{y} + \dot{X}y = X\dot{y} + A(t)Xy$$

from which $$X\dot{y} = \epsilon f(Xy,t)$$

or $$y = y(0) + \epsilon \int_0^t X^{-1}(s)f(Xy,s)\, ds$$

from which, since
$$x_0 = X(0)y(0) = y(0)$$

we have the variation-of-parameters integral equation

$$x(t) = X(t)x_0 + \epsilon \int_0^t X(t)X^{-1}(s)f[x(s),s]\,ds \qquad X(0) = I \quad (4\text{-}34)$$

which is equivalent to Eq. (4-33).

Remark: If $\epsilon f(x,t) = g(t)$, we have the solution of the general linear equation $\dot{x} = A(t)x + g(t)$.

Exercise 4-19   Verify the equivalence of (4-33) and (4-34).

Exercise 4-20   Show that when A is constant $X(t) = e^{At}$, and substitute for $X(t)$ and $X^{-1}(s)$ in Eq. (4-34).

Exercise 4-21   Using (4-34), solve $\ddot{x} + 2a\dot{x} + x = A \cos \omega t, -1 < a < 1$.

Exercise 4-22   Show that Duffing's equation with a forcing term,
$$\ddot{x} + ax + bx^3 = A \cos \omega t$$
when written in the form
$$\ddot{x} - \omega^2 x = A \cos \omega t - (\mu^2 x + bx^3)$$
or [by putting $s = t$ and writing $f(s)$ for the right side] in the form
$$\frac{d^2x}{ds^2} + \omega^2 x = f(s)$$
gives
$$x(t) = x(0) \cos \omega t + \dot{x}(0)\frac{\sin \omega t}{\omega} + \frac{1}{\omega}\int_0^t \sin \omega(t - s)f(s)\,ds \qquad (4\text{-}35)$$

4-5   NONLINEAR EQUATIONS—PERTURBATION METHOD

We now turn to the analysis of oscillations in nonlinear systems. The principle of superposing solutions to obtain the most general solution does not apply here. As it is usually not possible to write down the general solution of such a system or even to obtain an exact particular solution, one frequently resorts to careful approximations whose analysis reveals the characteristic oscillatory properties of the system containing a parameter (either inherent or introduced for convenience). By treating the case of a small parameter, one is sometimes able to deduce properties of the solution which hold in general. Unlike the linear case, in which the free and the forced oscillations were superimposed and analyzed separately, because of nonlinearity the restoring force and the damping force in nonlinear systems interact and cannot be studied separately. However, if the values of the damping constant are small, the free oscillations decay, and the forced oscillations become dominant in time and can then be analyzed alone.

The question is whether nonlinear systems, while subjected to periodic oscillations, reveal properties not shown by linear systems. The answer to this question is in the affirmative, and a characteristic property of such systems, as we shall see, is the presence of limit cycles. Our task for the present is to give a method for obtaining a periodic solution of a nonlinear equation with a small parameter.

A well-known method for obtaining such a solution is the method of perturbations; the solutions are valid for small values of the parameter. Consider the equation

$$\ddot{x} - \epsilon(1 - x^2)\dot{x} + ax = 0 \qquad a > 0 \tag{4-36}$$

We shall use a well-known perturbation method to obtain an approximate periodic solution, valid for small $\epsilon > 0$.

If $\epsilon = 0$, $x = A \cos a^{1/2}t$ is a periodic solution with A arbitrary. Therefore we may expect that, for $\epsilon \neq 0$,

$$x(t) = x_0(t) + x_1(t)\epsilon + x_2(t)\epsilon^2 + \cdots \tag{4-37}$$

with
$$x_0(t) = A \cos a_0^{1/2}t \tag{4-38}$$

for some A, with a_0 close to a. We write

$$a = a_0 + a_1\epsilon + a_2\epsilon^2 + \cdots \tag{4-39}$$

substitute the two series for x and a into (4-36), and equate like powers of ϵ. This results in a system of linear equations in x_i which can be solved, and which will yield $a_0, a_1, a_2, \ldots$. If we stop at second-order terms, we have

$$a \approx a_0 + a_1\epsilon + a_2\epsilon^2$$

from which a_0 can be obtained in terms of a and ϵ. If we then take $a = 1$, we have van der Pol's equation and its solution. In the following, we shall write

$$\omega_0 = a_0^{1/2}$$

Let the initial conditions be given by

$$x_0(0) = A \qquad \dot{x}_0(0) = 0 \qquad x_i(0) = \dot{x}_i(0) = 0 \qquad i \neq 0 \tag{4-40}$$

Since our purpose is to illustrate the use of a well-known perturbation technique, we shall construct an approximate periodic solution for small ϵ in the most direct manner. In a later section on periodic solutions, there will be a more detailed theoretical account of the method.

If we substitute the series expansion in Eq. (4-36) we get

$$(\ddot{x}_0 + \ddot{x}_1\epsilon + \ddot{x}_2\epsilon^2 + \cdots)$$
$$- \epsilon[1 - (x_0 + x_1\epsilon + x_2\epsilon^2 + \cdots)^2](\dot{x} + \dot{x}_1\epsilon + \dot{x}_2\epsilon^2 + \cdots)$$
$$+ (a_0 + a_1\epsilon + a_2\epsilon^2 + \cdots)(x_0 + x_1\epsilon + x_2\epsilon^2 + \cdots) = 0 \tag{4-41}$$

We now equate coefficients of like powers of ϵ. We have

$$\ddot{x}_0 + a_0 x_0 = 0 \tag{4-42}$$
and
$$\ddot{x}_1 + a_0 x_1 = (1 - x_0^2)\dot{x}_0 - a_1 x_0 \tag{4-43}$$
$$\ddot{x}_2 + a_0 x_2 = -2 x_0 x_1 \dot{x}_0 + (1 - x_0^2)\dot{x}_1 - a_1 x_1 - a_2 x_0 \tag{4-44}$$

The first equation has a solution of the form

$$x_0(t) = A \cos(\omega_0 t + \alpha)$$

where $\omega_0^2 = a_0$. Since α merely corresponds to a shift in t, we can put $\alpha = 0$ and use $x_0(t) = A \cos \omega_0 t$. If we substitute for x_0, we obtain

$$\ddot{x}_1 + \omega_0^2 x_1 = [-A\omega_0 \sin \omega_0 t + A^3 \omega_0 \sin \omega_0 t \cos^2 \omega_0 t - a_1 A \cos \omega_0 t] \tag{4-45}$$

By the application of the variation-of-parameters integral formula for the case of Duffing's equation given in Exercise 4-22, we obtain [recall $x_1(0) = \dot{x}_1(0) = 0$]:

$$x_1(t) = \frac{1}{\omega_0} \int_0^t \sin \omega_0 (t - s)[\quad] \, ds \tag{4-46}$$

where [ ] is the right side of the previous equation with s as variable. Now

$$\sin(\omega_0 t - \omega_0 s) = \sin \omega_0 t \cos \omega_0 s - \cos \omega_0 t \sin \omega_0 s$$

Thus we integrate $\cos \omega_0 s[\quad]$ and $\sin \omega_0 s[\quad]$. Let $u = \omega_0 s$, $du = \omega_0 \, ds$, or $ds = du/\omega_0$, and we have

$$x_1(t) = \sin \omega_0 t \int_0^{\omega_0 t} \cos u \, [-A\omega_0 \sin u + A^3 \omega_0 \sin u \cos^2 u - a_1 A \cos u] \, du$$
$$- \cos \omega_0 t \int_0^{\omega_0 t} \sin u \, [-A\omega_0 \sin u + A^3 \omega_0 \sin u \cos^2 u$$
$$- a_1 A \cos u] \, du$$
$$= \sin \omega_0 t \left[-\frac{A}{2} \omega_0 \sin^2 \omega_0 t - \frac{A^3}{4} \omega_0 \cos^4 \omega_0 t \right.$$
$$\left. - a_1 A \left(\frac{\omega_0 t}{2} + \frac{\sin 2\omega_0 t}{4} \right) \right] + \frac{A^3 \omega_0}{4} \sin \omega_0 t$$
$$+ \cos \omega_0 t \left[A\omega_0 \left(\frac{\omega_0 t}{2} - \frac{\sin^2 \omega_0 t}{4} \right) + \frac{A^3 \omega_0}{8} \left(\frac{\sin 4\omega_0 t}{4} - \omega_0 t \right) \right.$$
$$\left. + \frac{a_1}{2} A \sin^2 \omega_0 t \right] \tag{4-47}$$

The coefficients of all nonperiodic terms, often referred to as *secular* terms, must vanish (here we have terms of the form $t \sin \omega_0 t$ and $t \cos \omega_0 t$). This gives

$$-a_1 \omega_0 = 0$$

from which

$$a_1 = 0$$

since obviously we do not want $\omega_0 = 0$, and

$$\frac{A\omega_0^2}{2} - \frac{A^3\omega_0^2}{8} = 0 \qquad (4\text{-}48)$$

from which $A = 2$. (Note that $A = -2$ merely leads to a different phase.) This is called the *bifurcation* equation, about which we shall say more later.

We now have

$$x_1(t) = -\omega_0 \sin^3 \omega_0 t - 2\omega_0 \sin \omega_0 t \cos^4 \omega_0 t + 2\omega_0 \sin \omega_0 t$$
$$- \frac{\omega_0}{2} \cos \omega_0 t \sin 2\omega_0 t + \frac{\omega_0}{4} \cos \omega_0 t \sin 4\omega_0 t$$

Now use
$$\sin^3 \omega_0 t = \frac{3 \sin \omega_0 t - \sin 3\omega_0 t}{4}$$

$$\cos^4 \omega_0 t = \frac{\cos 4\omega_0 t + 8(1 - \sin^2 \omega_0 t) - 1}{8}$$

Hence

$$\sin \omega_0 t \cos^4 \omega_0 t = \frac{\sin \omega_0 t \cos 4\omega_0 t}{8} + \sin \omega_0 t - \frac{3 \sin \omega_0 t - \sin 3\omega_0 t}{4} - \frac{\sin \omega_0 t}{8}$$

Thus we have

$$x_1(t) = \omega_0 \sin \omega_0 t - \frac{\omega_0}{2} \cos \omega_0 t \sin 2\omega_0 t = \frac{\omega_0}{4} (3 \sin \omega_0 t - \sin 3\omega_0 t) \quad (4\text{-}49)$$

Then, up to first-order terms in ϵ, from Eq. (4-37) we have

$$x(t) = 2 \cos \omega_0 t + \frac{\omega_0}{4} (3 \sin \omega_0 t - \sin 3\omega_0 t)\epsilon \qquad (4\text{-}50)$$

Exercise 4-23 By equating to zero the sum of the secular terms in the equation in x_2, show that $a_2 = a_0/8$.

Remark: Now we can write $\omega_0 \simeq (1 + \epsilon^2/8)^{-1/2} \simeq 1 - \epsilon^2/16$ $(a = 1)$.

Exercise 4-24 Note in the equation of van der Pol that when $\epsilon = 0$ the resulting periodic solution has unit frequency. If ϵ is nonzero and small, then the frequency ω depends on ϵ and is near unity. If $x(t)$ has period $2\pi/\omega$, then, by setting $\theta = \omega t$, we obtain a new function $x(\theta)$ with period 2π. With primes for differentiation with respect to θ, our equation becomes

$$\omega^2 x'' - \epsilon(1 - x^2)x' + x = 0$$

which, on substituting

$$x = x_0 + x_1\epsilon + x_2\epsilon^2 + \cdots$$

and

$$\omega = \omega_0 + \omega_1\epsilon + \omega_2\epsilon^2 + \cdots$$

becomes

$$(x_0'' + x_1''\epsilon + x_2''\epsilon^2 + \cdots)[\omega_0^2 + 2\omega_0\omega_1\epsilon + (\omega_1^2 + 2\omega_0\omega_1)\epsilon^2 + \cdots]$$
$$- \epsilon[1 - (x_0 + x_1\epsilon + x_2\epsilon^2 + \cdots)^2(\omega_0 + \omega_1\epsilon + \omega_2\epsilon^2 + \cdots)]$$
$$(x_0' + x_1'\epsilon + x_2'\epsilon^2 + \cdots) + (x_0 + x_1\epsilon + x_2\epsilon^2 + \cdots) = 0$$

Carry out the computations, and obtain x_0, x_1, ω_0, and ω_1.

It has been pointed out by C. H. Murphy[64a] that the solution (4-50) may be expressed as

$$x(t) = 2 \cos \omega_0 t + \frac{\epsilon \omega_0}{4} [(3 - 8\alpha) \sin \omega_0 t - \sin 3\omega_0 t] \qquad (4\text{-}51)$$

where α is arbitrary; i.e., in the linear approximation the coefficient of $\sin \omega_0 t$ is completely arbitrary. Here the origin has been shifted along the time axis by a small amount in replacing ωt by $\omega t + \alpha \epsilon$.

The foregoing perturbation technique can also be applied to an equation with a periodic forcing function. To avoid resonance, the frequency of the forcing function must not be an integral multiple of the frequency of the unforced equation.

Note that the value of A can be determined without the elaborate use of perturbation theory. We use $x = A \cos (t + \alpha)$—to a first approximation ω_0 is close to 1—and determine A from the energy equation. This equation is obtained by multiplying the above force equation by $\dot{x} \, dt$ and integrating from zero to 2π. Since work (measured as energy) is force times distance and $\dot{x} \, dt = (dx/dt) \, dt = dx$, we have

$$\int_0^{2\pi} \{ -A^3 \cos^2 (t + \alpha) \sin (t + \alpha)$$
$$- \epsilon[1 - A^2 \cos^2 (t + \alpha)]A^2 \sin^2 (t + \alpha)$$
$$- A^2 \cos (t + \alpha) \sin (t + \alpha) \} \, dt = 0 \qquad (4\text{-}52)$$

Thus $\pi A^2 \epsilon (1 - A^2/4) = 0$, from which it follows that $A = 2$.

As the mechanics of constructing periodic solutions of nonlinear equations is tedious and takes up a considerable amount of space, it is suggested that the reader pursue the analysis of super- and subharmonic resonance through the references.

Stability

One is usually interested in the solution of a differential equation if the initial conditions or the right side of the equation is changed. This change corresponds to a real-life situation in which, for example, the differential equation may be an idealization which assumes a calm atmosphere, whereas the true state is described by a turbulent atmosphere which buffets a missile and hence comprises a disturbance of its trajectory. The simplest case is that in which the initial conditions are changed (disturbed). The existence theorem guarantees a unique solution of the differential equation for each choice of initial conditions. Thus for two initial conditions, i.e., the original and the disturbed, we obtain two solutions, and the following question arises: Will the difference between the trajectories remain bounded, tend to zero, oscillate, or grow without bound as time goes on? This is the fundamental concern of the Lyapunov theory of stability.   Another problem in this vein is stability under continuously acting disturbances, which we shall mention later, but only briefly.   Our concern will be mostly with disturbances acting at a particular moment, e.g., an error in the launch of a missile.

Since a system of differential equations is equivalent to a possibly larger system of first-order equations of the general form

$$\dot{y}_k = Y_k(y_1, \ldots, y_n; t) \qquad k = 1, \ldots, n \qquad (4\text{-}53)$$

where the y_k's may represent the state of the system at time t, for example, coordinates, velocities, or other physical quantities, stability theory concentrates on studying the behavior of first-order systems.

Any one particular solution of this system, $y_k = f_k(t)$, corresponds to a particular motion.   Such a motion may be considered to be an undisturbed motion.   Suppose now that position disturbances $x_k(t_0)$ are introduced at time t_0; then the position coordinates of the system at t_0 become

$$\bar{y}_k(t_0) = f_k(t_0) + x_k(t_0) \qquad (4\text{-}54)$$

Now, for $t > t_0$ we have the same system as before with the initial conditions (4-54), and we can denote by $\bar{y}_k(t)$ the solution of the new problem. We evidently have

$$\frac{d\bar{y}_k}{dt} = Y_k(\bar{y}_1, \ldots, \bar{y}_n; t) \qquad (4\text{-}55)$$

Our principal interest is in the difference

$$x_k(t) \equiv \bar{y}_k(t) - y_k(t) = \bar{y}_k(t) - f_k(t)$$

From the last equation we see that the disturbance $x_k(t)$ is the unique solution of

$$\frac{dx_k}{dt} = \frac{d\bar{y}_k}{dt} - \frac{df_k}{dt} = Y_k(x_1 + f_1, \ldots, x_n + f_n; t) - Y_k(f_1, \ldots, f_n; t)$$
$$\equiv X_k(x_1, \ldots, x_n; t) \qquad (4\text{-}56)$$

with initial values $x_k(t_0)$ at $t = t_0$.   The equations $\dot{x}_k = X_k(x_1, \ldots, x_n; t)$, $k = 1, \ldots, n$, are known as the *equations of the disturbed motion.* By means of Lyapunov's theory which we give later, the stability of a solution of a nonlinear equation may be examined through the associated variational equation, which is the linear part of the equations of disturbed motion.   For example, if we disturb a given solution $f(t)$ of the equation

$$\ddot{y} + a\dot{y} + y^2 = g(t) \qquad (4\text{-}57)$$

by $x(t)$ and hence substitute the new solution $y = x + f$, we obtain

$$\ddot{x} + \ddot{f} + a(\dot{x} + \dot{f}) + (x + f)^2 = g(t) \qquad (4\text{-}58)$$

Since f is a solution of (4-57), this becomes

$$\ddot{x} + a\dot{x} + 2xf + x^2 = 0 \tag{4-59}$$

Thus the forcing function drops out, and we need only the linear part to determine stability; i.e.,

$$\ddot{x} + a\dot{x} + 2xf = 0 \tag{4-60}$$

Thus the original solution is stable if the general solution of (4-60), which is linear in x, is stable. Equation (4-60) is known as the *variational* equation. It can be obtained directly by applying the variational operation to the differential equation. To see this, recall that in deriving Euler's equation we wrote

$$y(x,\epsilon) = \bar{y}(x) + \epsilon\eta(x)$$

If we form

$$\Delta f = f[x, y(x,\epsilon), y'(x,\epsilon)] - f[x,\bar{y}(x),\bar{y}'(x)] = \frac{\partial f}{\partial y}\,\epsilon\eta$$
$$+ \frac{\partial f}{\partial y'}\,\epsilon\eta' + \epsilon^2(\quad) + \cdots$$

by expanding the difference in powers of ϵ, then the variation of f is defined by the linear terms in ϵ. On putting $\delta y \equiv \epsilon\eta$, $\delta y' \equiv \epsilon\eta'$, we write, for the variation of f,

$$\delta f = \frac{\partial f}{\partial y}\,\delta y + \frac{\partial f}{\partial y'}\,\delta y'$$

which is analogous to the differential. It is easy to prove that variation and differentiation are commutative. Consequently, the variation operator can be directly applied to a differential equation. Thus

$$\delta\left(\frac{d^2y}{dt^2} + a\frac{dy}{dt} + y^2\right) = 0$$

or
$$\frac{d^2}{dt^2}\,\delta y + a\frac{d}{dt}\,\delta y + 2f\,\delta y = 0 \tag{4-61}$$

since $y = f$ is the solution to be tested for stability. This is a linear equation in δy, where δy is the same as the disturbance x in the previous interpretation. As we shall see later, the same method can be applied to a system of equations.

Consider van der Pol's equation:

$$\ddot{x} - \epsilon(1 - x^2)\dot{x} + x = 0$$

The variational equation with $\delta x = y$ is

$$\ddot{y} - \epsilon(1 - x^2)\dot{y} + (1 + 2\epsilon x\dot{x})y = 0$$

Given the approximate solution $x = A \sin t$, we wish to determine A for stability.   If we substitute in the last equation and then put

$$y = e^{v(t)} u(t)$$

where
$$v(t) = -\frac{\epsilon}{2}\left[\left(\frac{A^2}{2} - 1\right)t - \frac{A^2}{4}\sin 2t\right]$$

the term in $\dot{y}$ disappears, and we have, on neglecting terms in ϵ^2 and putting $t = s - \pi/2$,

$$\ddot{u} + \left(1 - \frac{\epsilon A^2}{2}\cos 2s\right)u = 0$$

which is Mathieu's equation.

Some of the simplest nonlinear problems lead to rather difficult linear equations with variable coefficients.   Many important examples lead to Hill's equation or Mathieu's equation during the analysis of their stability.   Because of the difficulty in analyzing such linear equations, the use of Lyapunov's theory of stability offers an encouraging alternative.

Exercise 4-25   Derive the variational equation of

$$\ddot{y} + y + \epsilon y^2 = 0$$

in order to test the stability of the approximate periodic solution $A \cos t$.

Remark: Through an interesting theorem given in Ref. 47, it is possible to determine the stability of $A \cos \omega t$ as an approximate solution of van der Pol's equation as $\epsilon \to 0$.   This fact hinges on the bifurcation equation

$$\frac{A}{2} - \frac{A^3}{8} = 0$$

Roughly, if the derivative of the left side is different from zero for $A = 2$, then as $\epsilon \to 0$ we have asymptotic stability of the solution if this derivative is negative for $A = 2$ (which is the case here), and instability if it is positive for this value.   There is a unique oscillatory solution which tends to this solution as $\epsilon \to 0$ and whose period tends to 2π.

4-6   PHASE - PLANE ANALYSIS—STABILITY BEHAVIOR IN THE SMALL (SYSTEMS OF TWO EQUATIONS)

We shall consider later an analytic approach to the study of stability. However, in this section we shall be concerned with the geometric concepts related to the stability of a system of two equations.   The generalization to the case of several equations is given in Refs. 58 and 65.

We start by considering the autonomous system of two equations

$$\frac{dx}{dt} = P(x,y) \qquad \frac{dy}{dt} = Q(x,y) \qquad (4\text{-}62)$$

There is a unique solution $x(t)$, $y(t)$ of this system which satisfies the initial conditions $x(t_0) = x_0$, $y(t_0) = y_0$. This solution defines a curve called a *characteristic curve* with a continuously turning tangent. Note that $x(t - \bar{t})$, $y(t - \bar{t})$ is also a solution for any $\bar{t}$ but represents the same characteristic; hence a characteristic may belong to more than one solution.

The study of stability involves investigating conditions under which characteristics stay near equilibrium points or tend to an equilibrium point as $t \to \infty$. At most one characteristic passes through any point of the domain of definition of the system. If a single point is a solution of the system, then clearly no characteristic passes through it, and it is said to be a *singularity* of the system. Singularities represent points of equilibrium of the system, whereas at any regular point nonzero components of both derivatives (indicating velocity and hence motion) exist, and they are therefore points of disequilibrium. Thus the characteristics represent the trajectories of motion.

The study of the stability of this autonomous system is based on the stability of the linear approximation obtained by expanding P and Q about a singular point (if the equations contain a linear part; if not, a theory based on the higher-order terms must be applied). The equations are assumed to have continuous partial derivatives of all orders everywhere. We have, expanding in series about the origin,

$$\begin{aligned}
P(x,y) &= ax + by + p(x,y) \\
Q(x,y) &= cx + dy + q(x,y)
\end{aligned} \qquad (4\text{-}63)$$

where p and q represent the nonlinear terms. Note that, by transforming the variables linearly, one can eliminate constant terms if they appear. A point at which both P and Q vanish is called an *equilibrium* or a *critical* point. Such a point is called a *singular* point of the expression obtained by dividing the two equations of the system, i.e., of the expression

$$\frac{dy}{dx} = \frac{Q(x,y)}{P(x,y)}$$

Critical points are assumed to be isolated, i.e., there is a neighborhood about each of them in which there are no other critical points. We have seen that there is a unique solution to the system for any point chosen as initial point. Later we shall see that solutions corresponding to all initial points reveal characteristic behavior with regard to the critical points, which from their definition can be obtained by solving $P(x,y) = 0$, $Q(x,y) = 0$ simultaneously. Since, by a translation of variables, a singular point can be brought to the origin, it suffices to examine systems having the latter as a critical point. A linear system of equations is obtained by suppressing the nonlinear terms p and q. In matrix notation it may be written as

$$\frac{dX}{dt} = AX$$

where
$$X = \begin{pmatrix} x \\ y \end{pmatrix} \qquad A = \begin{pmatrix} a & b \\ c & d \end{pmatrix}$$

As we have seen, this matrix equation has a solution in terms of exponentials. Thus, if the roots λ_1 and λ_2 of the characteristic equation $|A - \lambda I| = 0$ are distinct one has

$$x = c_1 e^{\lambda_1 t} + c_2 e^{\lambda_2 t} \qquad y = c_3 e^{\lambda_1 t} + c_4 e^{\lambda_2 t}$$

One may write similar expressions for multiple roots. As previously indicated, we shall study the behavior of solutions near the origin, which is a singular point of the linear system. The ideas developed can be applied to the nonlinear system in the neighborhood of the origin, since in that neighborhood the contribution of the higher-order terms is negligible.

If we solve for t above and then eliminate it, introducing the new variables

$$x_1 = c_1 y - c_3 x \qquad x_2 = c_4 x - c_2 y$$

we obtain the following expression, useful when λ_1 and λ_2 are real or pure imaginary:

$$x_2 = (c_1 c_4 - c_2 c_3)^{1 - \lambda_1/\lambda_2} \frac{x_1 \lambda_1}{\lambda_2}$$

If it happens that the constant coefficient is equal to unity, then the transformation to new variables corresponds to a rotation of the xy plane.

In general, if λ_1 and λ_2 are complex numbers with

$$\lambda_1 = \alpha + i\beta \qquad \lambda_2 = \alpha - i\beta$$

then by means of a real affine transformation the original system may be transformed to

$$\dot{x} = \alpha x - \beta y \qquad \dot{y} = \beta x + \alpha y$$

where
$$\dot{x} = \frac{dx}{dt} \qquad \dot{y} = \frac{dy}{dt}$$

which has a solution of the form

$$x = e^{\alpha t} \cos \beta \qquad y = e^{\alpha t} \sin \beta t$$

Remark: An affine transformation has the form

$$x' = a_1 x + b_1 y + c_1 \qquad y' = a_2 x + b_2 y + c_2$$

with a nonsingular coefficient matrix. An affine transformation can always be factored into a product of translation, rotation, stretching and contracting, reflections, and simple elongation and compression.

It is easily verified by varying the coefficients that a family of trajectories is obtained, and, depending on the nature of the roots λ_1 and λ_2,

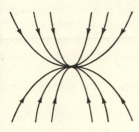

Stable nodal point
(reverse arrows for instability)

Saddle point, unstable equilibrium

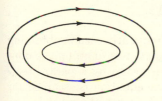

Center or vortex point,
a stable singularity

Stable spiral or focal point
(reverse arrows for instability)

Fig. 4-3

we have the following description of the trajectories near the origin (see Fig. 4-3) as t ranges from $-\infty$ to $+\infty$:

1. If λ_1 and λ_2 are real, and if

(a) $\lambda_1 < \lambda_2 < 0$, then all trajectories move towards the origin, forming a stable node there.

(b) $0 < \lambda_1 < \lambda_2$, the origin is an unstable nodal point, since the trajectories move away from it.

(c) $\lambda_1 < 0 < \lambda_2$, the trajectories tend asymptotically to the x_1 and x_2 axes: the origin forms a saddle point. Each axis in this case is called a *separatrix*, since it separates the plane into regions of trajectories.

2. If $\lambda_1 = i\beta$, $\lambda_2 = -i\beta$, that is, they are purely imaginary, then $\alpha = 0$ and $x = \cos\beta$, $y = \sin\beta$, and the trajectories form circles (ellipses before the affine transformation) with center at the origin. The origin is called a *vortex* point and (by definition) is a stable singularity.

Exercise 4-26 By examining the original system, show that, if clockwise direction is positive, then the direction of movement along the circles is that of the sign of c or the negative of the sign of b.

3. If $\lambda = \alpha + i\beta$, $\lambda = \alpha - i\beta$, for α, $\beta \neq 0$, and if $\alpha < 0$, the trajectories are spirals that wind around and move towards the origin, which is stable and is called a *focal* point. If $\alpha > 0$, the spirals unwind and move away from the origin, which is an unstable focal point (*focus* or *spiral* point).

Thus we have stability only when the real parts of the roots are zero or negative.

Exercise 4-27 Using the foregoing solution, obtain the trajectories, and show that the behavior remains the same if an affine transformation is applied to the system.

We now examine the nonautonomous system

$$\frac{dx}{dt} = P(x,y,t) \qquad \frac{dy}{dt} = Q(x,y,t) \tag{4-64}$$

As we have already seen in Sec. 4-5, a method frequently used in obtaining periodic solutions of a nonlinear differential equation is to group the nonlinear terms of the equation and multiply this group by a parameter ϵ which may be small. A series expansion of the equation is assumed for small values of ϵ. The coefficient of the zeroth power of ϵ must be the solution of the linear equation. The remaining coefficients are then determined by iterative procedures using the first coefficient. In this manner a solution of the original equation is obtained for small values of ϵ.

We usually take the linear approximation as the solution to the system, since higher powers of ϵ are negligible. Our goal here is to transform our nonautonomous system to an autonomous one by means of such an approach.

We assume that $y = \dot{x}$ is introduced in going from a second-order equation to a first-order system. The procedure of reduction is known as the *stroboscopic* method (see Minorsky[48]).

Let
$$\rho = r^2 = x^2 + y^2 \qquad \theta = \arctan\frac{y}{x}$$

Then
$$\frac{d\rho}{dt} = 2(x\dot{x} + y\dot{y}) \qquad \text{and} \qquad -\rho\frac{d\theta}{dt} = y\dot{x} - x\dot{y} \tag{4-65}$$

The system (4-65) can be written as

$$\frac{d\rho}{dt} = U_1(\rho,\theta,t) \qquad \frac{d\theta}{dt} = U_2(\rho,\theta,t) \tag{4-66}$$

The method assumes that U_1 and U_2 are periodic of the same period T, that the system is studied near a harmonic oscillator, where it becomes

$$\frac{d\rho}{dt} = 0 \qquad \frac{d\theta}{dt} = -1$$

By studying solutions for this special case, we hope to obtain the solution, under appropriate restrictions, to the general system given above. This method also assumes that the differential equations contain a small parameter ϵ, and hence the system may be expanded in powers of ϵ to linear terms; thus

$$\frac{d\rho}{dt} = \epsilon u_1(\rho,\theta,t) \qquad \frac{d\theta}{dt} = -1 + \epsilon u_2(\rho,\theta,t)$$

Note that for $\epsilon = 0$ we have the conditions for a harmonic oscillator. Now one can expand $\rho(t)$ and $\theta(t)$ in two series in ϵ with coefficients as functions of t. This gives

$$\begin{aligned}
\rho(t) &= \rho_0 + \rho_1(t)\epsilon + \cdots \\
\theta(t) &= \theta_0 - t + \theta_1(t)\epsilon + \cdots
\end{aligned} \tag{4-67}$$

The first-term coefficients are, respectively, a constant ρ_0 and $\theta_0 - t$, and hence differentiation gives, for $\epsilon = 0$, the harmonic-oscillator equations. The coefficients $\rho_1(t)$ and $\theta_1(t)$ of ϵ in both series are obtained using Picard's method:

$$\rho_1(t) = \int_0^t u_1(\rho_0, \theta_0 - z, z)\, dz \qquad \theta_1(t) = \int_0^t u_2(\rho_0, \theta_0 - z, z)\, dz \tag{4-68}$$

As we pointed out, it would be desirable to use the linear approximation to the solution for small ϵ. However, one cannot stop at the linear terms, since there may be secular terms of the form $t \cos t$ in the coefficients of higher powers of ϵ, which do not yield convergence to the true solution of the general system as $t \to \infty$. To avoid this problem of taking on large values, t is allowed to increase from 0 to T, and then from $2T$ to $4T$, and so on, from which the first and second coefficients are calculated, thus making possible the use of the linear approximation in ϵ through discrete values of t (stroboscopic flashes). It then becomes possible to apply a limiting operation.

Thus to determine $\rho_1(t)$ and $\theta_1(t)$, we evaluate the integrals on $[0,T]$, obtaining $\rho_1(T,\rho_0,\theta_0)$ and $\theta_1(T,\rho_0,\theta_0)$; since T is given, it is omitted in the argument.   We then have

$$\bar{\rho} = \rho_0 + \epsilon\rho_1(\rho_0,\theta_0) \qquad \bar{\theta} = \theta_0 - T + \epsilon\theta_1(\rho_0,\theta_0)$$

However, this is just one set of values of ρ and θ obtained at T.   We use this value to estimate a new variable:

$$\bar{\bar{\rho}} = \bar{\rho} + \epsilon\rho_1(\bar{\rho},\bar{\theta}) \qquad \bar{\bar{\theta}} = \bar{\theta} - T + \epsilon\theta_1(\bar{\rho},\bar{\theta})$$

where we have replaced the old initial values by the new ones $\bar{\rho}$ and $\bar{\theta}$, determined at time T.   Again $\bar{\rho}$ and $\bar{\bar{\theta}}$ are used as initial points to estimate new variables.   Note that the periodicity of u_1 and u_2 on $[0,T]$ makes this repetition meaningful.   Thus we form a difference equation, using $\Delta\rho$ and $\Delta\theta$ for the difference between two successive estimates of ρ and θ, and $\Delta\tau = \epsilon$.   Since θ is the angle and T is a period, we can drop T in the equation for θ.   We have, on simplifying a typical equation and using a tilde to avoid bars on the arguments ρ and θ,

$$\frac{\Delta\rho}{\Delta\tau} = \rho_1(\tilde{\rho},\tilde{\theta}) \qquad \frac{\Delta\theta}{\Delta\tau} = \theta_1(\tilde{\rho},\tilde{\theta})$$

Passing to the limit has been justified by the method of averaging due to Krylov, Bogoliubov, and Mitropolsky (see Sec. 4-12).   For example, an interval coincides with its initial point at which ρ and θ are calculated, and we have

$$\frac{d\rho}{dt} = \rho_1(\rho,\theta) \qquad \frac{d\theta}{dt} = \theta_1(\rho,\theta) \qquad (4\text{-}69)$$

which is an autonomous system.

It turns out in the analysis that if (4-69) has a stable singular point, then the original nonautonomous system has a stable periodic solution.

An Application: Brenner and Weisfeld,[55] using phase-plane methods, give a proof that the equation (analogous to one which arises in orbital flight)

$$\ddot{x} + x - 1 = a_1x^2 + a_2x^4$$

has, for small values of a_1 and a_2, bounded periodic solutions, and every bounded solution is periodic.   In the proof the equation is multiplied by $2\dot{x}$ and integrated.   This gives

$$\dot{x}^2 = x^2 - 2x + \frac{a_1}{3}x^3 + \frac{a_2}{5}x^5 + c \qquad (4\text{-}70)$$

where $c \equiv e^2 - 1$ is the integration constant.   Equation (4-70) is satisfied by every solution of the original equation for some value of c.   If $a_1 = a_2 = 0$, a plot of (4-70) in the $x\dot{x}$ plane yields a circle with center at $(1,0)$ and radius e.   The motion must follow this circle.

Exercise 4-28 Prove the last statement by examining the phase plane. Show by using the equation $\ddot{x} + x - 1 = 0$ that, unless $e = 0$, neither of the two values of x corresponding to $\dot{x} = 0$ is an equilibrium point.

If $a_1 \neq 0$, $a_2 \neq 0$ are small, there is a value of c such that two roots r_1, r_2 of (4-70) are close to $1 - e$ and $1 + e$. A plot of this case yields a possible periodic solution as indicated in Fig. 4-4.

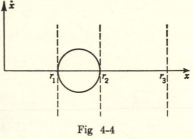

Fig 4-4

Since one can show that the graph of $y^2 = f(x)$ with $y = \dot{x}$, where $f(x)$ is a polynomial, contains both open and closed branches, and since every bounded branch is closed, it follows that every bounded solution of (4-70) is periodic.

Exercise 4-29 In the equation

$$\ddot{x} + 2a\dot{x} + x = 0$$

show that for $a^2 \geq 1$ the origin is a stable node if $a > 0$, and an unstable node if $a < 0$. If $a^2 < 1$, the origin is a spiral point that is stable for $a > 0$, and unstable for $a < 0$.

Exercise 4-30 Examine the nature of the singularity at the origin in Duffing's equation with different conditions on the damping constant.

Exercise 4-31 To examine the singular points of van der Pol's equation

$$\ddot{x} - \epsilon(1 - x^2)\dot{x} + x = 0$$

we write

$$\frac{dy}{dx} = -\frac{x - \epsilon(1 - x^2)y}{y}$$

If $|x| \ll 1$, we have, approximately,

$$\frac{dy}{dx} \simeq -\frac{x - \epsilon y}{y}$$

and hence $x = y = 0$ is singular. If $|x| \gg 1$,

$$\frac{dy}{dx} \simeq -\frac{x + \epsilon x^2 y}{y}$$

and the origin is also a singular point. Show that the origin is an unstable spiral point when $|x| \ll 1$. For $|x| \gg 1$, by showing that half the derivative of the square of the distance from the origin to the curve, i.e., $\frac{1}{2}\frac{d}{dt}[x^2(t) + y^2(t)]$, is negative, conclude that a point on the curve moves toward the origin as $t \to \infty$. It can be shown that this curve is also a spiral.

4-7 LIMIT CYCLES—STABILITY BEHAVIOR IN THE LARGE (SYSTEMS OF TWO EQUATIONS)

Let $x(t)$, $y(t)$ be a solution of

$$\dot{x} = P(x,y) \qquad \dot{y} = Q(x,y) \tag{4-71}$$

with initial conditions given for $t = t_0$.

If the curve corresponding to this solution is closed, then for some T

$$x(T) = x(0) \qquad y(T) = y(0)$$

Thus the solution is periodic. That is,

$$x(t) = x(t + T) \qquad y(t) = y(t + T)$$

To see this, note that the right sides are also a solution because they satisfy the system, have the same initial conditions, and must coincide with the solution given by the left sides. Clearly, every periodic non-equilibrium solution gives rise to a closed curve. The study of closed curves and their existence is equivalent to the search for periodic solutions. In nonlinear systems some closed curves are limiting sets for the trajectory corresponding to a solution as $t \to \pm \infty$. In this case they are known as *limit cycles*. In linear systems only equilibrium points are stable, whereas in a nonlinear system we can have additional stability at limit cycles. Thus limit cycles are stable closed curves independent of the initial conditions toward which the solutions tend as $t \to \pm \infty$, and they are either isolated periodic solutions or asymptotic limits of solutions. However, the converse need not be true. For example, $\dot{x} = y$, $\dot{y} = -x$ has periodic solutions corresponding to closed curves around a vortex point.

Exercise 4-32 Consider the system

$$\dot{x} = y + x(1 - x^2 - y^2) \qquad \dot{y} = -x + y(1 - x^2 - y^2) \tag{4-72}$$

Using the polar coordinates

$$r^2 = x^2 + y^2 \qquad \theta = \arctan \frac{y}{x}$$

$$r\dot{r} = x\dot{x} + y\dot{y} \qquad r^2\dot{\theta} = x\dot{y} - y\dot{x}$$

reduce the system to

$$\dot{r} = r(1 - r^2) \qquad \dot{\theta} = -1$$

and verify that the general solution is given by

$$r = \frac{1}{(1 + ce^{-2t})^{1/2}} \qquad \theta = -(t - t_0)$$

With $t_0 = 0$, we have

$$x = \frac{\cos t}{(1 + ce^{-2t})^{1/2}} \qquad y = -\frac{\sin t}{(1 + ce^{-2t})^{1/2}}$$

If $c > 0$, the solution is a spiral in the phase plane, starting inside the circle and tending to the circle as a limit as t varies from $-\infty$ to ∞. If $c < 0$, the spiral approaches the circle from the outside. The circle is an example of a limit cycle. Draw a diagram (see Fig. 4-5). See also Sec. 4-12, Geometric Methods, page 258.

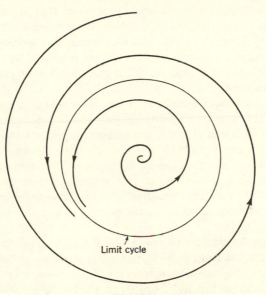

Fig. 4-5

An interesting illustration of the limit-cycle behavior of a physical system is demonstrated by a block resting on a large, slowly rotating drum and attached to a fixed spring whose axis is perpendicular to the rotation axis of the drum.[8] Experimental data give the relation between sliding friction and the relative velocity of oscillation of the spring and block with respect to the drum. This function has a positive value corresponding to static friction when the relative velocity is zero. As this velocity increases from zero, the value of the sliding friction decreases to a minimum along a parabolic curve and then symmetrically increases. Starting from rest, the block and spring are stretched and placed on the drum such that the static-friction force exactly balances the elastic force of the spring. Now the drum begins to rotate in the direction of expansion of the spring. Starting from rest, the relative velocity increases, breaking the static friction. The sliding friction first decreases, thus encouraging the contraction of the spring, but then increases to a point where the frictional force exceeds the contracting force of the spring. The spring tends to expand in the direction of motion of the drum, the relative velocity decreases, the sliding friction first decreases as one moves towards the origin and then again increases to the point where it exceeds the expansion of the spring, which now begins to contract, and the cycle is repeated. The spring oscillates back and forth, spirally approaching a limiting periodic cycle of oscillations as it moves away from the initial conditions to a steady state. Clauser points out that in this manner one can also explain the vibration of a violin string when stroked by a bow, the creaking of doors at hinges, the chattering and groaning of brakes, and the chirping of crickets. A musical note produced by rubbing a wet finger along the rim of a thin drinking glass is another example of this type of phenomenon.

Remark: An interesting question is: How many singular points of a given kind and how many limit cycles can an autonomous system of two equations (in which P and Q are quadratic polynomials) have? For example, Bautin and Petrovsky and Landis have shown in complicated proofs that the system can have at most three limit cycles. It can also be shown that there are at most two singular points that are vortex points. Another interesting problem is to establish the conditions under which the trajectories about a vortex point are closed. For proofs and references, see Davis.[11]

If there are trajectories of the system which spiral toward a limit cycle from inside and from outside, the cycle is called *orbitally stable*. If, for example, only the outside trajectories wind around a limit cycle while the inside trajectories unwind from it, it is *orbitally semistable*. Finally we have *neutral stability* if the periodic solution is no longer a limit cycle, as in the case of a sequence of concentric circles. The maximum deviations, from this solution, of nearby periodic solutions can be made arbitrarily small.

Consider the equation

$$\ddot{y} - (2\mu - c\dot{y}^2)y + ay = 0 \qquad a, c > 0$$

which represents a system with nonlinear damping coefficient $-(2\mu - c\dot{y}^2)$ but with linear control. If $y(0) > 0$ and $\dot{y}(0) = 0$, y is oscillatory, due to negative damping. If, in the neighborhood of the origin of the $y\dot{y}$ plane, $2\mu \gg c\dot{y}^2$, the origin is an unstable spiral point; if $c\dot{y}^2 \gg 2\mu$, the origin is a stable spiral point. Since these two spirals (one moving inward and the other outward) cannot intersect, they are asymptotic to a closed curve, i.e., a limit cycle, and hence the motion of the system is ultimately periodic.

If $c = 0$, $a > 0$, and $\mu < 0$, the motion is stable and periodic for any $y(0)$ as long as $\dot{y}(0) > a/2\mu$ (otherwise it is unstable). There is a focus at the origin $y = \dot{y} = 0$.

A sketch of the proof of the existence of a limit cycle for van der Pol's equation will now be given.   Using polar coordinates, with

$$r^2 = x^2 + y^2 \qquad y = \dot{x}$$

the equation can be written as

$$r \frac{dr}{dt} = (x + y)x = \epsilon(1 - x^2)y^2$$

from which
$$\frac{dr}{dt} = \frac{\epsilon(1 - x^2)y^2}{r}$$

If $y \neq 0$, then $\epsilon y^2/r > 0$, and dr/dt has the same sign as $1 - x^2$.   If $|x| < 1$, then $dr/dt > 0$, and r increases with t.   Thus a plot of the figure in the xy plane for small ϵ and $|x| \ll 1$ shows an outward spiral, whereas if $|x| \gg 1$, $dr/dt < 0$, and the plot shows an inward spiral.   These two curves cannot meet except at a singular point, and hence each is asymptotic to a closed curve.   This curve is the desired limit cycle.   A simple topological proof using Brouwer's fixed-point theorem is also possible.

4-8   TOPOLOGICAL CONSIDERATIONS: INDICES AND THE EXISTENCE OF LIMIT CYCLES (SYSTEMS OF TWO EQUATIONS)

Suppose we are given the system[11,33,37]

$$\frac{dx}{dt} = P(x,y) \qquad \frac{dy}{dt} = Q(x,y) \tag{4-73}$$

where P and Q have continuous first-order partial derivatives and a finite number of singularities.   Let $(\bar{x}, \bar{y})$ be a singular point, and suppose that C is a circle surrounding this point such that there are no other singular points inside or on C.   Associated with any point on C is a vector with components P and Q.   Let $\tan \alpha = Q/P$, where α measures the variation in this vector as the point moves counterclockwise on C.   After each circuit on C, the angle will increase or decrease by a multiple of 2π.   To relate the change in α to the coordinate system, we introduce polar coordinates with origin at $(\bar{x}, \bar{y})$ and angle θ, and we compute $d\alpha/d\theta$.   Then we compute the change in α in one circuit.   This gives the index of $(\bar{x}, \bar{y})$.   Thus the index is given by

$$I = \frac{1}{2\pi} \int_0^{2\pi} \frac{d\alpha}{d\theta}\, d\theta = \frac{1}{2\pi} \int_0^{2\pi} \frac{P(dQ/d\theta) - Q(dP/d\theta)}{P^2 + Q^2}\, d\theta \tag{4-74}$$

Exercise 4-33   Prove that $I = 0$ for a nonsingular point.   Recall that a singularity is a zero of the system $P = 0$, $Q = 0$.

Exercise 4-34 Show that for the case $\dot{x} = ax + by + \cdots$, $\dot{y} = cx + dy + \cdots$, $\begin{vmatrix} a & b \\ c & d \end{vmatrix} \neq 0$, the index of a singularity is 1, unless it is a saddle point, in which case the index is -1.

Exercise 4-35 Show that the index is unchanged if t is replaced by $-t$.

Exercise 4-36 If in the interior of a closed curve there is more than one of a finite number of singular points, then the total change in α in a single circuit is equal to 2π times the sum of the indices of the singularities. Prove this fact by dividing C into closed curves, each having one of the singularities in its interior. Show that the sum of the indices of the singularities enclosed by a closed trajectory is unity, and hence the index of the trajectory is 1. The enclosed singularities of index 1 exceed by unity those of index -1.

The following two theorems are concerned with the existence of limit cycles and, hence, periodic solutions. We shall not prove the first theorem, since it requires a considerable amount of background material. However, we prove the second theorem using the above ideas.[33]

Theorem (**Poincaré-Bendixon**) If a solution curve $C:[x(t),y(t)]$ of the system (4-73) is in and remains in a domain of the xy plane (enclosed by a simple closed curve) in which P and Q have continuous first partial derivatives for all $t \geq \bar{t}$, for some $\bar{t}$, without approaching singular points, then there exists a limit cycle in the domain, and either C is a limit cycle or it approaches a limit cycle as $t \to \infty$.

Remark: L is a limit point of C if there is a sequence $\{t_i\}$, $i = 1$, $2, \ldots$, such that $\lim\limits_{i \to \infty} [x(t_i),y(t_i)] = L$. The totality of limit points forms a limit set of C.

Theorem 4-2 (**Bendixon's nonexistence theorem**) If $\partial P/\partial x + \partial Q/\partial y \neq 0$ in a region of the (x,y) plane enclosed by a single closed curve, then no limit cycles of (4-73) exist in this domain.

Proof: $P_x + Q_y$ is continuous and maintains the same sign, e.g., positive, in the region. The integral over a limit cycle C, by Green's theorem, gives

$$\int_C P \, dy - Q \, dx = \iint_A \left(\frac{\partial P}{\partial x} + \frac{\partial Q}{\partial y} \right) dx \, dy \geq 0$$

where A is the region enclosed by C. However, because C is closed, the line integral must vanish; i.e.,

$$\int_C P \, dy - Q \, dx = \int_0^{2\pi} (P\dot{y} - Q\dot{x}) \, dt = \int_0^{2\pi} (QP - PQ) \, dt = 0$$

where 2π is the period of C. This implies that $P_x + Q_y$ vanishes or changes sign contrary to our assumption.

4-9 PERIODIC SOLUTIONS OF SYSTEMS WITH PERIODIC COEFFICIENTS

We now seek periodic solutions of the nonautonomous system

$$\frac{dx_k}{dt} = f_k(x_1, \ldots, x_n, t) \qquad k = 1, \ldots, n \qquad (4\text{-}75)$$

with the assumption that the f_k, as functions of t, are periodic with the same period T.

We shall first give conditions for the existence of periodic solutions of a linear system with periodic coefficients. A procedure for deriving such solutions will be examined. Then we shall study the construction of solutions for the general nonautonomous case with a small parameter. Some well-known theorems regarding the existence of periodic solutions of special equations will also be given.

As an illustration of an equation with periodic coefficients, we have, for the pendulum with an oscillating point of support,[54]

$$\frac{d^2\theta}{dt^2} = -\frac{g}{L}\left(1 + \frac{a\omega^3}{g}\sin \omega t\right)\sin \theta \qquad (4\text{-}76)$$

This equation contains the same quantities as the equation describing a simple pendulum except for a (small) and ω, the amplitude and frequency, respectively, of the vertical harmonic oscillations undergone by the point of support of the pendulum.

Linear Nonautonomous Systems—Floquet Theory

Assume that the linear part of our nonautonomous system is given by

$$\dot{x}_k = p_{k1}(t)x_1 + \cdots + p_{kn}(t)x_n \qquad k = 1, \ldots, n \qquad (4\text{-}77)$$

where the $p_{ki}(t)$, $i = 1, \ldots, n$, are periodic with period T. We shall simply write p_{ki}. (This study follows along lines developed by Floquet in studying periodic solutions of linear systems.)

The system of differential equations (4-77) has a fundamental system of solutions (a set of n linearly independent functions of which all other solutions are linear combinations) $(x_{k1}(t), \ldots, x_{kn}(t))$, $k = 1, \ldots, n$. Note that each solution is a vector. The first subscript denotes the function, and the second the solution to which it belongs. The functions $x_{ki}(t + T)$, $i = 1, \ldots, n$, are also solutions, and we can express them as linear combinations of the original set as follows:

$$x_{kj}(t + T) = \sum_{i=1}^{n} a_{ij}x_{ki}(t) \qquad j = 1, \ldots, n \qquad (4\text{-}78)$$

In matrix notation (4-77) may be written as

$$\dot{x} = P(t)x$$

Let $X(t)$ be the matrix whose columns are the fundamental solutions; thus $\dot{X} = PX$. Hence there exists a constant matrix A corresponding to the coefficients in (4-78) such that

$$X(t + T) = X(t)A \tag{4-78a}$$

If we assume that the roots of A are distinct, then there are a diagonal matrix S (with diagonal elements s_j) and a nonsingular matrix Q such that $A = QSQ^{-1}$. If we put $Z(t) = X(t)Q$, then (4-78a) becomes $Z(t + T) = Z(t)S$, which means that the jth column of $Z(t + T)$ is s_j times the jth column of $Z(t)$. Note that $Z(t)$ is again a fundamental solution matrix. The characteristic equation of A (which corresponds to the period T) is called the *characteristic equation* of the differential equations. It is given by

$$|A - sI| = \begin{vmatrix} a_{11} - s & a_{12} & \cdots & a_{1n} \\ a_{21} & a_{22} - s & \cdots & a_{2n} \\ \cdots\cdots\cdots\cdots\cdots\cdots\cdots\cdots \\ a_{n1} & a_{n2} & \cdots & a_{nn} - s \end{vmatrix} = 0 \tag{4-79}$$

Exercise 4-37 By choosing another fundamental system $y_{kj}(t)$, considering the $y_{kj}(t + T)$ as linear combinations of the $y_{kj}(t)$ with coefficient matrix C, and then expressing the $y_{kj}(t)$ as linear combinations of $x_{kj}(t)$ with coefficient matrix B, show that the characteristic equation does not depend on the fundamental system; i.e., show that $C = B^{-1}AB$, and hence $|C - sI| = |A - sI|$.

If we define

$$r_i = \frac{1}{T} \log s_i \qquad i = 1, \ldots, n \tag{4-80}$$

then the r_i are known as the *characteristic exponents* of the linear system of differential equations. There are periodic vectors $R_k(t)$, $k = 1, \ldots, n$, with period T such that for each k we can write

$$z_k(t) = e^{r_k t} R_k(t) \qquad k = 1, \ldots, n \tag{4-81}$$

where $z_k(t)$ is the kth column of $Z(t)$. This form of the solution satisfies

$$z_k(t + T) = s_k z_k(t) \tag{4-82}$$

In fact the two expressions are equivalent.

Exercise 4-38 Prove the statements made in the last paragraph.

It can be shown without great difficulty that to each root (the roots are assumed to be distinct) there corresponds one and only one solution of the above form.   Since there are n roots, we obtain n solutions which are clearly independent and hence form a fundamental system.   Note that by means of a general discussion we have deduced a form for our solutions without knowing the coefficients a_{ij}.   However, if, for example, we are given the initial conditions

$$x_{ki}(0) = \begin{cases} 1 & k = i \\ 0 & k \neq i \end{cases}$$

then

$$a_{kj} = x_{kj}(T) \qquad (4\text{-}83)$$

and we may replace the a_{kj} by $x_{kj}(T)$ in the characteristic equation, and the product of the roots becomes the wronskian evaluated at T and multiplied by $(-1)^n$.   We have used the initial conditions to arrive at this fact, as the reader may readily verify.

We conclude with the fact that no periodic solutions of a linear system whose coefficients are periodic exist if all the s_j have absolute values other than 1.   On the other hand, periodic solutions exist if any of the roots s_j are mth roots of unity, where m is a positive integer.   Indeed in this case we have $z_j(t + mT) = z_j(t)$; the period of the solution is mT (where m is the smallest integer such that $s_j^m = 1$), not T.   More generally, if s_j has unit absolute value, then s_j^m, for appropriate m, is as close to unity as desired, and $z_j(t + mT)$ is as close to $z_j(t)$ as desired, and $z_j(t)$ is almost periodic.

A continuous function $x(t)$ is *almost periodic* (a concept due to H. Bohr) if, given $\epsilon > 0$, there exists an $L(\epsilon)$ such that for every real number s there is a T which satisfies $s \leq T \leq s + L(\epsilon)$ and for which

$$|x(t + T) - x(t)| \leq \epsilon \qquad \text{for all } t$$

Any such T is referred to as a *translation number*.   Thus an almost-periodic function satisfies the periodic property within an arbitrarily small error.   Note that a continuous periodic function is almost periodic.   Investigations in differential equations have also examined conditions under which almost-periodic solutions exist.   These investigations have been concerned with equations such as

$$\dot{x} = A(t)x + g(t)$$

where x and g are n vectors, and $A(t)$ is an $n \times n$ matrix which is a function of t.   Both $A(t)$ and $g(t)$ are assumed almost periodic.   Sufficient conditions for the existence of almost-periodic solutions were derived.   The conditions and studies are too involved to present conveniently in this chapter.   We note that an almost-periodic function can be expanded in a series (as a special case of Dirichlet series) as follows:

$$x(t) = \sum_{n=0}^{\infty} a_n e^{i\lambda_n t}$$

where the series of coefficients $\Sigma|a_n|^2$ converges, and the λ_n are constants. Note that this series includes Fourier expansions as a special case, as one might expect.

Periodic solutions of the linear system (4-77) can be constructed using a small parameter ϵ, for example, by replacing $P(t)$ with $\epsilon P(t)$. Thus, suppose that the p_{ki} are also functions of a parameter ϵ (or of several parameters—but one suffices for our purposes). Assume that the period T does not depend on ϵ, nor do the initial conditions. We assume that the p_{ki} are analytic in ϵ and can therefore be expanded in series in ϵ. This fact also applies to a solution x_k which is now a function of ϵ. This enables us to calculate, approximately, the coefficients of the characteristic equation. Now let us write

$$p_{ki}(t,\epsilon) = c_{ki}(t) + c_{ki}^{(1)}(t)\epsilon + c_{ki}^{(2)}(t)\epsilon^2 + \cdots \qquad (4\text{-}84)$$

where the coefficients are continuous and periodic with period T, and the series converges for values of ϵ near the origin. If we are given

$$x_{ki}(0,\epsilon) = \begin{cases} 1 & k = i \\ 0 & k \neq i \end{cases} \qquad k, i = 1, \ldots, n$$

and if we expand in a convergent series for the same set of values of ϵ as for the coefficients defined by (4-84), we obtain

$$x_{ki}(t,\epsilon) = x_{ki}^{(0)}(t) + x_{ki}^{(1)}(t)\epsilon + x_{ki}^{(2)}(t)\epsilon^2 + \cdots \qquad (4\text{-}85)$$

Thus
$$x_{ki}^{(0)}(0) = \begin{cases} 1 & k = i \\ 0 & k \neq i \end{cases}$$

and all other coefficients are zero at $t = 0$. If the series expansions for the p_{ki} and for the x_{ki} are substituted in the system of differential equations (4-77) and the coefficients of the same powers of ϵ are equated to zero, we obtain

$$\frac{dx_{ki}^{(0)}}{dt} = c_{k1}x_{1i}^{(0)} + c_{k2}x_{2i}^{(0)} + \cdots + c_{kn}x_{ni}^{(0)}$$

$$\frac{dx_{ki}^{(1)}}{dt} = c_{k1}x_{1i}^{(1)} + c_{k2}x_{2i}^{(1)} + \cdots + c_{kn}x_{ni}^{(1)} + \sum_{p=1}^{n} c_{kp}^{(1)}x_{pi}^{(0)} \qquad (4\text{-}86)$$

$$\cdot \quad \cdot \quad \cdot \quad \cdot \quad \cdot \quad \cdot \quad \cdot$$

$$\frac{dx_{ki}^{(m)}}{dt} = c_{k1}x_{1i}^{(m)} + c_{k2}x_{1i}^{(m)} + \cdots + c_{kn}x_{ki}^{(m)} + \sum_{p=1}^{n}\sum_{q=0}^{m-1} c_{kp}^{(m-q)}x_{pi}^{(q)}$$

$$k = 1, \ldots, n; \, m = 0, 1, 2, \ldots$$

If we put $\epsilon = 0$ in the linear system of differential equations, we obtain

$$\frac{dx_k^{(0)}}{dt} = c_{k1}x_1^{(0)} + \cdots + c_{kn}x_n^{(0)} \qquad (4\text{-}87)$$

$$k = 1, \ldots, n$$

which gives the solution of $x_{ki}{}^{(m)}$ for $m = 0$.   This is then used to determine $x_{ki}{}^{(1)}$.   In this manner x_{ki} can be determined to any desired accuracy, and, by putting $t = T$, we obtain $a_{kj} = x_{kj}(T)$.

The question remains as to how one may introduce the parameter ϵ into a given linear system with periodic coefficients to make the calculations possible.   In general one introduces ϵ into the coefficients p_{ki} such that for some value of ϵ, for example, $\bar{\epsilon}$, in the region of convergence of the series for the coefficients (in ϵ) one has

$$p_{ki}(t,\bar{\epsilon}) = p_{ki}(t)$$

and such that for $\epsilon = 0$ the resulting system can be integrated in closed form (which, for example, would be the case if the coefficients were constants).   Note that if $\bar{\epsilon}$ is small, i.e., close to $\epsilon = 0$, the number of iterations is small.

Exercise 4-39   By examining the characteristic exponents, discuss the stability of solutions of a linear homogeneous system with periodic coefficients.

Example: We now apply the foregoing ideas to the well-known (Hill's) equation

$$\frac{d^2y}{dt^2} = p(t)y$$

where $p(t)$ is continuous and periodic with period T.   Let $x_1 = y$, $x_2 = dx_1/dt$.   We now have the system

$$\frac{dx_1}{dt} = x_2 \qquad \frac{dx_2}{dt} = p(t)x_1$$

Let us suppose that every solution is periodic with period T.

Our task, of course, is to determine the characteristic exponents which are related to the characteristic roots.   Let $X(t)$ be the fundamental matrix solution with $X(0) = I$.   Then $A = X(T)$.   Note that the wronskian of this system is constant.   Thus, using the initial conditions and our previous discussion, we find that the product of the characteristic roots is equal to unity, and the characteristic equation can be written in the form

$$s^2 - as + 1 = 0$$

We must determine a and, therefrom, the characteristic roots which enable us to write the condition for a periodic solution.   The characteristic equation may be written as

$$(a_{11} - s)(a_{22} - s) - a_{12}a_{21} = 0$$

from which $a = a_{11} + a_{22} = x_{11}(T) + x_{22}(T)$

If we substitute $\epsilon p(t)$ for $p(t)$, our system can be written in the form

$$\frac{dX}{dt} = \begin{pmatrix} 0 & 1 \\ \epsilon p & 0 \end{pmatrix} X = \begin{pmatrix} 0 & 1 \\ 0 & 0 \end{pmatrix} X + \epsilon \begin{pmatrix} 0 & 0 \\ p & 0 \end{pmatrix} X \equiv (N + \epsilon M)X$$
$$X(0) = I$$

(Note that $N^2 = 0$.)   Let

$$X(t) = \sum_{n=0}^{\infty} X_n(t)\epsilon^n \qquad X_n(0) = I\,\delta_{0n}$$

Substituting, for $X(t)$ in the expression for dX/dt, one obtains

$$\sum_{n=0}^{\infty} \dot{X}_n \epsilon^n = \sum_{n=1}^{\infty} \epsilon^n(NX_n + MX_{n-1}) + NX_0$$

Consequently, by equating powers of ϵ, we obtain

$$X_0(t) = e^{tN} = \begin{pmatrix} 1 & t \\ 0 & 1 \end{pmatrix}$$
$$X_n(t) = e^{tN} \int_0^t e^{-uN} M(u) X_{n-1}(u)\, du + e^{tN}\,\delta_{0n}$$

Writing out $x_{ij}{}^n$ in full for the ijth element of X_n, we have, for example,

$$x_{11}{}^n(t) = \int_0^t (t - u)p(u)x_{11}{}^{n-1}(u)\, du \qquad n \geq 1$$
$$x_{12}{}^n(t) = \int_0^t (t - u)p(u)x_{12}{}^{n-1}(u)\, du \qquad n \geq 1$$

with $\qquad\qquad x_{11}{}^0(t) = 1 \qquad x_{12}{}^0(t) = t$

Exercise 4-40   Obtain $x_{21}{}^n(t)$ and $x_{22}{}^n(t)$.

Thus

$$a = \sum_{n=1}^{\infty} x_{11}{}^{(n)}(T) + x_{22}{}^{(n)}(T) + 2$$

The proofs of periodicity for different possible forms of $p(t)$ are difficult. Some useful results are given by Cesari.[7]   A simple one is that if $p(t) \geq 0$, then for $t > 0$,

$$x_{11}{}^{(n)}(t), \qquad x_{12}{}^{(n)}(t), \qquad \text{and} \qquad x_{22}{}^{(n)}(t)$$

are positive; hence $a > 2$, the characteristic roots s_1 and s_2 are real with $0 < s_1 < 1 < s_2$, and there are no periodic solutions.   If $p(t) \leq 0$ and

$$\left| T \int_0^T p(t)\, dt \right| < 4$$

then s_1 and s_2 are conjugate imaginary with

$$|s_1| = |s_2| = 1$$

and the solution can be either periodic or almost periodic.

4-10 PERIODIC SOLUTIONS OF NONLINEAR SYSTEMS WITH PERIODIC COEFFICIENTS

We now consider periodic solutions of the simplest nonlinear system of n first-order differential equations

$$\dot{x}(t) = A(t)x(t) + \epsilon f[x(t),t] \tag{4-88}$$

where the elements of the matrix A and the components of the vector f are all periodic with period T.

For a solution of period T we must have $x(t + T) = x(t)$. This is equivalent to the condition that $x(0) = x(T)$, for the latter condition follows if we have periodicity and put $t = 0$. To show that this condition implies periodicity, we substitute $x(t + T)$ in the differential equation. Let $u = t + T$; since both $A(t)$ and f are periodic with period T, we have

$$\dot{x}(u) = A(t)x(u) + \epsilon f[x(t),t]$$

Thus $x(t + T)$ is also a solution. Since both $x(t)$ and $x(t + T)$ are solutions for the same initial conditions, by the uniqueness theorem they must be identical; hence $x(t)$ is periodic.

In the integral equation (4-34), we put $t = T$ to obtain

$$x(T) = X(T)x_0 + \epsilon \int_0^T X(T)X^{-1}(s)f[x(s),s]\,ds \qquad X(0) = I \tag{4-89}$$

If we impose the condition $x(T) = x(0) \equiv x_0$, our search for a periodic solution $x(t)$ is equivalent to finding an x_0 which satisfies Eq. (4-89), which we may write as

$$[I - X(T)]x_0 = \epsilon \int_0^T X(T)X^{-1}(s)f[x(s),s]\,ds \qquad X(0) = I \tag{4-90}$$

To obtain a solution of the linear part, it suffices to put $\epsilon = 0$. A nontrivial periodic solution x_0 (of period T) exists if and only if $[I - X(T)]$ has no inverse. We have two cases.

Case 1: Assume first that the linear part has no periodic solution of period T. Then $[I - X(T)]^{-1}$ exists, and there is a periodic solution of the entire equation, since

$$x_0 = \epsilon[I - X(T)]^{-1} \int_0^T X(T)X^{-1}(s)f[x(s),s]\,ds \tag{4-91}$$

or

$$x_0 = \epsilon \mathfrak{I}(x_0) \tag{4-92}$$

because the multiplier of ϵ in the integral equations is a function of x_0. The right side of Eq. (4-92) is a contraction operator (i.e., its norm can always be made less than unity for sufficiently small ϵ). Hence, by the fixed-point property of contraction operators, the equation has a unique solution which gives the desired x_0. Note here the advantage of introducing the perturbation parameter ϵ to yield the solution by iterative methods. An upper bound on ϵ can be determined within whose range one has periodic solutions.

The actual construction of the periodic solution $x(t)$ proceeds as follows: Since x is a function of ϵ, we expand $x(t,\epsilon)$ in powers of ϵ near $\epsilon = 0$, that is,

$$x(t,\epsilon) = x_0(t) + x_1(t)\epsilon + x_2(t)\epsilon^2 + \cdots \qquad (4\text{-}93)$$

where the coefficients are periodic with period T, and substitute in the differential equation, equating like powers of ϵ. Since the expansion of

$$\epsilon f[x_0(t) + \epsilon x_1(t) + \cdots, t]$$

in powers of ϵ near $\epsilon = 0$ gives $f[x_0(t),t]$ as the coefficient of ϵ, and $f[x_1(t),t]$ as the coefficient of ϵ^2, and so forth, we have

$$
\begin{aligned}
\dot{x}_0 &= A(t)x_0(t) \\
\dot{x}_1 &= A(t)x_1(t) + f[x_0(t),t] \\
&\cdots\cdots\cdots\cdots\cdots\cdots\cdots \\
\dot{x}_n &= A(t)x_n(t) + f[x_{n-1}(t),t] \\
&\cdots\cdots\cdots\cdots\cdots\cdots\cdots
\end{aligned}
\qquad (4\text{-}94)
$$

Here the integral equation associated with each of these equations gives rise to a periodic solution with period T since it is possible to invert $[I - X(T)]$ and determine the initial condition in each case. The solution is obtained by first solving $\dot{x}_0 = A(t)x_0(t)$. Since $x_0(t)$, the generating solution, is now known, the second of Eqs. (4-94) can be solved to yield $x_1(t)$, which now makes it possible to obtain $x_2(t)$ with the third equation, etc. In each case f remains periodic with period T. This gives the series expansion for $x(t,\epsilon)$ to the desired accuracy.

Case 2: Now suppose that the linear part has solutions of period T; then $[I - X(T)]$ has no inverse, and we are faced with the problem of finding a nontrivial x_0 in a problem similar to that encountered in solving a system of linear algebraic equations of the form $Ax = b$ when A is a singular matrix. The requirement for the existence of a solution in the latter case is obtained from the general condition for a solution for any rank of A, that is, that the augmented matrix (obtained by forming a matrix whose columns consist of all the columns of A and of the vector b) have the same rank as A. If we denote the columns of A by A_i, then

b is a linear combination of the A_i:

$$b = \Sigma a_i A_i$$

where some of the a_i may be zero. In addition b is orthogonal to all vectors that are orthogonal to A_i. This is equivalent to the statement that $y \cdot b = 0$ for all vectors y such that $yA = 0$. For our problem here, the requirement that a nontrivial x_0 exists is equivalent to asking that the right side of (4-90), which we denote by $F(x_0,\epsilon)$, be orthogonal to every nontrivial solution y of $y[I - X(T)] = 0$.

Because the direct method used to handle case 1 is relatively simple, it is desirable to reduce a problem which satisfies the assumption of case 2 to one which satisfies that of case 1, solve the latter, and in this manner hope to obtain a solution of the original problem. This is sometimes done by introducing a new known periodic function $g(t)$ multiplied by a new parameter μ which depends on ϵ and seeking a form of $\mu(\epsilon)$ such that the linear part is no longer periodic with period T. The development of this case leads to an equation encountered in a previous section (or to a set of equations) called the *bifurcation* equation.

Friedrichs[24] has studied the degenerate case in which the linear parts have solutions of period T; additional parameters are introduced, one for each solution. These parameters are combined with properly chosen functions so that the linear part of the modified equation has no solution of period T. The resulting equation is then solved, and the solution is required to satisfy the original (degenerate) equation. This leads to a set of conditions on the parameters, the bifurcation equations. The method used to examine the linear part is to form the variational equation which yields that part. We begin with the general system

$$\frac{dx}{dt} = f(x,t,\epsilon) \tag{4-95}$$

where $\qquad x = (x_1, \ldots, x_n) \qquad f = (f_1, \ldots, f_n)$

The f_i are periodic with period T, and ϵ is a small parameter. Suppose the linear part has a single periodic solution of period T which we denote by $x_0(t)$—the generating solution (a special case of putting $\epsilon = 0$ in f). The generating solution is the first term in the expansion

$$x(t,\epsilon) = x_0(t) + \sum_{n=1}^{\infty} x(t)\epsilon^n \tag{4-96}$$

Our object is to continue $x_0(t)$ to a periodic solution for $\epsilon > 0$.

Since differentiation and variation are commutative, we obtain, on applying the variational operator to our system,

$$\frac{d}{dt} \delta x = \frac{\delta f[x_0(t),t,0]}{\delta x} \delta x \qquad (4\text{-}97)$$

which is a linear homogeneous equation in δx.

Since both f and $x_0(t)$ are periodic, the first factor on the right is periodic. Note that f is evaluated at $\epsilon = 0$, and therefore at $x_0(t)$, after the variation is taken. We must ensure that it has no nonzero solution δx of period T, and then construct a periodic solution [a continuation of $x_0(t)$ for $\epsilon > 0$] of period T for the original equation.

The order of degeneracy is given by the number of nontrivial independent periodic solutions of period T of the linear part. Suppose that the degeneracy is of unit order. Let $\mu(\epsilon)$ be a new parameter depending on ϵ. The modified problem is then given by

$$\frac{dx}{dt} = f[x,t,\epsilon,\mu(\epsilon)] \qquad (4\text{-}98)$$

where the right side, which has period T, reduces to $f(x,t,\epsilon)$ at $\mu = 0$. Our object is to obtain a solution of period T for the modified equation near the solution of our original equation. Since, in addition to $x(t,\epsilon)$ we must also determine $\mu(\epsilon)$, it is necessary to have an additional condition, i.e., two equations, one of which is the above differential equation and another which requires that $x(t,\epsilon,\mu)$ has a prescribed value near $x_0(t)$ at $t = 0$. We can prescribe such a value since, so far, ϵ and μ are free parameters. In the case of a system of equations, one can impose this condition on a linear combination of the components of the vector x by a proper choice of coefficients (see later how they are chosen) and require that its value at $t = 0$ be near that of a linear combination (with the same coefficients) of the components of $x_0(t)$. Thus we let $\sum_{i=1}^{n} c_i x_i(0) = b$ where b is near $\sum_{i=1}^{n} c_i x_i^{(0)}(0)$, in which $x_i^{(0)}$ is the ith component of the vector x_0.

The variational equations of the two conditions are

$$\frac{d}{dt} \delta x = \frac{\delta}{\delta x} f[x_0(t),t,0] \delta x + \frac{\delta f}{\delta \mu} [x_0(t),t,0,0] \delta \mu \qquad (4\text{-}99)$$

$$\sum_{i=1}^{n} c_i \, \delta x_i(0) = 0$$

The new problem is nondegenerate if its only periodic solution of period T is $\delta x = 0$, $\delta \mu = 0$. For if $\delta \mu \neq 0$, then $\delta x/\delta \mu$ is a periodic solution with the same period T as $x_0(t)$, contrary to our desire to introduce μ so that no periodic solution of period T exists for the linear part.   Similarly one ensures that $\delta x = 0$ by a proper choice of the initial conditions (see below).   To see how one may introduce μ, we write

$$f(x,t,\epsilon,\mu) = f(x,t,\epsilon) - \mu g(t) \qquad (4\text{-}100)$$

whose variational equation is readily obtained from (4-71).   To ensure that the only periodic solution is $\delta\mu = 0$, we suppose that $\delta\mu \neq 0$; since μ does not depend on t, it suffices to find a periodic function $g(t)$ chosen (always possible if the original problem is degenerate) such that

$$\frac{dy}{dt} - \frac{\delta}{\delta x} f[x_0(t),t,0]y = g(t) \qquad y = \frac{\delta x}{\delta \mu} \qquad (4\text{-}101)$$

has no nonzero solution with period T.   To ensure that $\delta x = 0$, the coefficients c_i are chosen so that the only periodic solution $u(t)$ with period T for

$$\frac{du}{dt} - \frac{\delta}{\delta x} f[x_0(t),t,0]u = 0 \qquad (4\text{-}102)$$

with $$\sum_{i=1}^{n} c_i u_i(0) = 0$$

is the zero solution.

This procedure yields $x(t,\epsilon,b)$ and $\mu(\epsilon,b)$.   Finally, to obtain a periodic solution to our original problem, we require that $\mu(\epsilon,b) = 0$.   This is the bifurcation equation.   If it has a solution $b(\epsilon)$ which reduces to

$$\sum_{i=1}^{n} c_i x_i{}^{(0)}(0)$$ for $\epsilon = 0$, then $x[t,\epsilon,b(\epsilon)]$ is the desired periodic solution to our problem.

Let us consider the following simplified degenerate example studied by Friedrichs:

$$\frac{dx}{dt} - ix = \epsilon f(x,t) \qquad (4\text{-}103)$$

where $f(x,t)$ has period 2π.   For $\epsilon = 0$ we have the periodic solution $x_0(t) = x_0(0)e^{it}$, with period 2π.   The variational equation of the linear part is

$$\frac{d\delta x}{dt} - i\,\delta x = 0$$

which has the solution $\delta x = e^{it}$.   Thus the linear part has a solution of period 2π, which coincides with the solution for $\epsilon = 0$.   The modified problem is

$$\frac{dx}{dt} - ix = \epsilon f(x,t) - \mu e^{it} \qquad (4\text{-}104)$$

and we put $x(0) = b$, where $b = x_0(0)$ for $\epsilon = 0$. Recall that we assume a series expansion of $x(t,\epsilon,b)$ in powers of ϵ.

The associated variational equations are

$$\frac{d}{dt}\,\delta x - i\,\delta x = e^{it}\,\delta\mu \qquad x(0) = 0 \tag{4-105}$$

with $\delta x = 0$, $\delta\mu = 0$ as the only periodic solutions, since

$$\frac{dy}{dt} - iy = e^{it} \equiv g(t) \tag{4-106}$$

has no periodic solutions, and the only solution of

$$\frac{du}{dt} - iu = 0 \tag{4-107}$$

for which $u(0) = 0$ is zero.

The object now is to obtain the bifurcation equation.

For our series expansion we must have $b = x_0(0)$ when $\epsilon = 0$. Thus

$$x(t,\epsilon,\mu) = be^{it} + \epsilon \int_0^t e^{i(t-s)}\left[f(be^{is},s) - \frac{\mu e^{is}}{\epsilon} \right] ds + x_2(t)\epsilon^2 + \cdots \tag{4-108}$$

As a first-order approximation we are concerned with the coefficient of ϵ. We must ensure that it is periodic with period 2π by the proper determination of μ and b. Thus we truncate the series after the first power of ϵ. Since we wish x to be periodic, we must have

$$x(2\pi,\epsilon,\mu) - x(0,\epsilon,\mu) = 0 \tag{4-109}$$

Thus we form Eq. (4-109) from the truncated expansion of our solution. This gives

$$\epsilon \int_0^{2\pi} e^{-is}f(be^{is},s)\, ds - 2\pi\mu = 0 \tag{4-110}$$

Thus
$$\mu(\epsilon,b) = \frac{\epsilon}{2\pi} \int_0^{2\pi} e^{-is}f(be^{is},s)\, ds \tag{4-111}$$

which is a good approximation, since ϵ appears to the first power and is small. The bifurcation condition (seeking periodic solutions in the neighborhood of the periodic solution of our initial equation) becomes $\mu(\epsilon,b) = 0$. It has a solution $b(\epsilon)$ which satisfies $\mu[0,x_0(0)] = 0$ if the implicit-function theorem is satisfied. Otherwise an additional parameter is needed.

Hale[31] obtains the bifurcation equation in a manner related to the previous procedure. To illustrate his method, consider

$$\dot{x} = Ax + \epsilon f(t,x)$$

where f is periodic with period T (i.e., each component of f is periodic). The solutions of the linear part are obtained from e^{At}. Suppose every solution of the linear part is periodic with period T. We wish to determine $a(\epsilon)$ such that

$$x = e^{At}a + \epsilon u(t,\epsilon)$$

is a periodic solution of the above equation.

We now write

$$x = e^{At}y$$

and our equation becomes

$$\dot{y} = \epsilon e^{-At}f(t,e^{At}y) \equiv \epsilon F(t,y)$$

where F is periodic with period T.

We seek periodic solutions for this equation of the form

$$y = a + \epsilon v(t,\epsilon)$$

We fix a and determine a function $y(t,a,\epsilon)$ such that the mean value of $y(t,a,\epsilon) = a$, that is,

$$a = \frac{1}{T}\int_0^T y\, dt$$

and
$$y = \epsilon F[t,\, y(t,a,\epsilon)] - \text{mean value } F[t,\, y(t,a,\epsilon)]$$

We determine all periodic solutions $y(t,a,\epsilon)$ of this equation with mean value a.

The basic theorem here is as follows: A necessary and sufficient condition for a periodic solution $y(t,a,\epsilon)$ is that the bifurcation equation

$$\text{mean value } F[t,\, y(t,a,\epsilon)] \equiv g(a,\epsilon) = 0$$

has a solution $a(\epsilon)$.

Remark: Consider the undisturbed autonomous system

$$\frac{dy_k}{dt} = Y_k(y_1, \ldots ,y_n) \qquad k = 1, \ldots , n \qquad (4\text{-}112)$$

Let $y_k = f_k(t)$, $k = 1, \ldots , n$, be a periodic solution with period T. To obtain the system of disturbed motion, we let

$$x_k = y_k - f_k(t)$$

We have

$$\frac{dx_k}{dt} = p_{k1}x_1 + \cdots + p_{kn}x_n + X_n(t,x_1, \ldots ,x_n) \qquad k = 1, \ldots , n$$

$$(4\text{-}113)$$

where X_k can be expanded in series in the variables starting with terms of order greater than 1 with coefficients of period T. Of course, the p_{ki}

must be periodic with the same period T.   Note that $f_k(t + h)$ is also a solution of (4-112), and hence

$$x_k = f_k(t + h) - f_k(t)$$

is a solution of (4-113).   It has the series expansion

$$h\dot{f}_k(t) + \frac{h^2}{2}\ddot{f}_k(t) + \cdots$$

where dots indicate the order of differentiation with respect to t.   If we substitute in the second system and equate coefficients of like powers of h, we obtain

$$\frac{d\dot{f}_k}{dt} = \sum_{i=1}^{n} p_{ki}\dot{f}_i \tag{4-114}$$

and hence $\dot{f}_k$ (which, of course, is also periodic) is a solution of the linear part of the second system, and its characteristic equation has at least one root equal to unity.   It can be shown that the original system is orbitally stable if the remaining roots have moduli less than unity.

Periodic Solutions of a Second-order Equation

Consider the equation[55]

$$\frac{d^2y}{dt^2} + n^2y + g(t) = \epsilon F(t,y,\dot{y},\mu) \tag{4-114a}$$

with $g(t)$ periodic with period 2π, possessing a Fourier-series expansion, n a constant not an integer, F analytic in y and $\dot{y}$ and periodic of period 2π in t, and $\epsilon > 0$ a small parameter.   When $\epsilon = 0$, the equation describes a linear oscillation which has a periodic solution $f(t)$, the generating solution for the solution of the equation with $\epsilon \neq 0$.

If we have the Fourier expansion

$$g(t) = \frac{a_0}{2} + \sum_{k=1}^{\infty} a_k \cos kt + b_k \sin kt$$

then it is well known that the generating solution is given by

$$f(t) = -\frac{a_0}{2n^2} - \sum_{k=1}^{\infty} \frac{a_k \cos kt + b_k \sin kt}{n^2 - k^2}$$

Exercise 4-41   Verify this fact, and show that $f(t)$ is the only solution of period 2π.

The object now is to obtain a solution of period 2π for the case $\epsilon \neq 0$. If the two solutions differ in the initial values of y and $\dot{y}$ by β_1 and β_2, respectively, then we denote the periodic solution of the general case by $y(t,\beta_1,\beta_2,\epsilon)$; since the solution has period 2π, we must have

$$u(\beta_1,\beta_2,\epsilon) = y(2\pi,\beta_1,\beta_2,\epsilon) - y(0,\beta_1,\beta_2,\epsilon) = 0$$
$$v(\beta_1,\beta_2,\epsilon) = \dot{y}(2\pi,\beta_1,\beta_2,\epsilon) - \dot{y}(0,\beta_1,\beta_2,\epsilon) = 0$$

Exercise 4-42 Show that these conditions are necessary and sufficient for the periodicity of the solution which, assumed analytic in β_1, β_2, and ϵ and yielding $f(t)$ for $\epsilon = 0$, can be written in the form

$$y(t,\beta_1,\beta_2,\epsilon) = f(t) + a_1(t)\beta_1 + a_2(t)\beta_2 + a_3(t)\epsilon + \cdots$$

Substitute in the original equation, and obtain

$$\frac{d^2a_1}{dt^2} + n^2a_1 = 0 \qquad \frac{d^2a_2}{dt^2} + n^2a_2 = 0$$

with the initial conditions which $a_1(0)$, $\dot{a}_1(0)$, $a_2(0)$, and $\dot{a}_2(0)$ must satisfy. Show that $a_1 = \cos nt$, $a_2 = (\sin nt)/2$. Compute the linear terms of u and v, and show that they are identically satisfied for $\beta_1 = \beta_2 = \epsilon = 0$. Show that the jacobian

$$\left.\frac{\partial(u,v)}{\partial(\beta_1,\beta_2)}\right|_{\beta_1=\beta_2=\epsilon=0} = (\cos 2n\pi - 1)^2 + \sin^2 2n\pi \neq 0$$

and hence that for small ϵ there is a unique analytic solution $\beta_1(\epsilon)$, $\beta_2(\epsilon)$ which vanishes at $\epsilon = 0$. When this is substituted in $y(t,\beta_1,\beta_2,\epsilon)$, a unique periodic solution of the original equation is obtained. Thus y becomes a function of ϵ and t, and its series expansion is

$$y = f(t) + y_1(t)\epsilon + y_2(t)\epsilon^2 + \cdots$$

where the $y_i(t)$ are periodic with period 2π.

Substitute in Eq. (4-114a), and equate coefficients to obtain a set of differential equations satisfied by $y_i(t)$. If a_{ki} and b_{ki} are the Fourier coefficients of the right side of the resulting equations whose left side is $d^2y_i/dt^2 + n^2y_i$, $i = 1, 2, \ldots$, show that a unique periodic solution of the ith equation is given by

$$y_i = \frac{a_{0i}}{2n^2} + \sum_{k=1}^{\infty} \frac{a_{ki}\cos kt + b_{ki}\sin kt}{n^2 - k^2}$$

Note that for $i = 1$ the right side is $F(t,f,\dot{f},0)$, which is known. Thus for sufficiently small ϵ the series converges to a periodic solution.

Remark: The case in which n is an integer requires special treatment (see Malkin[55]).

Liénard proved the existence of periodic solutions of a second-order equation of the form

$$\frac{d^2x}{dt^2} + a(x)\frac{dx}{dt} + b(x) = 0$$

where $a(x)$ and $b(x)$ are continuous, $a(x) = a(-x)$, $b(-x) = -b(x)$, $b(x) > 0$ for $x > 0$, $b(x)$ satisfies the Lipschitz condition, and

$$A(x) = \int_0^x a(y)\, dy \to \infty \qquad \text{as} \qquad x \to \infty$$

One must also have $A(\bar{x}) = 0$, where $\bar{x} > 0$ is the single positive zero of A, $A(x) < 0$ for $0 < x < \bar{x}$, and $a(x) > 0$, $A(x) > 0$ for $x > \bar{x}$. Under these conditions the differential equation has a unique periodic solution to within a simple translation of t. Levinson and Smith have generalized the theorem to the case in which $a(x)$ is replaced by $a(x,\dot{x})$ [Ref. 11].

Conservative Systems

Let f be a force acting on a particle, and let r be the distance of the particle from a fixed origin at time t. Then f is conservative if there is a function g called the *potential function* such that $f \cdot \delta r = \delta g$. If the forces acting on a system during time $[a,b]$ are conservative, then Hamilton's principle gives $\delta \int_a^b (T - V)\, dt = 0$, where $V = -g$ is the potential-energy function, and T is the kinetic energy of the particle.[32]

An orbit is a locus of points (x,y). A trajectory describes motion in an orbit by giving $(x,y,\dot{x},\dot{y})$ as functions of time. If t is replaced by $-t$ and the corresponding trajectory has the same orbit, the motion is reversible.

G. D. Birkhoff[5] has shown that the most general conservative reversible two-variable second-order system is given by

$$\frac{d^2x}{dt^2} = \frac{\partial f}{\partial x} \qquad \frac{d^2y}{dt^2} = \frac{\partial f}{\partial y}$$

where $f(x,y,\epsilon)$ depends on a parameter ϵ. Multiplying the first equation by $2\dot{x}$ and the second by $2\dot{y}$ and adding yields

$$2\dot{x}\ddot{x} + 2\dot{y}\ddot{y} = 2\dot{x}f_x + 2\dot{y}f_y$$

Integration gives

$$\dot{x}^2 + \dot{y}^2 = 2f + C$$

where C is a constant.

Birkhoff also gave the normal form for the irreversible case in a restricted three-body problem describing the motion of an infinitesimal mass in the gravitational field of two masses which revolve about each other in circular orbits. Because of the rotation of the two masses we do not have reversibility. The equations are

$$\ddot{x} + g(x,y)\dot{y} = \frac{\partial f}{\partial x} \qquad \ddot{y} - g(x,y)\dot{x} = \frac{\partial f}{\partial y}$$

where $g \not\equiv 0$ and is continuous. The above system can also be reduced to

$$\dot{x}^2 + \dot{y}^2 = 2f + C$$

DeVogelaere[12] has studied periodic solutions of the reversible case by means of numerical methods, assuming that f is even in y. His method leads to an existence proof requiring only continuity arguments.

A two-variable system of the first order,

$$\frac{dx}{dt} = P(x,y) \qquad \frac{dy}{dt} = Q(x,y)$$

is conservative if $\partial P/\partial x + \partial Q/\partial y = 0$ or if there is a function $f(x,y)$ such that $\dot{x} = f_y$, $\dot{y} = -f_x$ from which we get $f(x,y) = $ const along a solution, so that the curves $f = C$ are invariant. If there is a C such that $f = C$ is a closed curve, then this curve is the trajectory of a periodic solution.

In the nonconservative case one may have periodic solutions which, when the linear part is taken, appear as limit cycles if the characteristic roots are real.

Exercise 4-43 For the pendulum $\dot{x} = y$, $\dot{y} = -\sin x$, $f = y^2/2 + (1 - \cos x)$ show that for C sufficiently small the curves $f = C$ are closed.

Remark: Considerable research has been applied to discontinuous systems. An example of such a system is the equation describing a vacuum tube with discontinuous characteristics,

$$\ddot{x} + a\dot{x} + bx = \begin{cases} +c & \text{for } \dot{x} > 0 \\ 0 & \text{for } \dot{x} < 0 \end{cases}$$

analogous to equations describing clock mechanisms. Another example is the multivibrator[48]

$$\dot{x} = \frac{y}{\varphi'(x) - \alpha^2} \qquad \dot{y} = \beta x + \frac{\gamma y}{\varphi'(x) - \alpha^2}$$

where $\varphi(x)$ is the characteristic of a pair of identical tubes, and α, β, and $\gamma > 0$ are constants. Let $\varphi(x) = x^3/3$; then the lines $x^2 = \alpha^2$ are critical, and the trajectories are outside the strip $|x| \leq \alpha$. Across the strip there is a jump; a unique discontinuous oscillation exists.

4-11 LYAPUNOV STABILITY

Lyapunov's second method examines the stability of a differential equation without the use of explicit solutions. It generally applies to a free system, i.e., an unforced system having the origin as a point of equilibrium. Stability itself is concerned with deviations about an equilibrium point. Thus stability means that if the initial conditions of the trajectory of an undisturbed motion are disturbed slightly from equilibrium at the origin, then subsequent motions remain in a small neighborhood of the origin. This is the case with harmonic oscillations.

Definitions (See Sec. 4-5)

An undisturbed motion $\dot{y}_k = Y_k(y_1, \ldots, y_n)$ corresponding to a particular solution $f_k(t)$ is called stable with respect to disturbances $x_k(t)$, $k = 1, \ldots, n$ (according to Sec. 4-5, these really are deviations resulting from the disturbances of the initial conditions), if, given $\epsilon > 0$, there exists a $\delta(\epsilon) > 0$ such that whenever the disturbances satisfy

$$|x_k(t_0)| \leq \delta \qquad k = 1, \ldots, n$$

then for all $t > t_0$,

$$|x_k(t)| < \epsilon \qquad k = 1, \ldots, n$$

The undisturbed motion is otherwise called unstable.   This occurs if no

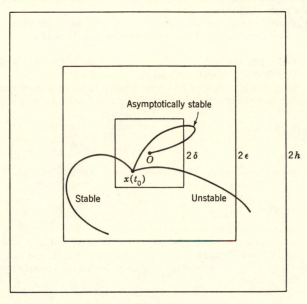

Fig. 4-6

$\delta(\epsilon)$ can be found to satisfy the conditions, and thus at least one inequality is violated.

An undisturbed motion is asymptotically stable if all disturbed motions satisfying $|x_k(t_0)| \leq \delta$ also satisfy $\lim_{t \to \infty} x_k(t) = 0$.

In the definition of stability both ϵ and $\delta(\epsilon)$ are assumed to be positive. However, they can, in addition and without loss of generality, be related to functions of the coordinates which must, of course, be positive.   Thus an undisturbed motion can be defined as stable if, for each number $\epsilon > 0$ a number $\delta(\epsilon)$ can be found such that for all disturbances obeying

$$\sum_{k=1}^{n} x_k{}^2(t_0) \leq \delta(\epsilon)$$

the disturbed motion satisfies the condition

$$\sum_{k=1}^{n} x_k{}^2(t) < \epsilon$$

or $t > t_0$. The undisturbed motion is otherwise unstable. When the last two inequalities are satisfied, the disturbed motion converges to the undisturbed motion.

The origin is said to be *completely stable* or *asymptotically stable in the large* if every solution tends to the origin as $t \to \infty$; this is equivalent to saying that it is asymptotically stable for every choice of initial conditions. It is evident that for complete stability the origin must be the only critical point. An important aspect of the subject of stability is the determination of the region (set of initial values) for which asymptotic stability holds. We shall examine this question later.

The origin is *equiasymptotically* stable if it is stable and if every motion starting sufficiently near the origin converges to it as $t \to \infty$ uniformly in the initial condition $x(t_0) \equiv x_0$.

A motion $x(t;x_0,t_0)$ is bounded for every x_0 and t_0 if for some $A(x_0,t_0)$ we have $\|x(t;x_0,t_0)\| \leq A$ for $t \geq t_0$. It is uniformly bounded if A does not depend on t_0. Note that in the definition of stability $\delta(\epsilon)$ also depends on t_0. If, on the other hand, δ does not depend on t_0, we have uniform stability. Asymptotic and equiasymptotic stability are the same if they are both uniform in t_0. The origin is *uniformly* asymptotically stable in the large if it is uniformly stable and uniformly bounded and every motion converges to it as $t \to \infty$ uniformly in t_0 and for all initial states x_0.

Lyapunov studied autonomous systems which can be put in the form $\dot{x} = Ax + f(x)$, where $x = (x_1, \ldots, x_n)$, and where $A \equiv [a_{ij}]$ is a constant $n \times n$ matrix and f has components f_i expressed in power series of degree at least 2 in the x_i. Note that by means of a simple change of variables, constant terms are eliminated and the origin becomes a singular point. At such a point the right sides of the equations vanish. Singular or critical points are zeros of $Ax + f(x)$ and are points of analyticity of coefficients and solutions. Lyapunov's first method uses perturbation and other procedures to study the solution and stability of a system. In his second method he introduces a special function V for studying stability. We shall be concerned mostly with the latter.

We shall hereafter study first-order systems in which the right side is assumed to be analytic and hence has a power-series expansion in a given region.

Lyapunov's Theorems on the Stability of Motion

We start this section by introducing definite and semidefinite functions, to be used in the analysis of stability.

A function is *definite* (positive or negative) in a region if it has the same sign throughout the region and vanishes only for zero values of the variables. It is *semidefinite* if it also vanishes at other points. Otherwise the function is called *indefinite* or *variable* with respect to sign; for example, $V(x_1,x_2) = x_1 + x_2$ is positive to the right of $x_1 = -x_2$, and negative to the left. On the other hand, $V(x_1,x_2) \doteq x_1^4$ is positive-semidefinite, since it can vanish for $x_1 = 0$ and x_2 arbitrary; finally,

$$V(x_1,x_2) = -(x_1^2 + x_2^2)$$

is negative-definite, and

$$V(x_1,x_2) \equiv \begin{cases} x_1^2 + x_2 & x_2 \geq 0 \\ x_1^2 + x_2^4 & x_2 < 0 \end{cases}$$

is positive-definite, which shows that V need not be analytic in order to have positive-definiteness. Note that if we express an analytic function $V(x_1, \ldots ,x_n)$ in a power series about the origin, the lowest-degree terms form a homogeneous polynomial (called a *homogeneous* form and having the degree of the polynomial) whose definiteness or indefiniteness near the origin will dominate the remaining terms (which also form homogeneous polynomials) and hence determine whether or not the function is definite near the origin. A necessary but not sufficient condition for definiteness of such a function is that the lowest-degree form be definite. If it is semidefinite, then the definiteness or indefiniteness of the terms of the next higher degree determine the definiteness or indefiniteness of the function. Thus our search for definite functions in order to study stability at the origin may be confined to a search for forms having the desired properties.

For a homogeneous form of order m we have

$$V(\lambda x_1, \ldots ,\lambda x_n) = \lambda^m V(x_1, \ldots ,x_n) \tag{4-115}$$

where λ is arbitrary. Hence if the form is definite (positive or negative) near the origin, it will remain so throughout the region, and similarly for indefiniteness. The reader may readily verify that V cannot be definite if m is odd; definite forms are among those of even order. The simplest homogeneous form of even order is a quadratic form,

$$V = \sum_{i,j=1}^{n} a_{ij}x_ix_j \tag{4-116}$$

which, according to a well-known theorem of Sylvester, is positive-definite if and only if all its principal minors are positive. By reducing to the canonical form

$$\sum_{i=1}^{n} \lambda_i x_i^2$$

one has definiteness, positive or negative, depending on whether the λ are all positive or all negative, respectively. If the signs are mixed, the form is indefinite.   It is semidefinite if the λ_i are all nonnegative or all nonpositive.

Consider the differential equations of disturbed motion of an autonomous system $\dot{x}_k = X_k(x)$ with a singularity at the origin and suppose that one has found a positive-definite function $V(x_1, \ldots, x_n)$ in a region R (about the origin) defined by $|x_k| \leq h, k = 1, \ldots, n$. (Throughout, V and its first-order partial derivatives are assumed to be continuous.)   We define $dV/dt = \sum_{i=1}^{n} (\partial V/\partial x_i) X_i(x) = \nabla V \cdot X$ where $x = (x_1, \ldots, x_n)$.

Theorem 4-3 If dV/dt is negative-semidefinite, then the undisturbed motion is stable at the origin.

Theorem 4-4 If dV/dt is negative-definite, then the undisturbed motion is asymptotically stable at the origin.

These theorems would also be true if V were negative-definite and dV/dt had the definiteness opposite to that indicated in the theorems.

Instability Theorems

Let $V(x_1, \ldots, x_n)$ be a function which assumes positive values arbitrarily close to the origin, and let $V(0, \ldots, 0) = 0$. Consider the equations of disturbed motion:

Theorem 4-5 If dV/dt is positive-definite, then the undisturbed motion is unstable at the origin.

Theorem 4-6 If $dV/dt = \lambda V + W(x_1, \ldots, x_n)$, where $\lambda > 0$ and W is nonnegative in R, the undisturbed motion is unstable at the origin.

Theorem (instability theorem of Chetaev) Considering the equations of disturbed motion, let $V(x_1, \ldots, x_n)$ be zero on the boundary of a region R which has the origin as a boundary point, and let both V and dV/dt be positive-definite in R; then the undisturbed motion is unstable at the origin.

The following proofs proceed along lines indicated by Malkin.[54]

Proof of Theorem 4-3: Let V be positive-definite, and let dV/dt be negative-semidefinite. Let $0 < \epsilon < h$, and consider the cube $|x| \leq \epsilon$. It is clear that a point x belongs to the boundary of this cube if and only if its maximum coordinate is ϵ, and hence in referring to such a point it is sufficient to write

$$\max(|x_1|, \ldots, |x_n|) = \epsilon \qquad (4\text{-}117)$$

Let α be the lower limit of V on the boundary of the cube; thus $V \geq \alpha > 0$, that is, $\alpha > 0$, since V is positive-definite and is zero only at the origin.   The value α exists since V is continuous. Let $x(t)$ be a solution whose initial value $x_1(t_0), \ldots, x_n(t_0)$ satisfies $|x_k(t_0)| < \delta$ for

$0 < \delta < \epsilon$. We choose δ small enough so that $V[x_1(t_0), \ldots ,x_n(t_0)] < \alpha$ is satisfied; this is possible because V is continuous and vanishes at the origin. Since $dV/dt \leq 0$, V, as a function of t, does not increase in the region R, and hence

$$V[x_1(t), \ldots ,x_n(t)] \leq V[x_1(t_0), \ldots ,x_n(t_0)] < \alpha$$

Thus for all $t > t_0$, $|x_k(t)| < \epsilon$; that is, the solution must remain in the cube of side 2ϵ, for otherwise $x(t)$ is equal to ϵ and V will have a value greater than or equal to α in contradiction to our choice of δ, which binds the choice of initial value and its corresponding unique solution for which $V < \alpha$ must hold.

Consequently, for all initial conditions within the region, the corresponding solutions lie within the ϵ region for $t > t_0$; hence the motion is stable, and the proof is complete.

Proof of Theorem 4-4: Since the hypotheses of the first theorem are satisfied, we have stability. To prove asymptotic stability, we show that

$$\lim_{t \to \infty} x_k(t) = 0 \qquad k = 1, \ldots, n \tag{4-118}$$

The origin is clearly a stable solution which is also asymptotically stable. Now for a solution which is not identically zero and which satisfies (4-118) we must have $dV/dt < 0$ for $t \geq t_0$. Otherwise, if we have $dV/dt = 0$ for some $t = t_a$, then, since dV/dt is negative-definite,

$$x(t_a) = [x_1(t_a), \ldots ,x_n(t_a)] = (0, \ldots ,0)$$

must hold, and, as the origin is a critical point, the solution $x(t)$ must coincide with it.

Since $dV/dt < 0$, V is monotone-decreasing, and, because it is bounded below, it has a limit α as $t \to \infty$. We may write $V[x_1(t), \ldots ,x_n(t)] > \alpha$ for all t. To show that the origin is asymptotically stable, we prove that $\alpha = 0$. If $\alpha \neq 0$, then $\alpha > 0$, since $V > 0$.

Because of the continuity of V and the fact that $\alpha \neq 0$, it follows that

$$\max [|x_1(t)|, \ldots , |x_n(t)|] > a$$

for some $a > 0$. Thus some $x_k(t)$ is bounded away from the origin for each instant t, and $dV/dt \leq -\beta$ for $\beta > 0$. Thus if t_0 is the initial time value, we have, for $t > t_0$,

$$V[x_1(t), \ldots ,x_n(t)] = V[x_1(t_0), \ldots ,x_n(t_0)] + \int_{t_0}^{t} \frac{dV}{dt}\, dt$$
$$\leq V[x_1(t_0), \ldots ,x_n(t_0)] - \beta(t - t_0) \tag{4-119}$$

But the right side can be negative for large t, which contradicts the positiveness of the left side; therefore, $\alpha = 0$, and

$$\lim_{t \to \infty} V = 0 \qquad (4\text{-}120)$$

Since V is positive-definite, it follows that

$$\lim_{t \to \infty} x_k(t) = 0 \qquad k = 1, \ldots, n \qquad (4\text{-}121)$$

Remark: If B is a symmetric positive-definite matrix and $V = x'Bx$, then $\dot{x} = Ax$ is asymptotically stable if and only if the equation

$$A'B + BA = -C$$

where C is positive-definite, has a solution B. Thus if C is positive-definite, then we have asymptotic stability, since

$$\dot{V} = x'(A'B + BA)x$$

is a negative-definite quadratic form. On the other hand, asymptotic stability implies that to each positive-definite C there corresponds a unique matrix B given by the solution:

$$B = \int_0^\infty \exp(A't)\, C \exp(At)\, dt$$

Interpretation: We now give very brief interpretations of a Lyapunov function.

If the Lyapunov function $V(x)$ measures the energy of an isolated system, then if the change of this energy in time is negative for every x except for $x = 0$, the energy will decrease until it assumes its minimum at the origin.

Lyapunov's theory also has a geometric interpretation. The function

$$V[x_1(t), \ldots, x_n(t)] = c$$

where c is a real parameter, defines a family of surfaces. The requirement that V be positive-definite implies that $c \geq 0$. The value $c = 0$ is satisfied only by $x_1 = \cdots = x_n = 0$ because of positive-definiteness. On the boundary of the cube $|x_k| < h$, V assumes a set of values with a lower bound. Since it is continuous, it assumes all intermediate values between $V = 0$ and this lower bound. Thus every curve from the origin to the boundary $|x_k| < h$ will meet V at any c assuming an intermediate value. The members of the family of surfaces about the origin, that is, $V = c$, are closed and do not intersect. The requirement that $\dot{V} \leq 0$ implies that a curve starting inside any of these closed surfaces must remain within it and move inwards towards surfaces closer to the origin, for otherwise V would increase as t increases, and its derivative would be positive, a contradiction.

Example: Consider the equation[44]

$$\ddot{x} + f(x)\dot{x} + g(x) = 0$$

with the equivalent system

$$\dot{x} = y \qquad \dot{y} = -g(x) - f(x)y$$

A natural way to choose the Lyapunov function is to let it represent the total energy; i.e.,

$$V(x,y) = \frac{y^2}{2} + G(x)$$

where

$$G(x) = \int_0^x g(u)\,du$$

If the damping is positive, that is, $f(x) > 0$ for $x \neq 0$, and the restoring force satisfies $xg(x) > 0$ for $x \neq 0$, $V(x,y)$ is positive-definite; yet

$$\dot{V}(x,y) = -f(x)y^2 \leq 0$$

is negative-semidefinite, and hence the motion is stable.

Proof of Theorem 4-5:   As in the proof of Theorem 4-3, let dV/dt be positive-definite in the cube with side $2h$.   It will be shown that no matter how small we choose δ such that $|x_k(t_0)| \leq \delta$, $k = 1, \ldots, n$, the solution $x_k(t)$ will leave the region R.   We choose δ so that

$$V[x_1(t_0), \ldots, x_n(t_0)] > 0$$

is also satisfied.   If $x_k(t)$ does not leave R for $t > t_0$, then dV/dt will be positive for all $t > t_0$, and thus V is monotone-increasing.   Hence $V(t) > V(t_0)$.   For some $\alpha > 0$ we always have max $[|x_1|, \ldots, |x_n|] \geq \alpha$, for otherwise, for small α, V cannot satisfy $V(t) > V(t_0)$.   Thus for the solution $x_k(t)$, $k = 1, \ldots, n$, we have $dV/dt \geq \beta$ for some $\beta > 0$ and for $t > t_0$.   Hence, finally,

$$V(t) = V(t_0) + \int_{t_0}^t \frac{dV}{dt}\,dt \geq V(t_0) + \beta(t - t_0) \tag{4-122}$$

which increases without bound as $t \to \infty$.   But this contradicts the fact that V is continuous and must be bounded because $x_k(t)$ remains in R.   Hence $x_k(t)$, $k = 1, \ldots, n$, must leave R.

Proof of Theorem 4-6:   We have $(d/dt)e^{-\lambda t}V = e^{-\lambda t}W \geq 0$.   Thus $e^{-\lambda t}V$ is nondecreasing.   Exactly as in the proof of Theorem 4-5, we arrive at our choice of δ for which $V(t_0) > 0$.   Now if $x_k(t)$ remains within R then V is bounded.   However, we have seen that

$$e^{-\lambda t}V \geq e^{-\lambda t_0}V(t_0) \tag{4-123}$$

from which

$$V \geq e^{\lambda(t - t_0)}V(t_0) \tag{4-124}$$

and V increases indefinitely, a contradiction. Thus $x_k(t), k = 1, \ldots, n$, must leave the region R.

Example: As an illustration of the effort involved in finding V, let us examine the stability of the linear part with constant coefficients

$$\dot{x}_k = a_{k1}x_1 + \cdots + a_{kn}x_n \qquad k = 1, \ldots, n$$

of an autonomous system. We seek a Lyapunov function V which will provide the desired stability proof.

We write

$$\frac{dV}{dt} = \sum_{k=1}^{n} \frac{\partial V}{\partial x_k} \sum_{j=1}^{n} a_{kj}x_j = \lambda V \qquad (4\text{-}125)$$

where λ is a constant. The reason for writing λV is that this choice enables us to use the theory of linear equations, a powerful and almost indispensable tool. We first assume that V is a linear form; i.e.,

$$V = \alpha_1 x_1 + \cdots + \alpha_n x_n$$

The discussion for forms of higher order follows similar lines. This form is of odd order and cannot be definite. However, it will serve to illustrate how one determines V, a procedure followed for any choice of the form of V, since the theory of systems of linear equations is also applicable to forms of higher order.

To determine the coefficients of this form, we substitute for V above and equate the coefficients of $x_k, k = 1, \ldots, n$, on both sides. We obtain the following homogeneous system of linear equations in the unknown coefficients:

$$\sum_{k=1}^{n} \alpha_k a_{kj} = \lambda \alpha_j \qquad j = 1, \ldots, n$$

As is well known, the necessary and sufficient condition for a nontrivial solution is that the determinant of the system vanishes. This condition leads to the characteristic equation (an nth-degree polynomial in λ) of the original system, that is, $|A - \lambda I| = 0$, where A is the matrix with coefficients a_{kj}. To each root corresponds a choice of a set of coefficients, and hence a linear form for V.

The determination of the roots λ of the characteristic equation is essential for finding V. However, we do not need the explicit form of V if we can conclude stability in the process of computing V. For example, we already know that a linear system is asymptotically stable if its characteristic roots have negative real parts. We shall prove below that this is also true for a nonlinear system.

It is clear that some practical method is needed for finding a form V by means of which the stability of an autonomous system can be studied.

Exercise 4-44 Show that the number of terms in a form $V(x_1, \ldots, x_n)$ o order $m \geq 1$ is given by

$$N = \frac{n(n+1) \cdots (n+m-1)}{m!}$$

Note that N is also equal to the number of ways in which the nonnegative integer m_k $(0 \leq m_k \leq m)$—giving the power to which x_k is raised—may be chosen, so that

$$\sum_{k=1}^{n} m_k = m$$

Each choice of m_k, $k = 1, \ldots, n$, with $\sum_{k=1}^{n} m_k = m$, gives one term of an mth-order form. Let β_i, $i = 1, \ldots, N$, be the coefficient of the ith term; then we have, as in the case $m = 1$ which we studied above,

$$A_{i1}\beta_1 + \cdots + A_{in}\beta_n = \lambda\beta_i \qquad i = 1, \ldots, N$$

Here A_{ij} corresponds to sums of a_{ij} which arise in equating sides in $dV/dt = \lambda V$. As before, this system has a solution if and only if the coefficient determinant vanishes. This condition yields an Nth-order polynomial in λ. Thus it is necessary and sufficient that λ be a root of this polynomial in order that the coefficients β_k, $k = 1, \ldots, N$, and hence a form V which satisfies the requirements $dV/dt = \lambda V$, be determinable. It turns out that the roots of the polynomial are obtainable from the relation $\lambda = m_1\lambda_1 + \cdots + m_n\lambda_n$, where $\lambda_1, \ldots, \lambda_n$ are the roots of the characteristic equation of the linear approximation.

Exercise 4-45 Let V_j, $j = 1, \ldots, n$, be the linear form corresponding to λ_j, $j = 1, \ldots, n$; then $dV_j/dt = \lambda_j V_j$. By writing $V = V_1^{m_1}V_2^{m_2} \cdots V_n^{m_n}$ and calculating dV/dt prove that the roots are obtainable as indicated above. For a proof that all the roots are obtainable in this manner, see Malkin.[54]

We now examine the problem of finding (when possible) an mth order form $V(x_1, \ldots, x_n)$ whose time derivative is a given mth-order form $U(x_1, \ldots, x_n)$. Let us denote by $c_1, \ldots, c_N$ the coefficient (unknown) of V, and by $b_1, \ldots, b_N$ the coefficients (known) of U. Equating sides of like power in $dV/dt = U$, we obtain

$$\sum_{j=1}^{N} A_{ij}c_j = b_i \qquad i = 1, \ldots, N$$

This equation admits a unique solution for the c_j (determined by Cramer's rule) if and only if the coefficient determinant does not vanish. Note that this coefficient determinant is obtained from that of the system $U = \lambda V$ by putting $\lambda = 0$. Thus $\lambda = 0$ must be excluded as a root o the system $U = \lambda V$, since, when it is a root, the coefficient determinant vanishes (in order that V be determinable for arbitrary U the coefficient determinant must not vanish). Thus we have:

Theorem 4-7 For any mth-order form U, there is a unique form V for which $dV/dt = U$, provided that the coefficient determinant (which is the product of the characteristic roots) does not vanish, i.e., provided that the expression

$$\lambda = m_1\lambda_1 + \cdots + m_n\lambda_n$$

is never zero for any choice of the m's.

Theorem 4-8 (*converse of asymptotic-stability theorem*) If the characteristic roots of the linear approximation have negative real parts, there exists a unique mth-order form $V(x_1, \ldots, x_n)$ which satisfies $dV/dt = U$ for any definite mth-order form $U(x_1, \ldots, x_n)$. The function V is also definite and of sign opposite to that of U.

Proof: By the previous theorem V exists and is unique. To show that, if U is definite, then V is definite with opposite sign, let U be negative-definite. Since the roots have negative real parts, the motion is asymptotically stable.

1. V cannot assume negative values near the origin, for otherwise, it and its derivative have the same sign, and, by Lyapunov's first stability theorem, the motion is unstable, a contradiction.

2. V cannot be positive-semidefinite, for then it would vanish at some point (given at a time t_a) other than the origin; that point may be used as the initial point of the solution of the system. Since the derivative dV/dt is negative, V must decrease after t_a and become negative, a contradiction. It follows that V must be positive-definite.

Note that, for an actual computation of V, the calculations are simplified if the lowest-order U is chosen. Since a form of odd order cannot be definite, a quadratic form with unit coefficients, i.e., of the form $x_1^2 + \cdots + x_n^2$, may be used for U.

To show that the stability of a system is determined by the stability of the linear approximation, we have:

Theorem 4-9 If all the roots λ_j of the linear approximation have negative real parts, then the undisturbed motion is asymptotically stable, regardless of the terms of higher order in the equations of disturbed motion.

Proof: We have shown in Theorem 4-8 that there is a positive-definite quadratic form V which satisfies

$$\sum_{k=1}^{n} \frac{\partial V}{\partial x_k} \sum_{j=1}^{n} a_{kj}x_j = -(x_1^2 + \cdots + x_n^2) \qquad (4\text{-}126)$$

For the general autonomous system we have

$$\frac{dV}{dt} = \sum_{k=1}^{n} \frac{\partial V}{\partial x_k} \left[\sum_{j=1}^{n} a_{kj}x_j + f_k(x_1, \ldots, x_n) \right]$$

$$= -(x_1^2 + \cdots + x_n^2) + \sum_{k=1}^{n} \frac{\partial V}{\partial x_k} f_k(x_1, \ldots, x_n) \quad (4\text{-}127)$$

where f_k represents the terms of order higher than unity in the kth equation of the system. Since definiteness of a form near the origin is determined only by the lowest-order terms, the last quantity on the right whose terms are all of order greater than 2 will not affect the sign definiteness of dV/dt. Thus the derivative of the positive-definite function V is always negative-definite, and the motion is asymptotically stable.

Exercise 4-46   Consider the system

$$\dot{x} = ax + by \qquad \dot{y} = cx + dy$$

and let $\qquad V = \frac{1}{2}(Ax^2 + 2Bxy + Cy^2)$

Determine the conditions on the coefficients a, b, c, and d for asymptotic stability. Also compute A, B, and C; that is, tentatively let $U = V = -(x^2 + y^2)$ for asymptotic stability. Reduce the study of the stability of

$$\ddot{x} + \left[1 - \frac{df(x)}{dx} \right] \dot{x} + 2x + f(x) = 0$$

where $\qquad \dfrac{df(x)}{dx} = kx + \cdots \qquad f(0) = 0$

to the above case.

The question now arises as to whether there is stability if some roots are zero or pure imaginary. These critical cases, though exceptional in treatment, often arise in practice and, therefore, require adequate investigation. So far, cases in which all roots have negative real parts except for a single zero root, a pair of imaginary roots, two pairs of imaginary roots, one zero root and one pair of purely imaginary roots, or two zero roots and a coefficient matrix which has rank smaller than the number of variables by two have been investigated, along with a second-order system, both roots of whose characteristic equations are zero but whose coefficient matrix has unit rank.

One Zero Root

We shall examine briefly the case in which one root is zero while the remaining roots are assumed to have negative real parts.

From a practical standpoint, a motion is unstable if the real part of a root is small and negative and the maximum deviations are larger than can be tolerated. A similar argument shows that if the real part of a root is small and positive and the maximum deviations of the system from the initial disturbances are small, the system is stable from a practical standpoint (though not according to Lyapunov's theory). For example, if α is small in the equation $\dot{x} = \alpha^2 x - x^3$, regardless of the initial value of x, $x(t) \to \pm \alpha$ (which is near the origin) as $t \to \infty$.

Consider now the autonomous disturbed system

$$\dot{x}_i = \sum_{j=1}^{n+1} a_{ij} x_j + f_i(x_1, \ldots, x_{n+1}) \qquad i = 1, \ldots, n+1 \quad (4\text{-}128)$$

where the f_i are analytic in the variables. (The reader will see below the reason for starting with $n + 1$ variables.)

Suppose that the characteristic equation of the system of linear approximations

$$\dot{x}_i = \sum_{j=1}^{n+1} a_{ij} x_j$$

has n roots with negative real parts and one zero root. Let us write

$$x = \sum_{i=1}^{n+1} b_i x_i$$

and determine b_i to satisfy $\dot{x} = 0$. This leads to

$$\dot{x} = \sum_{i=1}^{n+1} b_i \sum_{j=1}^{n+1} a_{ij} x_j = 0 \qquad (4\text{-}129)$$

Equating the coefficients of the x_j to zero gives

$$\sum_{i=1}^{n+1} a_{ij} b_i = 0 \qquad j = 1, \ldots, n+1$$

The coefficient determinant vanishes since there is a zero root, and hence this system has a nontrivial solution. Let one of the b_i, for example, b_{n+1}, be different from zero; then replace x_{n+1} with x. In addition, let

$$c_{ij} = a_{ij} - a_{i,n+1} \frac{b_j}{b_{n+1}} \qquad i, j = 1, \ldots, n$$

$$c_i = \frac{a_{i,n+1}}{b_{n+1}}$$

The linear part can then be written as

$$\dot{x} = 0 \qquad \dot{x}_i = \sum_{j=1}^{n} c_{ij} x_j + c_i x \qquad i = 1, \ldots, n$$

and the general system becomes

$$\dot{x} = X(x, x_1, \ldots, x_n)$$
$$\dot{x}_i = \sum_{j=1}^{n} c_{ij} x_j + c_i x + f_i(x, x_1, \ldots, x_n) \qquad i = 1, \ldots, n \qquad (4\text{-}130)$$

where X and f_i are analytic in the variables.

Since a linear transformation leaves the characteristic equation invariant, the characteristic equation of the new system has one zero root; its remaining roots have negative real parts. To study the stability of this system, let $R(x)$ and $R_i(x)$ be the collection of terms in X and f_i which do not contain $x_1, \ldots, x_n$, and let

$$R(x) = a_k x^k + a_{k+1} x^{k+1} + \cdots$$
$$R_i(x) = a_{k_i}{}^i x^{k_i} + a_{k_i+1}{}^i x^{k_i+1} + \cdots \qquad (4\text{-}131)$$

Theorem Suppose that $R(x)$ is not identically zero, $k_i \geq k$, and $c_i = 0$, $i = 1, \ldots, n$. Then the stability of the undisturbed motion is obtained from the stability of the single equation defining $R(x)$.

The proof of this theorem is long. For the stability of the equation defining $R(x)$, see Exercise 4-47 below.

Exercise 4-47 Consider the equation

$$\dot{x} = X(x) = a_k x^k + a_{k+1} x^{k+1} + \cdots \qquad k \geq 2 \qquad (4\text{-}132)$$

representing the case in which the characteristic equation of the linear approximation has a single zero root. Show that if k is even, then the undisturbed motion is unstable at the origin (it increases or decreases away from the origin). Show by arguments on values of x near the origin that if k is odd and if $a_k \geq 0$, then the motion is unstable. *Hint:* $\dot{x}$ is the velocity of this motion. Consider movement of a point along the x axis. Draw a diagram for small values of k odd and even. Study the stability with the Lyapunov functions

$$V = \tfrac{1}{2} a_k x^2 \qquad \text{for } k \text{ odd}$$
$$V = x \qquad \text{for } k \text{ even}$$

If one or more of the assumptions of the foregoing theorem are violated, we equate the right sides of the n equations in the x_i to zero in the linear system involving x, and solve for the x_i in terms of x. The left side is zero when $x = x_i = 0$, $i = 1, \ldots, n$; since the jacobian determinant of the right side with respect to $x_1, \ldots, x_n$ evaluated at the origin is equal to the determinant $|c_{ij}| \neq 0$, there exists a unique solution for each x_i in terms of x, with $x_i = 0$ for $x = 0$. Each solution can be expressed in a power series about the origin ;i.e.,

$$x_i(x) = A_{i1} x + A_{i2} x^2 + \cdots$$

The $x_i(x)$ are then substituted in the equation

$$\dot{x} = X(x, x_1, \ldots, x_n)$$

and the stability is studied on the basis of the stability of this equation, provided it does not vanish with the substitution. The problem is more complicated if the equation vanishes.[54]

Remark: Suppose that the last equation does not vanish with the substitution. Its stability is determined from the lowest-degree terms. Thus it is not necessary to compute more than the coefficients of the lowest-degree terms in the expansion of $x_i(x)$ or substitute more than the first terms in $\dot{x} = X$. To compute these coefficients $A_{i1}, i = 1, 2, \ldots, n$, assume that they are unknown. Then substitute the expansion of $X_i(x)$ (linear terms only) in the linear system involving x after equating the right sides of these equations to zero. Finally, equate to zero the coefficients of the first power of x, and thus obtain a set of equations from which the unknown coefficients may be determined.

Exercise 4-48   Consider the system

$$a \frac{d^2y}{dt^2} + a \left(\frac{dz}{dt} \right)^2 \sin y \cos y - b \frac{dz}{dt} \cos y = c \sin y \cos z$$

$$a \frac{d^2z}{dt^2} \cos y - 2a \frac{dy}{dt} \frac{dz}{dt} \sin y + b \frac{dy}{dt} = c \sin z$$

Show that if we only retain the linear terms, we have

$$a \frac{d^2y}{dt^2} - b \frac{dz}{dt} - cy = 0$$

$$a \frac{d^2z}{dt^2} + b \frac{dy}{dt} - cz = 0$$

and if we assume two particular solutions of exponential form $C_1 e^{\lambda t}$, $C_2 e^{\lambda t}$, substitute in the equations, and then solve for C_1 and C_2, the coefficient determinant must vanish. This condition gives rise to the charactristic equation

$$\lambda^2 b^2 + (a\lambda^2 - c)^2 = 0$$

Show that the four roots are given by

$$\frac{\pm b \sqrt{-1} \pm \sqrt{4ac - b^2}}{2a}$$

and hence if the discriminant is positive we have instability, and if it is negative we have purely imaginary roots—a critical case which is stable. To prove this one must use the original system; let

$$V = \tfrac{1}{2}[a\dot{z}^2 + (b\lambda - c)y^2 + 2a\lambda \dot{z}y] + \tfrac{1}{2}[a\dot{y}^2 + (b\lambda - c)z^2 - 2a\lambda \dot{y}z]$$

Show that the two quadratic forms involved, which can be written in general as

$$au^2 + (b\lambda - c)v^2 \pm 2a\lambda uv$$

are positive-definite if $a > 0$.

In the case of nonautonomous systems whose disturbed equations are

$$\frac{dx_k}{dt} = X_k(t, x_1, \ldots, x_n)$$

the function $V(t, x_1, \ldots, x_n)$ and its derivative

$$\frac{dV}{dt} = \frac{\partial V}{\partial t} + \sum_{k=1}^{n} \frac{\partial V}{\partial x_k} X_k$$

must satisfy requirements similar to those in the autonomous case. Here, for stability V must, for example, be positive-definite, while dV/dt must be negative-semidefinite or identically zero. The proofs are very similar to those given for the autonomous case. One starts by using an auxiliary positive-definite form $W(x_1, \ldots, x_n)$ which is majorized by V for large values of t and sufficiently small values of x_i. Then one produces α for W as in Theorem 4-3. The constant α is also dominated by V for $x = \epsilon$. If the initial values of x_i are sufficiently small, then V is dominated by α at the initial value of t and the initial values of the x_i.

Exercise 4-49   Prove the above condition for stability.

Remark:[54]   Previously, the definition of stability was given with respect to momentary disturbances at a single point t_0. This, however, need not be the case, as it often happens in practice that there are constantly acting forces which cause continuous disturbances for all t. These forces may be indicated in the equations of the usual system of disturbed motion by $D_k(x_1, \ldots, x_n; t)$. Then we have

$$\dot{x}_k = X_k(x_1, \ldots, x_n; t) + D_k(x_1, \ldots, x_n; t) \qquad k = 1, \ldots, n$$

although in practice D_k would not be distinguished from X_k. An undisturbed motion is stable with respect to constantly acting disturbances if, given $\epsilon > 0$, there exist $\delta_1(\epsilon) > 0$, $\delta_2(\epsilon) > 0$ such that each solution $x_k(t)$ at a given initial t_0 for which $|x_k(t_0)| \leq \delta_1(\epsilon)$, and for arbitrary D_k which satisfy $|D_k(x_1, \ldots, x_k; t)| \leq \delta_2(\epsilon)$ for $t \geq t_0$ and for $|x_k| \leq \epsilon$, we have $|x_k| < \epsilon$ for $t > t_0$.

The condition for the stability of the undisturbed motion for constantly acting disturbances is that a positive-definite Lyapunov function V exists (which is difficult to prove) for the system $\dot{x}_k = X_k$ with $\dot{V}$ negative-definite and $\partial V/\partial x_k$, $k = 1, \ldots, n$, finite in the region.

Stability of Systems with Time-dependent Coefficients

We shall briefly introduce the terminology necessary to the presentation of the main stability theorem for systems with time-dependent coefficients.

The function $f(t)$ is bounded for $t > 0$ if $|f(t)| < A$ for sufficiently large A, and unbounded if $|f(t)| > A$ for some t, no matter how large A may be.[54]   A function for which $\lim_{t \to \infty} f(t) = 0$ holds is called *vanishing*. One can show that if there exist two numbers a and b such that $f(t)e^{at}$ is unbounded and $f(t)e^{bt}$ is vanishing, then there is a real λ such that, for any $\epsilon > 0$, $f(t)e^{(\lambda+\epsilon)t}$ is unbounded and $f(t)e^{(\lambda-\epsilon)t}$ is vanishing.   Lyapunov has called the number λ the *characteristic number* of $f(t)$.   One can also show that

$$\lambda = - \varlimsup_{t \to \infty} \frac{\ln |f(t)|}{t} \tag{4-133}$$

The characteristic number of a sum of two functions is the smaller of the two characteristic numbers if the latter are different; it may be smaller than either when they are equal.   The characteristic number of a product of two functions is not less than the sum of their characteristic numbers.   The sum of the characteristic numbers of f and $1/f$ is not greater than zero; it is zero if and only if $(\ln |f(t)|)/t$ approaches a finite limit as $t \to \infty$.   The characteristic number of the product of f and some function g is equal to the sum of their characteristic numbers if the last-named condition holds for both.   The characteristic number of an integral is not less than that of the integrand.   Every nonzero solution of the linear part $\dot{x} = A(t)x$ of the system

$$\dot{x} = A(t)x + f(t,x) \tag{4-134}$$

where the $a_{ij}(t)$ are finite for $t \geq 0$ and are continuous, has a finite characteristic number.   Here $a_{ij}(t)$ are the coefficients of A, and f_k are the components of f.

This linear system is called *correct* if the sum of the characteristic numbers of its fundamental system of solutions is equal to the characteristic number of the function

$$\exp \int_0^t \sum_{j=1}^n a_{jj}(t) \, dt$$

If the sum of this number and the characteristic number of

$$\exp \left[- \int_0^t \sum_{j=1}^n a_{jj}(t) \, dt \right]$$

is equal to zero, and if the $a_{ij}(t)$ are periodic, the linear system is correct.

The main theorem here is as follows: If the linear approximation $\dot{x} = A(t)x$ is correct, and all its characteristic numbers are positive, then the undisturbed motion for the general system is asymptotically stable for arbitrary $f_k(t,x)$ which satisfy

$$|f_k(t,x)| < A \left(\sum_{k=1}^{n} |x_k| \right)^m$$

where A and m are positive constants and $m > 1$.

Asymptotic Stability in the Large (*Complete Stability*)

The subject of complete stability is essential to the theory of automatic control discussed in the next chapter. There, it will be desirable to verify whether a system whose only singularity is the origin (a condition necessary for complete stability) is completely stable for all admissible choices of initial conditions. The theorems which follow provide criteria for this type of stability.

Theorem (***Barbashin-Krassovskii***)[2] The origin of the autonomous system $\dot{x} = X(x)$ is asymptotically stable in the large if it is asymptotically stable and its Lyapunov function satisfies the condition

$$V(x) \to \infty \quad \text{as} \quad \sum_{i=1}^{n} x_i^2 \to \infty \tag{4-135}$$

In spite of the fact that this theorem is not satisfied in the case of the equation $\ddot{x} + f(x)\dot{x} + g(x) = 0$ of a previous example, this equation, with $V(x,y)$ as chosen in that example, is completely stable. This fact is a consequence of the following.

Theorem (***LaSalle***)[44] Assume $V(x) \geq 0$ and $\dot{V}(x) \leq 0$ for all x. Let E be the locus $\dot{V}(x) = 0$, and let M be the union of all trajectories that remain in E for all t. Then all solutions of $\dot{x} = X(x)$ that are bounded for $t > t_1$, for some t_1, approach M as $t \to \infty$ (Lagrange stability).

LaSalle also assumes, for the equation of the example referred to above, that $xg(x) > 0$, $f(x) > 0$ for $x \neq 0$, and

$$|F(x)| = \left| \int_0^x f(u) \, du \right| \to \infty \quad \text{as} \quad |x| \to \infty$$

It follows that $\dot{V} = 0$ only on $y = 0$ and, possibly, $x = 0$. Therefore, if the origin is excluded, no solution remains on the x and y axes. Thus M is the origin, and by the foregoing theorem all solutions bounded for $t > t_1$, for some t_1, approach M as $t \to \infty$.

To prove asymptotic stability, define the region $R(A,a)$ for A and a satisfying

$$V(x,y) \leq A \qquad [y + F(x)]^2 \leq a^2$$

Because $\dot{V} \leq 0$, a solution starting in R cannot cross $V = A$. By choosing a sufficiently large, we can ensure that for a part of the boundary of

R, that is, for $y + F(x) = \pm a$, we have

$$\frac{d}{dt}[y + F(x)]^2 = -2|g(x)| < 0$$

Thus for any A and for sufficiently large a, every solution in R remains in R. Since R is bounded and can be extended to the entire plane, all solutions are bounded, and we have complete stability.

If we consider van der Pol's equation as given by the second system of Exercise 4-10, and if we take the Lyapunov function

$$V(x,y) = \frac{x^2 + y^2}{2}$$

then
$$\dot{V}(x,y) = \epsilon y^2\left(\frac{y^2}{3} - 1\right)$$

which remains negative for $y^2 < 3$. However it vanishes along the x axis (i.e., for $y = 0$). By the LaSalle theorem, the set along which it vanishes is the x axis; the only invariant set on the x axis is the origin, and thus every solution inside $x^2 + y^2 < 3$ converges to the origin. The limit cycle must lie outside this circle.

Hartman has shown, for the system $\dot{x} = F(x), F(0) = 0$, that by taking $V = x'Bx$ [and hence $\dot{V} = F'(x)Bx + x'BF(x)$] and noting that

$$F(x) = \int_0^1 J(sx)x \, ds$$

where $J(x)$ is the jacobian matrix of $F(x)$ obtained from $dF(sx)/ds$, we have

$$\dot{V} = \int_0^1 x'[J'(sx)B + BJ(sx)]x \, ds$$

Hence if B is positive-definite, $J'(x)B + BJ(x)$ is negative-definite for all $x \neq 0$, and the system is asymptotically stable in the large since $V(x) \to \infty$ as $\|x\| \to \infty$. This has also been shown by Krassovskii in a more complicated proof.

Region of Stability

The regions in which it is possible to construct a definite Lyapunov function to determine stability are variable. Some research has concentrated on determining the regions of stability. One well-known idea is contained in Aizerman's conjecture.[1]
Let

$$\dot{x}_1 = \sum_{j=1}^n a_{ij}x_j + f(x_m) \qquad 1 \le m \le n$$

$$\dot{x}_k = \sum_{j=1}^n a_{kj}x_j \qquad k = 2, \ldots, n$$

$$(4\text{-}136)$$

where $f(x_m)$ is a nonlinear function of the variable x_m, and let $f(0) = 0$. Let α_0 be a constant, and suppose the system is asymptotically stable in the large for $f(x_m) = \alpha x_m$ for $\alpha = \alpha_0$. Then there are constants (obtained from Lyapunov's functions) $\alpha_1 > 0$ and $\alpha_2 > 0$ such that the system is asymptotically stable in the large for all f which satisfy

$$(\alpha_0 - \alpha_1)x_m{}^2 < x_m f(x_m) < (\alpha_0 + \alpha_2)x_m{}^2$$

for all x_m; that is, $f(x_m)$ lies between the lines

$$(\alpha_0 - \alpha_1)x_m \quad \text{and} \quad (\alpha_0 + \alpha_2)x_m$$

Clearly all values of α for which the real parts of the characteristic roots are negative will render the system asymptotically stable. These values of α lie in an open interval; i.e., there exist β_1, β_2 real such that if α satisfies $\beta_1 < \alpha < \beta_2$, then the roots of the above system have negative real parts.

Aizerman's conjecture is that one can take $\beta_1 = \alpha_0 - \alpha_1$ and $\beta_2 = \alpha_0 - \alpha_2$, which is true for the two-variable case. Pliss[69] considered a number of special values of coefficients in a three-variable system and found the conjecture valid only in some of the cases. The system he studied is

$$\frac{dx_1}{dt} = x_2 - f(x_1) \qquad \frac{dx_2}{dt} = x_3 - cx_1 - ef(x_1) \qquad \frac{dx_3}{dt} = -ax_1 - bf(x_1) \quad (4\text{-}137)$$

In the single-variable case, note that $\dot{x}_1 = -ax_1 - f(x_1)$ with $V(x_1) = x_1{}^2/2$, $\dot{V}(x_1) = -x_1{}^2[a + f(x_1)/x_1]$ gives, for asymptotic stability in the large, $f(x_1)/x_1 > -a$. The two possible systems of two equations to consider for verifying the conjecture are:

$$\dot{x}_1 = ax_1 + bx_2 \qquad \dot{x}_2 = f(x_1) + cx_2$$
and
$$\dot{x}_1 = f(x_1) + cx_2 \qquad \dot{x}_2 = ax_1 + bx_2$$

For the first system, one can use the Lyapunov function

$$V(x_1,x_2) = (cx_1 - ax_2)^2 + 2\int_0^{x_1}\left[bc - a\frac{f(u)}{u}\right]u\,du$$

Exercise 4-50 Apply the transformation $y_1 = -x_1$, $y_2 = ex_1 - x_2$, $y_3 = bx_1 - x_3$ to bring the system (4-137) to the form (4-136).

Structural stability, i.e., the stability of a system of differential equations when the coefficients are changed by small amounts, is currently an important area of research into ideas initiated by Andronov and Pontryagin. The approach is to study systems whose phase-plane analysis remains unchanged under small changes in the coefficients. The size of this chapter does not permit inclusion of this material, a good summary of which is found in Refs. 47 and 66a.

4-12   GENERAL METHODS OF SOLUTION

The construction of a solution for a differential equation which does not belong to a class whose solutions are known can be aided by a qualitative study of the physical phenomenon which the equation describes and by quantitative information in the nature of physical data. This information is particularly useful when the equation has no unique solution and when it is difficult to establish from the model the necessary auxiliary conditions which lead to uniqueness. The methods to be given here comprise the majority of useful techniques for solving nonlinear equations, particularly in the pursuit of periodic solutions.

The reader interested in the construction of general solutions of a nonlinear equation may find it useful to consult the references. For example, by using Fuchs' method,[34] one can show that a linear differential equation whose coefficients are analytic except for poles has an analytic general solution except for poles if and only if the integrals are analytic in the neighborhood of the singularities, the roots of the characteristic equation relative to the singularities are integers, and logarithmic terms in the general solution disappear near the singularities.

We shall not treat all the methods mentioned in the outline in Sec. 4-1. For example, Picard's method has already been mentioned. Perturbation methods will be mentioned only briefly, since they occurred previously in the study of periodic solutions.

The First-order Equation[11]

Even a first-order nonlinear equation can be difficult to solve. For example, the degree of the derivative may be such that it is difficult to resolve the resulting algebraic equation in the derivative in order to obtain an equation of the first degree. Even after such a resolution, the problem may not be easily integrable, as the well-known cases treated below will indicate.

Let $F(x,y) = c$ be a solution of the first-order equation

$$\frac{dy}{dx} = f(x,y) \equiv -\frac{P(x,y)}{Q(x,y)} \tag{4-138}$$

that is,
$$P\,dx + Q\,dy = 0 \tag{4-139}$$

We have
$$\frac{\partial F}{\partial x}\,dx + \frac{\partial F}{\partial y}\,dy = 0 \tag{4-140}$$

Thus, if
$$\frac{\partial F}{\partial x} = P(x,y) \qquad \frac{\partial F}{\partial y} = Q(x,y) \tag{4-141}$$

and if in addition (because of the relation between these two conditions, that is, $\partial^2 F/\partial y\,\partial x = \partial^2 F/\partial x\,\partial y$) we have the necessary condition

$$\frac{\partial P}{\partial y} = \frac{\partial Q}{\partial x} \tag{4-142}$$

then Eq. (4-139) is said to be exact, and the solution $F(x,y) = c$ can be derived from (4-142).

We have, for example, from the first equation in (4-141),

$$F(x,y) = \int P(x,y)\,dx + R(y)$$

where $R(y)$ is an arbitrary function of y. Using the second of Eqs. (4-141), we have

$$\frac{dR(y)}{dy} = Q(x,y) - \frac{\partial}{\partial y}\int P(x,y)\,dx \tag{4-143}$$

Hence
$$R(y) = \int\left[Q(x,y) - \frac{\partial}{\partial y}\int P(x,y)\,dx\right]dy \tag{4-144}$$

No arbitrary function of x appears on the right of (4-144), since in (4-143) the left side is independent of x, and hence so is the right side; thus the derivative of the right side of (4-144) with respect to x is zero. Note that applying $\partial/\partial x$ to (4-143) yields (4-142).

A similar procedure is used if the second equation in (4-141) is used. Thus we have a solution of equation (4-138).

If equation (4-139) is not exact, then one introduces an integrating factor, i.e., a function $k(x,y)$ which is to be determined from the condition

$$\frac{\partial(kP)}{\partial y} = \frac{\partial(kQ)}{\partial x}$$

Exercise 4-51   Prove that if $k(x,y)$ is the integrating factor, then so is $k(x,y)h(F)$, where h is an arbitrary differentiable function, and $F(x,y) = 0$ is a solution of equation (4-139).   Note that

$$\frac{\partial F/\partial x}{\partial F/\partial y} = \frac{P}{Q} \tag{4-145}$$

Exercise 4-52   Prove that if P and Q are both homogeneous functions of degree n [for example, $P(\lambda x, \lambda y) = \lambda^n P(x,y)$, and thus by differentiating with respect to λ and putting $\lambda = 1$ we obtain $nP = x\,\partial P/\partial x + y\,\partial P/\partial y$], then $k(x,y) = 1/(xP + yQ)$ is an integrating factor for (4-139).   By substituting $y = xz$ in that equation, using the homogeneity property of P and Q and $dy = x\,dz + z\,dx$, and simplifying, obtain

$$\log cx = -\int \frac{Q(1,z)}{P(1,z) + zQ(1,z)}\,dz \tag{4-146}$$

where c is a constant.

Exercise 4-53 (Davis)[11]   Apply the ideas of the previous exercise to

$$\frac{dz}{dw} = \frac{\alpha z + \beta w + \gamma}{az + bw + c} \equiv \frac{A}{B} \qquad \alpha b - \beta a \neq 0$$

after using the transformation

$$z = x + \epsilon \qquad w = y + \delta$$

and determining ϵ and δ such that the resulting constants are zero.   If $\alpha b = \beta a$, then $\alpha z + \beta w = \alpha z + \alpha b w/a$.   Let $t = \alpha z + \beta w$; then, by writing dt/dz, obtain an equation in which the variables are separable, and integrate it.   Note that this problem is also solvable by introducing a variable u and writing

$$\frac{du}{u} = \frac{dz}{A} = \frac{dw}{B} = \frac{r\,dz + s\,dy}{rA + sB}$$

where r and s are to be determined such that

$$\frac{du}{u} = \frac{r\,dz + s\,dw}{\lambda(rz + sw) + K}$$

where $$ra + s\alpha = \lambda r \qquad r\beta + sb = \lambda s \qquad K = rc + s\gamma$$

Here the first two equations yield a homogeneous system in r and s.   It has a solution for those λ for which the coefficient determinant is zero.   Obtain the two roots λ and λ_2 and their corresponding r and s, substitute in the last differential equation, and solve.   If $\alpha b - \beta a = 0$, then one root is zero.   Provide the solution for this case.

Exercise 4-54   Show that $f(y/x)\,dx - dy = 0$ has the integrating factor $1/[y - xf(y/x)]$.

Exercise 4-55 Using the transformation $z = y^{1-n}$, reduce Bernoulli's equation, $dy/dx = f(x)y + g(x)y^n$, to the form $dz/dx = (1 - n)f(x)z + g(x)(1 - n)$, and, by a single quadrature, write down its solution.

Remark: The first-order equation given in the form $f(x,y,y') = 0$ may be differentiated with respect to y' to yield $f_{y'}(x,y,y') = 0$, and with respect to x to yield (because of the last relation) $f_x + f_y \, dy/dx = 0$. A necessary condition for a singular solution of $f = 0$ is that these three equations are satisfied by a continuous function of x and y. The condition is sufficient if for this continuous function $f_y \neq 0$.

Exercise 4-56 Clairaut's equation is given by $y = xy' + f(y')$, where $y' = dy/dx$. By differentiating this equation once with respect to x, conclude that $y''(x) = 0$, solve this equation, and obtain $y = cx + f(c)$ as the general solution. Also conclude that $x + f'(y') = 0$, substitute $y' = c$, and eliminate c between the last two equations to obtain the singular solution $x + f'(c) = 0$. Note that this solution can be obtained from the theory of envelopes by eliminating c.

Exercise 4-57 Reduce the generalized form of Riccati's equation (usually known as $dy/dx + by^2 = cx^n$) given by

$$\frac{dy}{dx} + a(x)y + b(x)y^2 = c(x)$$

to a second-order linear differential equation by means of the transformation $y = [1/b(x)z] \, dz/dx$. Reduce the linear homogeneous differential equation of the second order

$$\alpha(x) \frac{d^2z}{dx^2} + \beta(x) \frac{dz}{dx} + \gamma(x)z = 0$$

to a Riccati equation by means of the substitution $dz/dx = b(x)yz$.

Exercise 4-58 Let $y = F(x)$ be a solution of the general Riccati equation, and let $y = 1/z + F(x)$. Reduce the equation to a linear first-order equation, and obtain its general integral. If $G(x)$ is another solution, then again let $y = 1/w + G(x)$, and solve as above. Obtain the relation

$$\frac{z}{w} = \frac{y - F}{y - G} \equiv KH(x)$$

Solve for y.

Exercise 4-59 Let

$$y = \frac{A(x) + \alpha B(x)}{C(x) + \alpha D(x)}$$

Solve for the constant α. Differentiate y with respect to x, and solve for α. Equate the two expressions for α, and obtain the general Riccati equation, thus showing that the above expression is the general solution of the Riccati equation. Assume four different values of α, that is, α_i, $i = 1, \ldots, 4$, obtain four particular solutions y_i, and show that the following cross-ratio (theorem) for four linearly independent solutions is constant:

$$\frac{(y_1 - y_2)(y_2 - y_4)}{(y_1 - y_4)(y_2 - y_2)} = \frac{(\alpha_1 - \alpha_3)(\alpha_2 - \alpha_4)}{(\alpha_1 - \alpha_4)(\alpha_2 - \alpha_3)}$$

Exercise 4-60   Assume that three particular solutions of Riccati's equation are known; obtain the fourth using the cross-ratio theorem.

Exercise 4-61   Abel's equation is given by $dy/dx = f(x,y)$, where $f(x,y)$ is a cubic polynomial in y with coefficients that are functions of x.   With constant coefficients, the roots y_1, y_2, y_3 of this polynomial are solutions.   Show that if $a_3(x)$ is the coefficient of y^3, a, b, c are fixed constants, and K is an arbitrary constant, then the general solution is given in the form

$$(y - y_1)^a(y - y_2)^b(y - y_3)^c = Ke^{a_3(x)x}$$

Remark: A partial differential equation of the first order in two independent and one dependent variables has the form

$$F\left(x,y,z,\frac{\partial z}{\partial x},\frac{\partial z}{\partial y}\right) = 0$$

This equation is linear if it has the form

$$f_1(x,y)z + f_2(x,y)\frac{\partial z}{\partial x} + f_3(x,y)\frac{\partial z}{\partial y} = f(x,y)$$

homogeneous if $f = 0$, and quasi-linear if it has the form

$$f_1(x,y,z)\frac{\partial z}{\partial x} + f_2(x,y,z)\frac{\partial z}{\partial y} = f(x,y,z)$$

which in general cannot be reduced to the linear form.   Note that the system

$$\dot{x} = Ax + \epsilon f(x,t)$$

where the matrix A is constant, is quasi-linear.

Remark: G. Temple[84] has pointed out that one reason why it is simpler to solve a linear differential equation than it is to solve a nonlinear one is the existence of a principle of superposition of solutions in the linear case.   This principle is the usual way of expressing the general solution as a function of a certain finite number of particular solutions.   The function has the same form whatever the particular solutions.   Riccati's equation is a nonlinear equation which, as we have indicated, has a general solution satisfying the cross-ratio formula for any three independent solutions; here we have a function expressing the general solution in terms of particular ones.

Temple points out that Sophus Lie's exhaustive classification of all transitive groups up to three variables (which can be extended to more variables) enables one to make a complete classification of nonlinear equations soluble by composition. However, it can be shown that there are nonlinear equations for which such compositions involving a finite number of particular solutions do not exist.

Geometric Methods

The method of isoclines applied to $dy/dx = f(x,y)$ consists in drawing curves known as *isoclines* in the plane for which $f(x,y) = $ const.   Thus isoclines are curves along which the derivatives of an integral of the equation is constant.   By varying the constant, a family of isoclines is obtained.   At each of a carefully chosen set of points along an isocline, a short line segment is drawn whose midpoint coincides with the chosen point and whose slope is equal to the constant corresponding to the isocline. Thus we have a family of parallel short segments on each isocline.   If the isoclines are sufficiently close and the segments from one isocline to a neighboring one are drawn to indicate a gradual turn of slope, the diagram will give an idea of how to draw an integral curve having this gradually turning tangent. This process is simplified if the second derivative is used, for then it is possible to calculate the radius of curvature

$$R = \frac{(1 + y'^2)^{3/2}}{y''}$$

at a given point (x,y), from which the center of curvature is

$$\left(x - \frac{y'R}{(1 + y'^2)^{1/2}}, \; y + \frac{R}{(1 + y'^2)^{1/2}} \right)$$

An arc of the circle of curvature is drawn to pass through the given point. The process is continued for nearby points, and an integral curve is plotted.

It is also possible to generate integral curves by substituting for nearby chosen points in $f(x,y)$—thus obtaining values of the derivative—and then drawing a short line segment through each point with the same slope as the tangent to the integral curve. Nearby points with gradually turning segments are connected so that the segments comprise tangents to the connecting curve. This procedure can also be pursued if the slopes, or "fields of tangents," are given as observational data rather than by means of a differential equation. Both methods are useful in searching for limit cycles and for the study of behavior in the large.

Exercise 4-62 Apply the method of isoclines to the equation of van der Pol in the phase plane.

Remark: K. O. Friedrichs[22] has analyzed the problem of the existence of a periodic solution to a third-order autonomous equation by resorting to three-dimensional graphic constructions and obtaining a solid invariant torus. Thus a solution which begins in the torus remains in it.

Perturbation Theory

We have already used this method in a previous section.

Consider the nonlinear differential equation

$$\sum_{i=0}^{n} a_i x^{(i)} = f_1(t) + \epsilon f_2(x^{(n-1)}, x^{(n-2)}, \; \ldots \; ,x,t) \tag{4-147}$$

where the a_i are constants, $x^{(i)} \equiv d^i x/dt^i$, $\epsilon > 0$ is a small constant, and f_1 and f_2 are functions of the indicated variables.

The method proceeds by assuming an expansion of the form

$$x(t) = \sum_{k=0}^{\infty} x_k(t)\epsilon^k \tag{4-148}$$

substituting in Eq. (4-147), and equating coefficients of like powers of ϵ, which gives, for example, for the coefficients of the zeroth power of ϵ, an nth-order linear equation in $x_0(t)$. This equation is solved for $x_0(t)$ subject to the given initial conditions. Then equating the coefficients of ϵ yields an equation in $x_1(t)$ and in $x_0(t)$, which is now known. It is solved subject to the condition that it vanishes at the initial condition.

An nth-order equation in $x_2(t)$ and $x_1(t)$ (now known) is obtained by equating the coefficients of ϵ^2 and is solved for $x_2(t)$, etc.   If the system is oscillatory with frequency ω, then it in turn must be expanded with coefficients that are functions of the amplitude A, that is,

$$\omega = \omega_0 + \sum_{j=1}^{\infty} \alpha_j(A)\epsilon^j \tag{4-149}$$

and introduced into the various steps of the above procedure to determine $\alpha_j(A)$ such that powers of t (or other functions of t called *secular terms*) which prevent convergence of the solution as $t \to \infty$ can be eliminated.

To have a periodic solution $x(t)$ without expanding ω as in (4-149), the coefficients in the series are assumed to be periodic of the same period.   However, determining these unknown coefficients by substituting and equating like coefficients of ϵ does not always lead to linear differential equations with periodic solutions, since secular terms may appear in the solution of one of these equations.   Linstedt's method consists in dropping the term which gives rise to a secular expression.   However, Poincaré has shown that the series in $x(t)$ need not converge after such manipulation.

The Method of Lighthill and Temple

This method is concerned with the continuation of the solutions of

$$\frac{dx}{dt} = f(x,t,\epsilon) \tag{4-150}$$

from $\epsilon = 0$ to $\epsilon > 0$, where ϵ is a small parameter.   If $f(x,t,\epsilon)$ has a Taylor-series expansion at (x,t) in ϵ near $\epsilon = 0$, the perturbed equation (4-150) is said to be *regular* at (x,t); it is regular in a domain if it is regular at every point of the domain.   Thus

$$f(x,t,\epsilon) \equiv \sum_{n=0}^{\infty} a_n(x,t)\epsilon^n \quad \text{with} \quad a_0(x,t) = f(x,t,0)$$

It is *irregular* at (x,t) if no such expansion exists.   In the regular case one seeks a solution of the form

$$x(t) = x_0(t) + \sum_{n=1}^{\infty} x_n(t)\epsilon^n$$

where $x_0(t)$ is the solution as $\epsilon \to 0$.

Lighthill gives a method of reducing the irregular case to a regular perturbation equation which is often linear.   The object is to produce a uniformly convergent series, in the neighborhood of $\epsilon = 0$, converging to the solution as $\epsilon \to 0$.   This is achieved by expanding both x and t in series in ϵ with coefficients depending on a third variable y as follows:

$$x = \sum_{n=0}^{\infty} x_n(y)\epsilon^n$$

$$t = \sum_{n=0}^{\infty} t_n(y)\epsilon^n \qquad t_0(y) = y$$

$x_0(y)$ is the solution for $\epsilon = 0$. By substituting in Eq. (4-150) and equating coefficients of like powers of ϵ, one obtains an equation in $x_n(y)$ and $t_n(y)$, from which two equations are obtained by requiring that the above power-series expansions be uniformly convergent for a region of values of ϵ. This method can fail to produce a solution, as illustrated by an example due to Carrier which is given in Temple's notes. Temple splits the original equation into a pair of equations, each having a perturbation. The method is applicable to a system of equations of any order in one parameter.

Equivalent or Harmonic Linearization (the Method of Krylov-Bogoliubov and van der Pol)

This method will be illustrated with a special case, from which one also obtains the solution to van der Pol's equation:

$$\ddot{x} - \epsilon(1 - x^2)\dot{x} + x = 0$$

This equation has one limit cycle and, hence, a periodic solution.

Assume that the solution is given by

$$x = a(t) \sin [bt + w(t)]$$

where $a(t)$ tends to a constant limit independent of ϵ as $t \to \infty$, $w(t)$ varies slowly with t, and the period tends to a constant.

We have, on putting $u = bt + w(t)$,

$$\dot{x} = \dot{a}(t) \sin u + a(t)b \cos u + a(t)\dot{w}(t) \cos u$$

If we assume that

$$\dot{a}(t) \sin u + a(t)\dot{w}(t) \cos u = 0 \tag{4-151}$$

we have

$$\ddot{x} = \dot{a}(t)b \cos u - a(t)b^2 \sin u - a(t)b\dot{w}(t) \sin u$$

If we consider a more general equation of the form

$$\ddot{x} + b^2 x + \epsilon f(x,\dot{x}) = 0$$

and substitute in it for x, $\dot{x}$, and $\ddot{x}$, we obtain

$$\dot{a}(t)b \cos u - a(t)b\dot{w}(t) \sin u + \epsilon f[a(t) \sin u, a(t)b \cos u] = 0 \tag{4-152}$$

If we solve simultaneously for $\dot{a}$ and $\dot{w}$ in (4-151) and (4-152), we obtain

$$\dot{a}(t) = -\frac{\epsilon}{b} f[a(t) \sin u, a(t)b \cos u] \cos u$$

$$\dot{w}(t) = \frac{\epsilon}{a(t)b} f[a(t) \sin u, a(t)b \cos u] \sin u$$

It is generally difficult to solve these equations and thus determine, without approximation, $a(t)$ and $w(t)$. Hence we introduce the simplifying assumption that for sufficiently small ϵ both $\dot{a}(t)$ and $\dot{w}(t)$ will be small in the last expressions, and $a(t)$ and $w(t)$ will change slowly with time t. The assumption is made that $a(t)$ will be constant during an entire period; its value may be corrected after the lapse of the period.

With this assumption over a period T of $\sin u$ and $\cos u$ we have the expansion in Fourier series over T:

$$f[a(t) \sin u, a(t)b \cos u] \cos u = \tfrac{1}{2}a_0 + \sum_{n=1}^{\infty} a_n \cos nu + b_n \sin nu$$

$$f[a(t) \sin u, a(t)b \cos u] \sin u = \tfrac{1}{2}c_0 + \sum_{n=1}^{\infty} c_n \cos nu + d_n \sin nu$$

where

$$a_n = \frac{2}{T} \int_0^T f(a \sin u, ab \cos u) \cos u \cos nu \, du$$

$$b_n = \frac{2}{T} \int_0^T f(a \sin u, ab \cos u) \cos u \sin nu \, du$$

(4-153)

There are similar expressions for c_n and d_n.

Integrating the expression for $\dot{a}$ over $(t, t + T)$, we have

$$\int_t^{t+T} \dot{a}(t) \, dt = a(t + T) - a(t) = Ta'(t + hT) \sim Ta'(t) \qquad 0 < h < 1 \quad (4\text{-}154)$$

On substituting the Fourier expansion in the expression for $\dot{a}(t)$ and integrating from t to $t + T$, we obtain

$$T\dot{a}(t) = -\frac{\epsilon}{b} \frac{T}{2} a_0$$

or

$$\dot{a}(t) = -\frac{\epsilon}{2b} a_0 \qquad (4\text{-}155)$$

Similarly, we have

$$\dot{w}(t) = \frac{\epsilon}{2a(t)b} c_0$$

Exercise 4-63   Let $f(x,\dot{x}) = -(1 - x^2)\dot{x}$, let $T = 2\pi$, and show that for van der Pol's equation,

$$\dot{a}(t) = \frac{\epsilon}{2} a(t) \left[1 - \frac{a^2(t)}{4} \right]$$

or that

$$a(t) = \frac{2ge^{\epsilon t/2}}{(1 + g^2 e^{\epsilon t})^{1/2}}$$

where $g = a^2(t)/[4 - a^2(0)]$.   Show that w is a constant and that, for small ϵ, $a(t) \to 2$ as $t \to \infty$.   Note from the derivative $\dot{x}(t)$ of $x = a(t) \sin (t + w)$ that, as ϵ increases, the assumptions made to develop this linearization method would be untenable.

The functions $a(t)$ and $w(t)$, together with their derivatives, may be used in the expression for $\ddot{x}$ given above [that is, $\dot{x}/a(t)$ replaces $\cos u$, and x replaces $a(t) \sin u$] to yield the "equivalent" second-order linear equation

$$\ddot{x} - b\dot{a}(t) \frac{\dot{x}}{a(t)} + [b^2 + b\dot{w}(t)]x = 0$$

T. R. Goodman and T. P. Sargent[27] have applied the method of equivalent linearization to the equation

$$\ddot{x} + f(\dot{x}) + g(x) = h(t)$$

where $h(t)$ is a random gaussian force.

The equivalent linear form is

$$\ddot{x} + q\dot{x} + bx = h(t)$$

and the object is to determine the coefficients a and b. This is accomplished by minimizing the mean-square error

$$\lim_{T \to \infty} \frac{1}{2T} \int_{-T}^{T} [a\dot{x} + bx - f(\dot{x}) - g(x)]^2 \, dt$$

which in the ergodic case (in which the time average is equal to the space average) may be replaced by the ensemble average

$$\iint_{-\infty}^{\infty} [a\dot{x} + bx - f(\dot{x}) - g(x)]^2 F(\dot{x},x) \, d\dot{x} \, dx$$

where $F(\dot{x},x)$ is the joint density function of the responses $\dot{x}$ and x.

If $F(\dot{x},x)$ is taken as the bivariate normal distribution, one can show that the minimizing values of a and b are given by

$$a = \frac{1}{\sigma_{\dot{x}}{}^2} \int_{-\infty}^{\infty} x f(x) \frac{1}{\sigma_{\dot{x}} \sqrt{2\pi}} \exp - \frac{\dot{x}^2}{2\sigma_{\dot{x}}{}^2} \, d\dot{x}$$

where

$$\sigma_{\dot{x}}{}^2 = \int_{-\infty}^{\infty} \int_{-\infty}^{\infty} \dot{x}^2 F(\dot{x},x) \, d\dot{x} \, dx$$

If, in the expression for a, we replace $\dot{x}$ by x and $\sigma_{\dot{x}}$ by σ_x, where the latter is obtained from the expression for $\sigma_{\dot{x}}{}^2$ with $\dot{x}^2$ replaced by x^2, we obtain b. Note that $F(\dot{x},x)$ involves a correlation factor ρ such that $\rho\sigma_{\dot{x}}\sigma_x$ is defined as $\sigma_{\dot{x}}$ where $\dot{x}^2$ is replaced by $x\dot{x}$.

Exercise 4-64 Work out the details of the Goodman-Sargent method.

The Method of Collocation

Consider the linear nth-order equation

$$\sum_{i=0}^{\infty} a_i(t) x^{(i)}(t) = f(t) \qquad a \le t \le b \qquad (4\text{-}156)$$

where $x^{(i)} = d^i x/dt^i$. Suppose that the boundary conditions are in the form

$$\sum_{i=0}^{\infty} a_i(t^j) x^{(i)} \Big|_{t=t^j} = A_j \qquad j = 1, \ldots, n \qquad (4\text{-}157)$$

where $a \le t^j \le b$, $j = 1, \ldots, n$, are prescribed points in the interval, and the A_j are constants that are not all zero. The method of collocation proceeds as follows: Let

$$x(t) = x_0(t) + \sum_{k=1}^{m} c_k x_k(t) \qquad m \le n$$

where the $x_k(t)$, $k = 0, \ldots, m$, are linearly independent arbitrary functions (determined from knowledge of the physics of the problem), the c_k are constants to be determined so that the differential equation is satisfied approximately, $x_0(t)$ satisfies

(4-157), and the $x_k(t)$, $k = 1, \ldots, m$, also satisfy (4-157) but all A_j are replaced by zeros. Thus $x(t)$ itself satisfies (4-157). To determine the c_k, we define

$$f_k(t) \equiv \sum_{i=0}^{n} a_i(t) x_k^{(i)}(t) \qquad k = 0, \ldots, m \tag{4-158}$$

On substituting the expression for $x(t)$ in the original equation, we obtain

$$f_0(x) + \sum_{k=1}^{m} c_k f_k(x) = f(t) \tag{4-159}$$

We then use m of the initial t^j, $j = 1, \ldots, n$, to determine c_j so that the left side coincides with $f(t)$ at m of these points, which implies that our approximate solution will coincide (collocate) with the boundary conditions at m points. Thus we solve for c_j in

$$f_0(t^j) + \sum_{k=1}^{m} c_k f_k(x^j) = f(t^j) \tag{4-160}$$

for m values of j.

The method in its generality is applicable to a nonlinear differential equation.

The Method of Ritz-Galerkin[32]

This method differs from the previous one in the manner in which the c_k are found. Let

$$e(t) \equiv \sum_{i=0}^{n} a_i(t) x^{(i)}(t) - f(t) \tag{4-161}$$

and, as before, let

$$x(t) = x_0(t) + \sum_{k=1}^{m} c_k x_k(t)$$

Here we determine the c_k which minimize

$$I = \int_a^b [e(t)]^2 \, dt \tag{4-162}$$

i.e., we require that

$$\frac{\partial I}{\partial c_k} = 0 \qquad k = 1, \ldots, m$$

Because of the last relation and the fact that the variation δI is given by

$$\delta I = \sum_{k=1}^{m} \frac{\partial I}{\partial c_k} \delta c_k = 0 \tag{4-163}$$

we obtain, on substituting for $e(t)$ and applying (4-163),

$$0 = \delta I = \sum_{k=1}^{m} \int_a^b 2 \left[\sum_{i=0}^{m} a_i(t) \frac{d^i}{dt^i} \left(x_0 + \sum_{k=1}^{m} c_k x_k \right) - f(x) \right] x_k \, \delta c_k \, dt$$

or

$$2 \sum_{k=1}^{m} \int_a^b e(x) x_k(t) \, dt \, \delta c_k = 0$$

But the δc_k are arbitrary, since the c_k are the quantities being varied. Therefore, the coefficients of each δc_k must vanish, and we have the Ritz-Galerkin equations

$$\int_a^b e(t)x_k(t)\,dt = 0 \qquad k = 1,\,\ldots\,,\,m \tag{4-164}$$

This yields m linear equations in the c_k, $k = 1,\,\ldots\,,\,m$, whose determination minimizes the error of the approximating solution.

A more general approach to this method of approximation is to enlarge gradually the sequence x_k, thus determine a larger number of c_k, and hence improve the approximation. This method gives one an idea of the degree of accuracy of the successive iterations. For convenience, polynomials or trigonometric functions are often used in these approximations.

Klotter[38] has used the ideas of the Ritz-Galerkin method to study the stability of

$$\ddot{x} + ag(\dot{x}) + bf(x) = p\cos\omega t$$

where a and $b > 0$ are constants, p and ω are parameters, and both f and g are odd functions of the arguments.

We assume a solution of the form

$$x = A\cos(\omega t - \epsilon)$$

and determine A and ϵ by the Ritz-Galerkin method. Note that a solution of the form $\alpha\sin\omega t + \beta\cos\omega t$ (the first two terms of an expansion in a Fourier series) can be reduced to the above by putting $A = \alpha^2 + \beta^2$ and $\alpha = A\cos\epsilon$, $\beta = A\sin\epsilon$. If

$$H = \frac{4}{\pi A}\int_0^{\pi/2} f(A\cos s)\cos s\,ds$$

$$J = \frac{4}{\pi A\sqrt{b}}\int_0^{\pi/2} g(\omega A\sin s)\sin s\,ds$$

we have, from the Ritz-Galerkin method,

$$\left(H - \frac{\omega^2}{b}\right)^2 + \frac{a^2}{b}J^2 = \left(\frac{p}{Ab}\right)^2 \qquad \tan\epsilon = \frac{(a/\sqrt{b})J}{H - \omega^2/b}$$

Exercise 4-65 Obtain the foregoing expressions, and show that if $a = 0$ (and hence $\tan\epsilon = 0$, that is, ϵ is equal to zero or π), then from the first expression $\epsilon = 0$ corresponds to the positive sign of

$$H - \frac{\omega^2}{b} = \pm\frac{p}{Ab}$$

and $\epsilon = \pi$ to the negative sign. Study the stability of the approximate solution by dividing the $(A,\omega^2/b)$ plane into three regions with boundaries obtained from the foregoing equation. Values in only one of these three regions yield instability.

The Method of Lie Series[29a]

A Lie series has the form

$$\sum_{i=0}^{\infty} \frac{t^i}{i!}D^i f(z) = f(z) + t\,Df(z) + \cdots \tag{4-165}$$

where

$$D = g_1(z)\frac{\partial}{\partial z_1} + \cdots + g_n(z)\frac{\partial}{\partial z_n} \tag{4-166}$$

and the $g_k(z)$ are holomorphic functions of $z = (z_1, \ldots, z_n)$ in a neighborhood of one and the same point, and where t is a new complex variable independent of z. A Lie series can be shown by the method of Cauchy majorants to converge absolutely as a power series in t for all z for which $f(z)$ and $g_k(z)$ are holomorphic. Thus, a Lie series is a holomorphic function in z and t.

To illustrate with a simple application, the equation $\ddot{x} + \omega^2 x = 0$ may be transformed to

$$\dot{x} = y \qquad \dot{y} = -\omega^2 x \tag{4-167}$$

The Lie operator is

$$D = y\frac{\partial}{\partial x} - \omega^2 x\frac{\partial}{\partial y} \tag{4-168}$$

and if x_0, y_0 is the initial condition, then

$$x(t,x_0) = \sum_{i=0}^{\infty} \frac{t^i}{i!} D^i x \tag{4-169}$$

$$y(t,y_0) = \sum_{i=0}^{\infty} \frac{t^i}{i!} D^i x \tag{4-170}$$

If $x_0 = 1$ for $t = 0$, we have on expanding,

$$x(t) = ty_0 - \frac{t^3}{3}\omega^2 y_0 + \cdots = y_0\frac{\sin \omega t}{\omega} \tag{4-171}$$

$$y(t) = y_0 - \frac{t^2}{2}\omega^2 y_0 + \cdots = y_0 \cos \omega t \tag{4-172}$$

Exercise 4-66 Show that if the initial condition is $(x_0,0)$ for $t = 0$, then $x(t) = x_0 \cos \omega t$, $y(t) = -\omega x_0 \sin \omega t$.

Exercise 4-67 The operator corresponding to the equation $\ddot{y} + ky - \epsilon y^2 = 0$, $k \geq 0$, is

$$D = \dot{y}\frac{\partial}{\partial y} - (ky - \epsilon y^2)\frac{\partial}{\partial \dot{y}}$$

Obtain

$$y(t) = y_0 + t\dot{y}_0 + \frac{t^2}{2}(\epsilon y_0^2 - ky_0) + \frac{t^3}{6}(2\epsilon y_0\dot{y}_0 - k\dot{y}_0) +$$

$$\frac{t^4}{24}(2\epsilon\dot{y}_0^2 + 2\epsilon^2 y_0^3 - 3\epsilon ky_0^2 + k^2 y_0) + \cdots \tag{4-173}$$

A series of the form

$$y = y_0 + \epsilon y_1 + \epsilon^2 y_2 + \cdots$$

can be obtained for this equation by writing $D = D_1 + D_2$, with $D_1 = y(\partial/\partial y) - ky(\partial/\partial\dot{y})$ and $D_2 = \epsilon y^2(\partial/\partial\dot{y})$, and specializing the iteration formula

$$f_{i+1}(t,z) = g(t,z) + \int_0^t D_2 f_i(t - \tau; z)\Big|_{z=g} d\tau \tag{4-174}$$

which applies to equations of the form

$$\ddot{y} + f(y,\dot{y}) = g(t)$$

where

$$f_0(t,z) = \sum_0^\infty \frac{t^i}{i!} D_1{}^i(z)\Big|_0 \tag{4-175}$$

Asymptotic-series Method[7]

This method (well known in the theory of linear differential equations) is usually applied in the nonlinear case to a differential equation with a parameter ϵ when it is desired to study the behavior of the solution for large values of the parameter. Asymptotic expansions are also useful in analyzing the behavior of solutions near a singular point at infinity.

An asymptotic series expansion $f(t)$ is a series of the form

$$\sum_{n=0}^\infty a_n t^{-n} \tag{4-176}$$

where, if $R_n(t) \equiv f(t) - S_n(t)$, and $S_n(t)$ is the sum of the first $n+1$ terms, then

$$\lim_{t \to \infty} t^n R_n(t) = 0 \tag{4-177}$$

The sum and product of two asymptotic series are asymptotic series; integrating such a series term by term yields an asymptotic series. Such a series can be the expansion of more than one function.

One assumes asymptotic series expansions for the solution of a differential equation and substitutes to obtain relations among the coefficients. The latter are used to determine the coefficients. When a differential equation has a parameter which assumes large values, the expansion is made in terms of the parameter, with coefficients of the expansion depending on t.

Exercise 4-68 By reference to Minorsky,[62] page 679, obtain asymptotic expansions for the equation of van der Pol.

The Taylor-Cauchy Transform

This method, developed by Ku, Wolf, and Dietz,[42] applies to equations of the form

$$\sum_{i=0}^n a_i x^{(i)} + f(x,\dot{x}, \ldots ,x^{(n-1)}) = g(t)$$

where the a_i are constants, and f may be nonlinear.

The Taylor-Cauchy transform of a function $W(\lambda)$ is defined by

$$T_c[W(\lambda)] = \frac{1}{2\pi \sqrt{-1}} \int_C \frac{W(\lambda)}{\lambda^{m+1}} d\lambda \qquad m = 0, 1, 2, \ldots \tag{4-178}$$

where λ is a complex variable, and C is a closed contour in the λ plane enclosing the singularities of $W(\lambda)$.

In the differential equation one replaces t with λ and $x(t)$ with $W(\lambda)$. Then a given-order derivative of x with respect to t becomes the same-order derivative of W with respect to λ. Then one applies T_c to the equation and assumes that the highest-order derivative $W^{(n)}(\lambda)$ is analytic. Thus

$$W^{(n)}(\lambda) = \sum_{m=0}^{\infty} w_{m,n}\lambda^m \tag{4-179}$$

from which we have

$$T_c[W^{(n)}(\lambda)] = w_{m,n} \tag{4-180}$$

The inverse transform gives back the derivative; i.e.,

$$T_c^{-1}(w_{m,n}) = \sum_{m=0}^{\infty} w_{m,n}\lambda^m = W^{(n)}(\lambda) \tag{4-181}$$

To illustrate, we use the example given by Ku:[42]

$$\frac{dx}{dt} + x^2 = u(t) \equiv \text{unit step function}$$

Substituting $W(\lambda)$ and applying T_c yields

$$w_m + C_{m-2,0}^{(2)} = \delta_m$$

where

$$C_{m-2,0}^{(2)} = \sum_{k=0}^{m-2} \frac{w_k}{k+1} \frac{w_{m-k-2}}{m-k-2}$$

If the last two equations are solved recursively for w_m, we obtain

$$w_0 = 1 \quad w_1 = 0 \quad w_2 = -1 \quad w_3 = 0 \quad w_4 = \tfrac{2}{3} \quad w_5 = 0 \quad w_6 = -\tfrac{17}{45} \quad \cdots$$

Note that

$$W^{(1)}(\lambda) = \sum_{m=0}^{\infty} w_{m,1}\lambda^m$$

and, instead of $w_{m,1}$, we have used w_m. This gives, by integration,

$$W(\lambda) = \sum_{m=0}^{\infty} \frac{w_m\lambda^m}{m+1} + A$$

where A is a constant. If $x(0) = 0$, then $A = 0$, and we have

$$W(\lambda) = \lambda - \frac{\lambda^3}{3} + \frac{2}{15}\lambda^5 \cdots = \tanh\lambda$$

and the solution is given by $x(t) = \tanh t$.

The Method of Complex Convolution

Our brief outline of this method follows the presentation of E. Weber.[92]

We define the Laplace transform of a function $x(t)$ by

$$\bar{x}(s) = \int_0^{\infty} e^{-st}x(t)\,dt$$

Exercise 4-69 Evaluate $\int_0^{\infty} e^{-st}(dx/dt)\,dt$.

The Laplace transform of a product of real functions $x(t)$, $y(t)$ is the complex convolution:

$$\bar{x}(s) * \bar{y}(s) = \frac{1}{2\pi i} \int_{a-i\infty}^{a+i\infty} \bar{x}(s-u)\bar{y}(u)\, du$$

where a is a constant, $i = \sqrt{-1}$, and $\bar{x}$ and $\bar{y}$ are the Laplace transforms of x and y. Note that because of the foregoing we have the n-fold convolution $\bar{x}^{(n)}(s)$ giving the Laplace transform of $x^n(t)$, the nth power of x. It is frequently possible to apply the Laplace transform to a differential equation (assuming existence throughout) and to consider the singularities of the resulting complex integrals.

If, for example, we are given a nonlinear equation with constant coefficients

$$\frac{d^2x}{dt^2} + a_1 \frac{dx}{dt} + a_0(x + kx^n) = f(t) \tag{4-182}$$

we obtain, after taking the Laplace transform [assuming $\bar{x}(s)$ exists],

$$(s^2 + a_1 s + a_0)\bar{x}(s) + a_0 k\bar{x}^{(n)}(s) = \bar{f}(s) + (s + a_1)x\Big|_{t=0} + \frac{dx}{dt}\Big|_{t=0}$$

The procedure then assumes that $\bar{x}(s)$ is meromorphic (i.e., all its singularities are poles in the finite plane), and hence, if the s_α are its poles, we may write

$$\bar{x}(s) = \sum_\alpha \frac{A_\alpha}{s - s_\alpha}$$

It follows that the Laplace transform of any integral power of x is also meromorphic, and if, for example, $n = 3$, then

$$\bar{x}^{(3)}(s) = \sum_\alpha \sum_\beta \sum_\gamma \frac{A_\alpha A_\beta A_\gamma}{s - (s_\alpha + s_\beta + s_\gamma)}$$

These expressions are substituted in the transformed equation, after division by the coefficient $s^2 + a_1 s + a_0$, which is separated into its two factors (with roots s' and s''). If none of the s_α is equal to either s' or s'', then, if other assumptions are made on the roots, it is possible to determine a few of the roots and some of the coefficients A_α by comparing parts of the equation and computing the inverse transform of $\bar{x}(s)$.

Exercise 4-70 Make at least one useful observation (e.g., phase-plane analysis, stability, periodicity, method of solution, etc.) on each of the following equations:

Emden's equation:
$$\frac{d^2y}{dx^2} + \frac{2}{x}\frac{dy}{dx_n} + y^n = 0$$

which arises in the study of the thermal behavior of a spherical cloud of gas acting under molecular attraction subject to the laws of thermodynamics.

Duffing's equation:
$$\ddot{y} + n^2 y - \beta y^3 = \beta F_0 \cos \lambda t$$

(mass on a nonlinear spring) which arises in approximating the equation of the pendulum to cubic terms and then using a forcing function.

Thomas-Fermi equation: $\quad \ddot{y} = \left(\dfrac{y^3}{t}\right)^{\frac{1}{2}}.$

Rayleigh's equation: $\quad \dfrac{d^2y}{dx^2} + a\dfrac{dy}{dx} + m\left(\dfrac{dy}{dx}\right)^3 + n^2y = 0$

Volterra's system: $\quad \dot{x} = A(x - xy) \qquad \dot{y} = -B(y - xy)$

This system arises in the study of the struggle for existence, in a closed environment, between two species of animals, one of which preys exclusively on the other. Let $N_1(t)$ be the population of the prey, and $N_2(t)$ of the predator, at time t. If N_2 is small, N_1 increases, entailing an increase in N_2. This increase in N_2 is followed by a decrease in N_1, which causes starvation and hence a decrease, in N_2. The number of encounters of the two species is proportional to N_1N_2. In an encounter, one species decreases while the other increases. Thus $\dot{N}_1 = aN_1 - bN_1N_2$, and $\dot{N}_2 = -cN_2 + dN_1N_2$, from which we obtain Volterra's system.

REFERENCES

1. Aizerman, M. A.: On a Problem Concerning the In-the-large Stability of Dynamic Systems (in Russian), *Uspehi Mat. Nauk*, vol. 4, no. 4, pp. 187–188, 1949.

1a. Antosiewicz, H. A.: A Survey of Lyapunov's Second Method, in "Contributions to the Theory of Nonlinear Oscillations," vol. IV, Annals of Mathematics Studies, no. 41, pp. 141–166, Princeton University Press, Princeton, N.J., 1958.

2. Barbashin, E. A., and N. N. Krassovskii: On Stability of Motion in the Large, *Dokl. Akad. Nauk SSSR*, vol. 86, pp. 453–456, 1952.

2a. Bellman, Richard: "Stability Theory of Differential Equations," McGraw-Hill Book Company, Inc., New York, 1953.

2b. ———: "Introduction to Matrix Analysis," McGraw-Hill Book Company, Inc., New York, 1960.

3. Bierberbach, L.: "Theorie der Differentialgleichungen," Berlin, 1926 (reprinted by Dover Publications, Inc., New York, 1944).

4. ———: "Theorie der Gewohnlichen Differentialgleichungena uf Funktionentheoretischer Grundlage Dargestellt," Springer-Verlag OHG, Berlin, 1953.

5. Birkhoff, G. D.: "Collected Works," vols. I and II, American Mathematical Society, Providence, R.I., 1950.

5a. Brauer, F., and S. Sternberg: Local Uniqueness, Existence in the Large, and the Convergence of Successive Approximations, *Am. J. Math.*, vol. 80, pp. 421–430, 1958.

5b. Brenner, J. L., and M. Weisfeld: Special Satellite Orbits; 1. Equatorial Orbits, 2. Polar Orbits, Stanford Research Institute, December, 1960.

6. Cartwright, M. L., and J. E. Littlewood: On Non-linear Differential Equations of the Second Order, *Annals of Math.* vol. 48, pp. 472–494, 1947.

7. Cesari, Lamberto: "Asymptotic Behavior and Stability Problems in Ordinary Differential Equations," Springer-Verlag OHG, Berlin, 1959.

8. Clauser, Francis H.: The Behavior of Nonlinear Systems, *J. Aeronaut. Sci.*, pp. 411–434, May, 1956.

9. Coddington, Earl A., and Norman Levinson: "Theory of Ordinary Differential Equations," McGraw-Hill Book Company, Inc., New York, 1955.

10. Coddington, Earl A.: "An Introduction to Ordinary Differential Equations," Prentice Hall, Inc., Englewood Cliffs, N.J., 1961.

11. Davis, Harold T.: "Introduction to Nonlinear Differential and Integral Equations," U.S. Atomic Energy Commission, September, 1960.

12. DeVogelaere, René: On the Structure of Symmetric Periodic Solutions of Conservative Systems, with Applications, in "Contributions to the Theory of Nonlinear Oscillations," vol. 4, pp. 53–84, Princeton University Press, Princeton, N.J., 1958.

13. Diliberto, S. P.: On Systems of Ordinary Differential Equations, in "Contributions to the Theory of Nonlinear Oscillations," vol, 1, pp. 1–38, Princeton University Press, Princeton, N.J., 1950.

14. ———: An Application of Periodic Surfaces (Solution of a Small Division Problem), in "Contributions to the Theory of Nonlinear Oscillations," vol. 3, pp. 257–260, Princeton University Press, Princeton, N.J., 1956.

15. ———: Bounds for Periods of Periodic Solutions, in "Contributions to the Theory of Nonlinear Oscillations, vol. 3, pp. 269–276, Princeton University Press, Princeton, N.J., 1956.

16. ——— and M. D. Marcus: A Note on the Existence of Periodic Solutions of Differential Equations, in "Contributions to the Theory of Nonlinear Oscillations," vol. 3, pp. 237–242, Princeton University Press, Princeton, N.J., 1956.

17. ——— and G. Hufford: Perturbation Theory for Nonlinear Ordinary Differential Equations, in "Contributions to the Theory of Nonlinear Oscillations," vol. 3, pp. 207–236, Princeton University Press, Princeton, N.J., 1956.

17a. Doherty and Keller: "Mathematics of Modern Engineering," John Wiley & Sons, Inc., New York, 1936.

18. Dunford, Nelson, and Jacob T. Schwartz: "Linear Operators," Part I: General Theory, Interscience Publishers, Inc., New York, 1958.

19. El'Sgol'ts, L. E.: "Differential Equations," Hindustan Publishing Corp., Delhi, India, 1961.

20. Ford, Lester R.: "Differential Equations," 2d ed., McGraw-Hill Book Company, Inc., New York, 1955.

21. Frazer, R. A., W. J. Duncan, and A. R. Collar: "Elementary Matrices and Some Applications to Dynamics and Differential Equations," Cambridge University Press, New York, 1955.

22. Friedrichs, K. O.: On Non-linear Vibrations of the Third Order, pp. 65–103, in "Studies in Nonlinear Vibration Theory," New York University Press, New York, 1946.

23. ———: "Studies in Nonlinear Vibration Theory," New York University, Institute of Mathematical Sciences, 1946.

24. ———: Fundamentals of Poincaré's Theory, in "Nonlinear Circuit Analysis," pp. 56–67, Polytechnic Press of the Polytechnic Institute of Brooklyn, April, 1953.

25. Fuchs, B. A., and V. I. Levin: "Functions of a Complex Variable and Some of Their Applications," vol. II, Pergamon Press, New York, 1961.

26. Gantmacher, F. R.: "The Theory of Matrices" (translated from Russian), 2 volumes, Chelsea Publishing Company, New York, 1959.

27. Goodman, T. R., and T. P. Sargent: Launching of Airborne Missiles Underwater, part XI of "Effect of Nonlinear Submarine Roll Damping on Missile Response in Confused Seas," Allied Research Associates, Inc., Boston, Mass., September, 1961.

28. Goursat, Edouard (translated by E. R. Hedrick and Otto Dunkel): "Differential Equations, A Course in Mathematical Analysis," vol. 2, part 2, Dover Publications, Inc., New York, 1945.

29. Graffi, Dario: Forced Oscillations for Several Nonlinear Circuits, *Ann. Math.*, vol. 54, pp. 262–271, 1951.

29a. Gröbner, W., and F. Cap: A New Method to Solve Differential Equations: Lie Series and Their Applications in Physics and Engineering, Report, Office of Naval Research contract N62558-2992, August 31, 1962.

30. Hahn, Wolfgang: "Theorie und Anwendung der Direkten Methoden von Liapunov," Springer-Verlag OHG, Berlin, 1959.

31. Hale, Jack K.: "Nonlinear Oscillations," McGraw-Hill Book Company, Inc., New York, 1963.

31a. Hartman, P.: On Stability in the Large for Systems of Ordinary Differential Equations, *Can. J. Math.*, vol. 13, pp. 480–492, 1961.

31b. ——— and C. Olech: On Global Asymptotic Stability of Solutions of Differential Equations, *Trans. Am. Math. Soc.*, vol. 104, no. 1, pp. 154–178, July, 1962.

32. Hildebrand, F. B.: "Methods of Applied Mathematics," Prentice-Hall, Inc., Englewood Cliffs, N.J., 1961.

33. Hurewicz, Witold: "Lectures on Ordinary Differential Equations," The Technology Press of the Massachusetts Institute of Technology, Cambridge, Mass., and John Wiley & Sons, Inc., New York, 1958.

34. Ince, E. L.: "Ordinary Differential Equations," Dover Publications, Inc., New York, 1926.

35. Kamke, E.: "Differentialgleichungen, Reeller Funktionen," Chelsea Publishing Company, New York, 1947.

36. Kakutani, S., and L. Markus: On the Non-linear Difference-Differential Equations $y'(t) = [A - By(t - \tau)]y(t)$, in "Contributions to the Theory of Nonlinear Oscillations," vol. 4, Princeton University Press, Princeton, N.J., 1958.

37. Kaplan, Wilfred: "Ordinary Differential Equations," Addison-Welsey Publishing Company, Inc., Cambridge, Mass., 1961.

38. Klotter, K.: Nonlinear Vibration Problems Treated by the Averaging Method of W. Ritz, *Proc. First National Congress of Appl. Mech.*, pp. 125–131, ASME, New York, 1951.

39. Krasnosel'skii, M. A.: "Some Problems of Nonlinear Analysis," American Mathematical Society Translation Series 2, vol. 10.

40. Krassovskii, N. N.: Sufficient Conditions for Stability of Solutions of a System of Non-linear Differential Equations, *Dokl. Akad. Nauk SSSR*, vol. 98, pp. 901–904, 1954.

41. ———: On Stability with Large Initial Disturbances, *Dokl. Akad. Nauk SSSR*, vol. 21, pp. 309–319, 1957.

41a. ———: On the Stability of Solutions of Certain Differential Equations, *Prikl. Mat. Meh.*, vol. 18, pp. 735–737, 1954.

41b. Krylov, N. M., and N. N. Bogoliubov (translated by S. Lefschetz): "Introduction to Nonlinear Mechanics," Princeton University Press, Princeton, N.J., 1943.

42. Ku, Y. H.: Theory of Nonlinear Control, *J. Franklin Inst.*, vol. 271, no. 2, February, 1961.

43. LaSalle, J. P.: Some Extensions of Liapunov's Second Method, *Air Force Rept.* AFOSR TN 60-22, 1960.

44. ———: Asymptotic Stability Criteria, in "Proceedings of Symposia in Applied Mathematics," vol. 13, American Mathematical Society, 1962.

45. ——— and Solomon Lefschetz: "Stability by Liapunov's Direct Method with Applications," vol. 4 of "Mathematics in Science and Engineering," Academic Press Inc., New York, 1961.

45a. ——— and ——— (eds.): "Nonlinear Differential Equations and Nonlinear Mechanics," Academic Press Inc., New York, 1963.

46. Lefschetz, Solomon: Russian Contributions to Differential Equations, pp. 68–74 in "Nonlinear Circuit Analysis," Polytechnic Press of the Polytechnic Institute of Brooklyn, April, 1953.

47. ———: "Differential Equations: Geometric Theory," Interscience Publishers, Inc., New York, 1957.

48. Leimanis, E., and N. Minorsky: "Dynamics and Nonlinear Mechanics," John Wiley & Sons, Inc., New York, 1958.

49. Lewis, D. C.: Differential Equations Referred to a Variable Metric, *Am. J. Math.*, vol. 73, pp. 48–58, 1951.

50. ———: Periodic Solutions of Differential Equations Containing a Parameter, *Duke Math. J.*, vol. 22, no. 1, pp 39–56, March, 1955.

51. ———: On the Perturbation of a Periodic Solution When the Variational System Has Non-trivial Periodic Solutions, *J. Rational Mech. and Anal.*, vol. 4, no. 5, pp. 795–815, September, 1955.

52. ———: On the Role of First Integrals in the Perturbation of Periodic Solutions, *Ann. Math.*, vol. 63, no. 3, pp. 535–548, May, 1956.

53. Lyapunov, A. A.: "Problème Général de la Stabilité du Mouvement," photoreproduction as Annals of Mathematics Studies No. 17, Princeton University Press, Princeton, N.J., 1907.

54. Malkin, I. G.: "Theory of Stability of Motion," translated from a publication of the State Publishing House of Technical-Theoretical Literature, Moscow, by U.S. Atomic Energy Commission, Office of Technical Information, 1952.

55. ———: "Some Problems in the Theory of Nonlinear Oscillations," books 1 and 2, translated from a publication of the State Publishing House of Technical-Theoretical Literature, Moscow, by U.S. Atomic Energy Commission, 1956.

56. Markus, L.: Escape Time for Ordinary Differential Equations, *Rend. Sem. Mat., Univ. Torino*, vol. 11, pp. 271–276, 1951–1952.

56a. ———: Global Structure of Ordinary Differential Equations in the Plane, *Trans. Am. Math. Soc.*, vol. 76, pp. 127–148, 1954.

56b. ——— and H. Yamabe: Global Stability Criteria for Differential Systems, University of Minnesota mimeographed report, 1960.

57. McLachlan, N. W.: "Ordinary Non-linear Differential Equations in Engineering and Physical Sciences," 2d ed., Oxford University Press, Fair Lawn, N.J., 1955.

58. Mendelson, Pinchas: On Phase Portraits of Critical Points in n-Space, in "Contributions to the Theory of Nonlinear Oscillations," vol, 4, pp. 167–199, Annals of Mathematics Studies No. 41, Princeton University Press, Princeton, N.J., 1958.

59. Minorsky, N.: "Introduction to Non-linear Mechanics," J. W. Edwards, Publisher, Incorporated, Ann Arbor, Mich., 1947.

60. ———: Stationary Solutions of Certain Non-linear Differential Equations, *J. Franklin Inst.*, vol. 254, pp. 21–42, 1952.

61. ———: On Interaction of Non-linear Oscillations, *J. Franklin Inst.*, vol. 256, pp. 147–167, 1953.

62. ———: "Nonlinear Oscillations," D. Van Nostrand Company, Inc., Princeton, N.J., 1962.

63. Mitropol'skiy, Yu. A.: Nonstationary Processes in Nonlinear Oscillatory Systems, Kiev, 1955 (*Air Technical Intelligence Translations* AD 162846 and AD 162847).

64. Moser, Jurgen: A New Technique for the Construction of Solutions of Nonlinear Differential Equations, *Proc. National Acad. Sci.*, vol. 47, no. 11, pp. 1824–1831, November, 1961.

64a. Murphy, C. H.: On Dissimilar Solutions of a Differential Equation, *Am. Math. Monthly*, vol. 62, no. 6, p. 438, 1955.

65. Nemytskii, V. V., and V. V. Stepanov: "Qualitative Theory of Differential Equations" (translation), Princeton University Press, Princeton, N.J., 1949.

66.   Olech, Z: Sur la Stabilité Asymptotique des Solutions d'un Système d'Équations Différentielles, *Ann. Polon. Math.*, vol. 7, pp. 259–267, 1960.

66a. Peixoto, M. M.: Structural Stability on Two-dimensional Manifolds, *Topology*, vol. 1, pp. 101–120, Pergamon Press, New York, 1962.

67.   Perlis, S.: "Theory of Matrices," Addison-Welsey Publishing Company, Inc., Cambridge, Mass., 1956.

68.   Picard, Emile: "Traité d'Analyse," 3 vols., 3d ed., Gauthier-Villars, Paris, 1928.

69.   Pliss, V. A.: "Some Problems in the Theory of Stability of Motion in the Large, Leningrad University Press, 1958.

70.   Poincaré, Henri: Sur les Propriétés des Fonctions Définies par les Équations aux Différences Partielles, thesis, Gauthier-Villars, Paris, 1879.

71.   ———: Mémoire sur les Courbes Définies par une Équation Diffréentielle, *J. Math. Pures Appl.*, vol. 7, no. 3, pp. 375–422, 1881; vol. 8, pp. 251–296, 1882; vol. 1, no. 4, pp. 167–244, 1885; vol. 2, pp. 151–217, 1886; Oeuvres, t.1, pp. 3–84; 90–161; 167–221.

72.   ———: Sur le Problème des Trois Corps et les Équations de la Dynamique, *Acta Math.*, vol. 13, pp. 1–270, 1890.

73.   ———: "Les Méthodes Nouvelles de la Mécanique Céleste," 3 vols., Gauthier-Villars, Paris, 1892–1899.

74.   Pontryagin, L. S.: "Ordinary Differential Equations," Addison-Wesley Publishing Company, Inc., Reading, Mass., 1962.

75.   Rauch, L. L.: Oscillations of a Third Order Nonlinear Autonomous System, in "Contributions to the Theory of Nonlinear Oscillations," Annals of Mathematics Studies No. 20, Princeton University Press, Princeton, N.J., 1950.

76.   Reddick, H. W., and F. H. Miller: "Advanced Mathematics for Engineers," John Wiley & Sons, Inc., New York, Chapman & Hall, Ltd., London, 1947.

77.   Reuter, G. E. H.: Subharmonics in a Nonlinear System with Unsymmetrical Restoring Force, *Quart. J. Mech. Appl. Math.*, vol. 2, pp. 198–207, 1949.

78.   ———: A Boundedness Theorem for Non-linear Differential Equations of Second Order, *J. Proc. Cambridge Phil. Soc.*, vol. 47, pp. 49–54, 1951.

79.   ———: Boundedness Theorems for Nonlinear Differential Equations of the Second Order II, *J. London Math. Soc.*, vol. 27, pp. 48–58, 1952.

80.   Southwell, R. V.: "Relaxation Methods in Theoretical Physics," Oxford University Press, Fair Lawn, N.J., 1956.

81.   Stoker, J. J.: "Nonlinear Vibrations in Mechanical and Electrical Systems," Interscience Publishers, Inc., New York, 1950.

82.   Stokes, A. A.: Floquet Theory for Functional Differential Equations, *Proc. National Acad. Sci.*, vol. 48, no. 8, pp. 1330–1334, August, 1962.

83.   Struble, Raimond A.: "Nonlinear Differential Equations," McGraw-Hill Book Company, Inc., New York, 1961.

84.   Temple, G.: "The 'P-L-K' Method," lecture notes from David Taylor Model Basin, 1960.

85.   ———: "A Superposition for Ordinary Non-linear Differential Equations," lecture notes from David Taylor Model Basin, 1960.

86.   Urabe, Kojuro: On the Existence of Periodic Solutions for Certain Nonlinear Differential Equations, *Math. Japon.*, vol. 2, pp. 23–26, 1949.

87.   van der Pol, B.: On Oscillation Hysteresis in a Triode Generator with Two Degrees of Freedom, *Phil Mag.*, vol. 6, pp. 700–719, 1922.

88.   van Kampen, E. R.: Remarks on Systems of Ordinary Differential Equations, *Am. J. Math.*, vol. 59, pp. 144–152, 1937.

89. Volosov, V. M.: On the Theory of Nonlinear Differential Equations of Higher Orders with a Small Parameter in the Highest Derivative, *Am. Math. Soc. Translations*, ser. 2, vol. 8, 1958.

90. Von Karman, Theodore: The Engineer Grapples with Nonlinear Problems, *Bull. Am. Math. Soc.*, vol. 46, pp. 615–683, 1940.

91. Wasow, Wolfgang R.: Singular Perturbation Methods for Nonlinear Oscillations, in "Nonlinear Circuit Analysis," pp. 75–97, Polytechnic Press of the Polytechnic Institute of Brooklyn, April, 1953.

92. Weber, Ernst: Complex Convolution Applied to Nonlinear Problems, in "Proceedings of the Symposium on Nonlinear Circuit Analysis," pp. 409–427, Polytechnic Press of the Polytechnic Institute of Brooklyn, April, 1956.

93. Zubov, V. I.: "Methods of A. M. Lyapunov and Their Application," translated from a publication of the Publishing House of Leningrad University by U.S. Atomic Energy Commission, Division of Technical Information, AEC-tr-4439, 1957.

FIVE

INTRODUCTION TO AUTOMATIC CONTROL AND THE PONTRYAGIN PRINCIPLE

5-1 INTRODUCTION

In this and the next chapter we shall develop ideas which have recently flourished in theory and application. The mathemati cal models used are deterministic through differential equation and the calculus of variations, and probabilistic through predic tion theory and related stochastic processes.

Let us describe briefly the physical ideas underlying the theories of control and prediction. An interesting illustration is found in a guided missile launched at time t_0 to intercept an airplane at a later time T. There are a number of trajectorie which the missile can follow; each trajectory leads to a differen angle of impact. The fuel consumption of the rocket and the angle between the direction of thrust and the direction of motion of the vehicle are two examples of quantities which can be con trolled during the transit time of the rocket to obtain, for exam ple, a desired maximum speed at the point of impact, or in genera

276

to optimize some cost function. Let us develop the differential equations for a simplified problem of this kind.

We consider an example of rocket flight under two major simplifying assumptions:[47]

1. The earth is "flat."
2. The atmospheric factors (drag, lift, etc.) are negligible.

The "flat earth" assumption merely means that the acceleration due to gravity is constant, but implies nothing further about the shape of the earth. The approximation is valid for short ranges and altitude changes (compared to the earth's radius).

The force of the rocket in the direction of motion is given by the thrust of the rocket engine, which we denote by $P(t)$ since it varies with the fuel available and, hence, with the time t. We write $u_1(t) = P(t)/P_{max}$ to indicate the dimensionless thrust. From its definition, $u_1(t)$ is subject to the constraint $0 \leq u_1(t) \leq 1$. Let $x_1(t)$ be the horizontal projection of the velocity, and let $x_2(t)$ be its vertical projection. Let $u_2(t)$ be the angle between the direction of thrust and the horizontal. Then $\dot{x}_1(t)$ is the horizontal component of the acceleration, and $\dot{x}_2(t)$ is its vertical component.

Let $m(t)$ denote the variable mass; then for convenience we write $m(t) = m(0)x_3(t)$, where $0 \leq x_3(t) \leq 1$ is a dimensionless factor with $x_3(0) = 1$. Newton's law gives

$$m(t)\dot{x}_1(t) = P(t) \cos u_2(t) = P_{max}u_1(t) \cos u_2(t)$$

If we define $A = P_{max}/m(0)$, our equation becomes

$$\dot{x}_1(t) = \frac{Au_1(t) \cos u_2(t)}{x_3(t)} \tag{5-1}$$

Similarly,
$$\dot{x}_2(t) = \frac{Au_1(t) \sin u_2(t)}{x_3(t)} - g \tag{5-2}$$

where g is the acceleration due to gravity.

Since the rate of change of the mass is proportional to the thrust, i.e., the thrust is proportional to the amount of fuel used (which equals a negative change in mass), we have, in dimensionless form,

$$\dot{x}_3(t) = -\alpha u_1(t) \tag{5-3}$$

where $\alpha = IP_{max}/m(0)$ is a constant, and I, called the *specific impulse*, is a constant characteristic of the rocket and is measured in pounds per second per pound of thrust. Hence, one can write directly $\dot{m}(t) = -IP(t)$.

We also have, for the range $x_4(t)$ and the altitude $x_5(t)$,

$$\dot{x}_4(t) = x_1(t) \qquad \dot{x}_5(t) = x_2(t) \tag{5-4}$$

Note that, in the operation of the rocket, one can control $u_1(t)$ and $u_2(t)$, and hence by an appropriate "optimal choice" one is able, for example, to send the rocket to a maximal distance.   In so-called *rendezvous* problems, it may be desired to select the control variables $u_1(t)$ and $u_2(t)$ such that the rocket, starting from initial conditions $x_i(0)$ at $t = 0$, yields a maximum value to

$$\sum_{i=1}^{5} c_i x_i(T)$$

which may be regarded as a payoff function at the terminal time T.

The c_i are arbitrary constants; all except one may be equated to zero. The exceptional c_i is equated to unity.   For example, to maximize the range, we set $c_i = 0$, $i \neq 4$, and $c_4 = 1$.   It may also be possible to formulate a problem in which more than one c_i is nonzero.

It is frequently essential to determine the stability of a control system. Although this cannot be done in general, we shall examine later a few systems which have been investigated.

Even if stability has been established, there are several difficulties which must be taken into account.   One is that the differential equations apply to ideal conditions and do not include the possibility of environmental disturbances which cannot be simply described because of their randomness and variability.   Aerodynamics, irregularities in the earth's gravitational field, and various other factors change the missile trajectory. What is the mathematical description of this new situation, and how should the control variables be determined?   Once the control variables are determined, how are they actually stored and put into effect?   It is well known to engineers that even the most practically perfect electrical network through which information is transmitted and received is susceptible to spurious signals or to noise arising in the process of converting and amplifying signals.   The problem would be simpler if all conditions were ideal, but since they are not, and since the trajectory *will* be disturbed, there must be a meaningful input which predicts a path for the rocket based on analysis of past information on the effects of disturbances. Thus, for the input we use a mixture of a deterministic control function and a stochastic variable.   Some process of filtering in the electrical network is necessary to minimize the effect of the noise and produce a useful output for guiding the missile.   Of course, the missile will not be in the predicted position, but in a neighborhood tolerated by the prediction technique which considers deviations about the predicted value.

As further interpretation, if the differential equations describing the problem are formulated so as to contain most of the significant physical factors of missile flight, and if they can be solved for the state variables after the determination of the control variables by methods to be discussed later, we have what may be called an *ideal* trajectory for the missile. Even after this effort, the disturbances will perturb the missile path. The prediction mechanism can then compare the present position of the missile with the ideal and predict a position which considers the past disturbances and minimizes the difference between the actual and the ideal future positions. In spite of this approach and its possible ramifications, in general the missile will not be exactly on the target at time T. In fact, depending on the prediction technique and on how the ideal trajectory is determined, the final position of the missile may be quite far from the target. Thus, a more general problem to be considered is to deliver the missile within a tolerated neighborhood of the target, or even to find a predicted trajectory which minimizes the size of this neighborhood, i.e., the difference error at this point.

A part of the input signal to the missile may be information received through an antenna tracking the moving target, which may have its own evasive strategy. Accordingly, this type of problem may be divided as follows: A missile may be "manned" or "unmanned," in modern terminology. In either case the target may be cooperative by sending signals which are used to improve the tracking, or it may be uncooperative. In the latter case noncooperation can be active (the target may use chaff and evasive action to mislead the tracking mechanism of the missile) or passive (the target simply does not indicate its presence). Servomechanisms are generally used to control the missile. In such a mechanism, for example, a combination of mechanical devices and electrical circuits may be employed to convert the input signals to a filtered signal which is then changed to mechanical energy and compared with the input, enabling the missile to assume its correct course. (An example of a device which converts mechanical energy to electrical energy is an electric motor. We shall refer to the use of motors in writing some well-investigated equations of automatic control.) Such usage of the output process by servomechanisms is a *feedback* process. Thus one employs the present behavior of a system to automatically influence its future behavior. At any instant its state is determined by its past history.

Note that the values predicted for a given time may be used to compute the control variables at that time in order to minimize errors. It has been pointed out by rocket-flight experts that the most general problem as it is encountered in practice has yet to be formulated.

A control problem is often described by a differential equation which when reduced to a first-order system, acquires the vector form

$$\dot{x} = f(x(t), u(t), t) \qquad x(0) \equiv x_0 \qquad (5\text{-}5)$$

in the nonautonomous case or

$$\dot{x} = f(x(t), u(t)) \qquad x(0) \equiv x_0 \qquad (5\text{-}6)$$

in the autonomous case, where

$$x(t) = (x_1(t), \ldots, x_n(t))$$

is called the *state-variables* vector and

$$u(t) = (u_1(t), \ldots, u_m(t))$$

is called the *control* or *steering-function* vector. As we shall see, various conditions are imposed on $u(t)$, and sometimes on $x(t)$.

The object of this chapter is to analyze the stability of special cases of such a system and to investigate methods yielding appropriate steering functions, mainly through Pontryagin's maximum principle, which can be derived using the calculus of variations. A functional-analysis approach to the optimum control problem will also be used.

A control system is stable if, starting from any prescribed initial condition, its actual state and its desired state ultimately coincide, i.e., its error becomes zero. In reality this is an ambitious goal, as one usually minimizes the error within a tolerated range.

Thus it is futile to seek $u(t)$ for an unstable system. However, as we have seen in Chap. 4, the problem is not so simple, since for many problems there are regions of stability bordered by regions of instability.

It is usually assumed for a practically conceived system that stability has been affirmed, and one proceeds to determine the steering function $u(t)$. The problem of determining $u(t)$ is known as a *synthesis* problem. It is generally difficult to set up and derive the steering function for any physical problem. Some information is available on linear systems of the form

$$\dot{x} = A(t)x + B(t)u + f(t) \tag{5-7}$$

where x and f are n vectors, u is an m vector, and $A(t)$ and $B(t)$ are $n \times n$ and $n \times m$ matrices, with the constraints

$$|u_i(t)| \leq 1 \qquad i = 1, \ldots, m \tag{5-8}$$

In addition, the class of steering functions is required to be measurable, and A, B, and f are required to be continuous for $t \geq 0$. We shall also discuss nonlinear systems of this type.

Although we shall be concerned mostly with the continuous case, a control problem may also be given in discrete form, in which the system can occupy any one of a finite number of states. Gaffney[33] has given the following illustration: Suppose that there are six positions, or states, that a system can be in, denoted by $s_1, s_2, \ldots, s_6$, with s_6 being the "target" position (in a more complicated example these might be positions of a missile seeking a target, or any of various other interpretations).

At each position there is available a collection of loaded dice—three dice, to be specific. If the system starts (for example) in position s_2, the mechanism for making the first move is to select one of the three dice available at s_2 and throw it; the system is then moved to the position given by the value of the throw. Thus one moves to s_1 with probability p_1, stays at s_2 with probability p_2, etc., where the probabilities depend on the particular choice of a die. The procedure is then repeated at the new position. This process continues until the target s_6 is reached. If the die to be used has been determined in advance for each position, one can compute, on the basis of probabilities, the expected number of moves it will take to reach s_6. The selection of a die at each position is called a *strategy* or *policy*.

The problem of optimal control in this context is to find an optimal policy, i.e., a policy for which, from each of the five possible starting positions, the expected number of moves is less than (or at most equal to) that for any other policy. It is not a priori evident that there is one policy which will simultaneously achieve this minimization for all starting positions. However, in a much more general context one can establish the existence of an optimal policy (given certain mild assumptions) and develop various of its properties. Procedures for computing both the optimal policy and the associated cost function from each starting position are available (in the example the cost function is the expected number of moves).[98]

5-2 STABILITY AND A CLASS OF CONTROL EQUATIONS

Among the most investigated control equations are the Lur'e-Letov systems. According to A. I. Lur'e,[70] feedback control systems are either direct or indirect. In the former case the error of Sec. 5-1 is used to displace the input control vector directly (i.e., control can be applied through solenoids with springs or through the relay action of a galvanometer, both of which are displaced in proportion to the signal). In the indirect case the error signal is used to obtain a rate of displacement of the input control vector. Here a motor controls the system, and its rate of rotation is proportional to the signal. The equations of indirect control are a special case of the direct-control equations. The basic method of constructing the Lyapunov functions for control systems is given by Lur'e and Postnikov.

Consider the direct control system with a single nonlinear element f:

$$\dot{y} = Ay + bf(\sigma) \qquad \sigma = c'y \qquad (5\text{-}9)$$

with $\quad \sigma f(\sigma) > 0$ for $\sigma \neq 0$ $\quad f(0) = 0$ $\quad \int_0^\sigma f(s)\,ds \to \infty$ as $|\sigma| \to \infty$

(By the theorem of Barbashin-Krassovskii, the last condition is needed below to define a Lyapunov function for complete stability.) Here y is an n vector, A is an $n \times n$ constant-coefficient matrix, b and c are constant vectors with $c'b < 0$, and $f(\sigma)$ is a continuous scalar function, a characteristic of the servomotor mentioned above depending on the feedback control signal σ. The vector $bf(\sigma)$ is the control function. Let the λ_i be the characteristic roots of A with negative real parts, all distinct and none of which is zero, and let T be a matrix such that

$$T^{-1}AT = \begin{pmatrix} \lambda_1 & 0 & \cdots & 0 \\ 0 & \lambda_2 & \cdots & 0 \\ \cdots\cdots\cdots\cdots\cdots \\ 0 & 0 & \cdots & \lambda_n \end{pmatrix} \equiv \Lambda \qquad (5\text{-}10)$$

If we write $z = T^{-1}y$, our system assumes the canonical form

$$\dot{z} = \Lambda z + T^{-1}bf(\sigma) \qquad \sigma = c'Tz \qquad (5\text{-}11)$$

In expanded notation this system, after differentiation of the second equation and substituting $x_i = z_i/s_i$ where s_i is the ith component of $T^{-1}b$, here all assumed nonzero, and with $\alpha_i = -\lambda_i$, becomes

$$\dot{x}_i = -\alpha_i x_i + f(\sigma)$$
$$\dot{\sigma} = \sum_{i=1}^{n} \beta_i x_i - rf(\sigma) \qquad r = -c'b \qquad (5\text{-}12)$$
$$\sigma f(\sigma) > 0 \text{ for } \sigma \neq 0 \qquad f(\sigma) = 0 \text{ for } \sigma = 0$$

where $r > 0$, $\alpha_i > 0$, and the β_i are real constants. The transformation from (5-11) to the canonical form (5-12) is possible if no $\lambda_i = 0$. This is the condition for controllability; otherwise, if $\lambda_i = 0$, the control has no effect on x_i.

The problem before us is to determine the conditions which α_i and β_i must satisfy in order that the origin

$$x_1 = x_2 = \cdots = x_n = \sigma = 0$$

should be asymptotically stable in the large.

Let us assume that α_i and β_i are real and define a Lyapunov function for the foregoing system as follows:[72]

$$V(x_1, \ldots, x_n, \sigma) = \int_0^\sigma f(s)\, ds + F(a_1x_1, \ldots, a_nx_n)$$
$$+ \tfrac{1}{2}(A_1x_1^2 + \cdots + A_nx_n^2) \qquad (5\text{-}13)$$

where both $A_i > 0$ and a_i, $i = 1, \ldots, n$, are arbitrary constants, and where

$$F(x_1, \ldots, x_n) \equiv \sum_{i,j=1}^{n} \frac{1}{\alpha_i + \alpha_j} x_i x_j$$

$$= \sum_{i,j=1}^{n} \left[\int_0^\infty e^{-(\alpha_i + \alpha_j)w} \, dw \right] x_i x_j$$

$$= \int_0^\infty \left(\sum_{i=1}^{n} x_i e^{-\alpha_i w} \right)^2 dw \qquad (5\text{-}14)$$

Since we assume that $\alpha_i \neq \alpha_j$ for $i \neq j$, Eq. (5-14) is zero if and only if all the x_i are zero. Thus F is positive-definite, from which V is also positive-definite. Yacubovic showed that the condition

$$\int_0^\sigma f(s) \, ds \to \infty \qquad \text{as} \qquad |\sigma| \to \infty$$

was not necessary. Lefschetz added the condition to ensure that $V(x,\sigma) \to \infty$ as

$$\sum_{i=1}^{n} x_i^2 + \sigma^2 \to \infty$$

validating the argument for all values of the variables.

We also have

$$-\frac{dV}{dt} = \sum_{i=1}^{n} A_i \alpha_i x_i^2 + 2 \sum_{i,j=1}^{n} \frac{\alpha_i a_i a_j}{\alpha_i + \alpha_j} x_i x_j + rf^2(\sigma)$$

$$- f(\sigma) \sum_{i=1}^{n} x_i \left(A_i + \beta_i + 2a_i \sum_{j=1}^{n} \frac{a_j}{\alpha_i + \alpha_j} \right)$$

$$= \left[\sum_{i=1}^{n} a_i x_i + \sqrt{r} f(\sigma) \right]^2 + \sum_{i=1}^{n} A_i \alpha_i x_i^2$$

$$- f(\sigma) \sum_{i=1}^{n} x_i \left(A_i + \beta_i + 2\sqrt{r} \, a_i + 2a_i \sum_{j=1}^{n} \frac{a_j}{\alpha_i + \alpha_j} \right) \qquad (5\text{-}15)$$

since

$$2 \sum_{i,j=1}^{n} \frac{\alpha_i a_i a_j}{\alpha_i + \alpha_j} x_i x_j = \left(\sum_{i=1}^{n} a_i x_i \right)^2 \qquad (5\text{-}16)$$

Hence, to achieve asymptotic stability, i.e., in order that $\dot{V}$ be negative-definite, it is sufficient that

$$A_i + \beta_i + 2a_i \left(\sqrt{r} + \sum_{j=1}^{n} \frac{a_j}{\alpha_i + \alpha_j} \right) = 0 \qquad (5\text{-}17)$$

which is a quadratic equation in a_i.

The problem reduces to one of finding $A_i > 0$ such that real a_i are obtained from (5-17).

On dividing (5-17) through by α_i and summing over i, the sufficient condition for asymptotic stability becomes

$$\sum_{i=1}^{n} \frac{A_i}{\alpha_i} + \left(\sum_{i=1}^{n} \frac{a_i}{c_i} + 1 \right)^2 = 1 - \sum_{i=1}^{n} \frac{\beta_i}{\alpha_i} \tag{5-18}$$

from which we have

$$1 - \sum_{i=1}^{n} \frac{\beta_i}{\alpha_i} > 0 \tag{5-19}$$

as a sufficient condition for asymptotic stability. If $f(\sigma)$ is analytic such that

$$f(\sigma) = a_k \sigma^k + a_{k+1} \sigma^{k+1} + \cdots$$

where $k \geq 2$, then the condition (5-19) is also necessary. To see this note that, because of the requirement $\sigma f(\sigma) > 0$, k must be odd, and $a_k > 0$. Consider the linear approximation to the original system. It has nonnegative roots α_i and one zero root, and hence, from our study of the critical case with a single zero root, we define

$$x = \sigma + \frac{\beta_1}{\alpha_1} x_1 + \cdots + \frac{\beta_n}{\alpha_n} x_n$$

If we solve for σ and substitute in the system of equations (5-12), we obtain

$$\dot{x}_i = -\alpha_i x_i + f\left(x - \frac{\beta_1}{\alpha_1} x_1 - \cdots - \frac{\beta_n}{\alpha_n} x_n \right)$$

$$\dot{x} = \left(\frac{\beta_1}{\alpha_1} + \cdots + \frac{\beta_n}{\alpha_n} - 1 \right) f\left(x - \frac{\beta_1}{\alpha_1} x - \cdots - \frac{\beta_n}{\alpha_n} x_n \right) \tag{5-20}$$

Since the study of this critical case gives

$$R(x) = \left(\frac{\beta_1}{\alpha_1} + \cdots + \frac{\beta_n}{\alpha_n} - 1 \right) f(x)$$

$$R_i(x) = f(x) \qquad \text{and} \qquad c_i = 0$$

the stability of the system depends on that of $R(x)$, which is stable only if

$$\frac{\beta_1}{\alpha_1} + \cdots + \frac{\beta_n}{\alpha_n} - 1 < 0 \tag{5-21}$$

thus establishing necessity.

We return to the original system (our previous proof was confined to α_i and β_i real),

$$\dot{y} = Ay + bf(\sigma) \qquad \dot{\sigma} = c'y - rf(\sigma) \qquad (5\text{-}22)$$

with $\qquad\qquad \sigma f(\sigma) > 0 \qquad \text{for} \qquad \sigma \neq 0$

and define $\qquad\quad V(x,\sigma) = \int_0^\sigma f(s)\, ds + x'Bx \qquad (5\text{-}23)$

where B is positive-definite.   We have

$$\dot{V} = -x'Cx + f(\sigma)(2b'B + c')x - rf^2(\sigma) \qquad (5\text{-}24)$$

where $C = -(A'B + BA)$, and hence $\dot{V}$ is quadratic in x and in $f(\sigma)$. Lefschetz has shown that $\dot{V}$ is negative-definite for all x and σ (and hence sufficient for complete stability of the origin) if and only if the following "fundamental control" inequality holds:

$$C > 0 \qquad r > \left(Bb + \frac{c}{2}\right)' C^{-1} \left(Bb + \frac{c}{2}\right) \qquad (5\text{-}25)$$

Thus the matrix B must be determined to satisfy these conditions. LaSalle has shown that if V is defined as above, and if $\sigma f(\sigma) > 0$ for $\sigma \neq 0$, then the origin is completely stable.   He also shows that $r + c'A^{-1}b > 0$ follows from the fundamental-control inequality.   This condition arises from the fact that, for complete stability, the origin must be the only solution of

$$Ay + b\xi = 0 \qquad c'y - r\xi = 0 \qquad \xi = f(\sigma)$$

whose coefficient determinant is related to the last inequality.   One refers to the complete stability of the above system, satisfied for $f(\sigma)$ as described here, as *absolute* stability.

Lefschetz[65] has examined the more general form of the control equations proposed by Letov[68] in which we have

$$\ddot{\zeta} + \beta\dot{\zeta} + \gamma\zeta = f(\sigma) \qquad (5\text{-}26)$$

where $\beta > 0$ and $\gamma > 0$ are constants.   On writing $\eta = \dot{\zeta} + \beta\zeta$, the system becomes

$$\dot{\zeta} = -\beta\zeta + \eta \qquad \dot{\eta} = -\gamma\zeta + f(\sigma)$$

The full system is taken in the form

$$\dot{x} = Ax + \zeta b + \dot{\zeta}b^0 \qquad \sigma = c'x - \rho\zeta - \rho^0\dot{\zeta} \qquad (5\text{-}27)$$

which, on introducing η, becomes

$$\begin{aligned} \dot{x} &= Ax + \zeta b_1 + \eta b_2 & \dot{\zeta} &= -\beta\zeta + \eta \\ \dot{\eta} &= -\gamma\zeta + f(\sigma) & \sigma &= c'x - \rho_1\zeta - \rho_2\eta \end{aligned} \qquad (5\text{-}28)$$

Exercise 5-1   Assuming that A is stable, form the matrix of the first-degree terms, and show that its characteristic roots are the same as those of A and of the quadratic equation

$$r^2 + \beta r + \gamma = 0 \tag{5-29}$$

and hence show that it, too, is stable.

Remark: Lefschetz also studied the stability of the generalized multiple feedback system proposed by Letov:

$$\dot{x} = Ax + Gu \qquad \dot{u} = f(v) \qquad v = Hx - Ru \tag{5-30}$$

where R is a diagonal matrix with positive coefficients, f, x, u, v are vectors, A is a stable matrix, and G and H are constant matrices.

Exercise 5-2   Give the condition under which the origin is the only critical point of the system (5-30).

Let $\qquad\qquad\qquad y = \dot{x}$
The system (5-30) becomes

$$\dot{y} = Ay + Gf(v) \qquad \dot{v} = Hy - Rf(v) \tag{5-31}$$

Exercise 5-3   Give conditions for the global stability of the system

$$
\begin{aligned}
y_1 &= a_{11}y_1 + a_{12}x \\
y_2 &= a_{21}y_1 + a_{22}x \\
\dot{x} &= f(c_1y_1 + c_2y_2)
\end{aligned}
\tag{5-32}
$$

$$a_{11} < 0 \qquad a_{21} > 0 \qquad f(0) = 0 \qquad \sigma f(\sigma) > 0$$

using

$$V = \int_0^\sigma f(s)\, ds + b_1y_1{}^2 + b_2y_1y_2 + b_3y_2{}^2 \tag{5-33}$$

5-3   PONTRYAGIN'S MAXIMUM PRINCIPLE

Pontryagin's principle, which gives the conditions necessary for a stationary value of an integral arising in control theory, subject to restrictive assumptions, is a special case of a general variational approach. In recent papers, Berkovitz has unified a variational approach to the optimum control principle used by a number of investigators. He has enlarged the principle to include constraints on the state and control variables. Pontryagin initially developed the principle for a free system without such constraints and without consideration of corner and transversality conditions and other needed refinements which are included in the calculus of variations.

There are two general classes of problems to consider. *Time-optimal* problems are those in which one seeks the trajectory which yields the minimum time of transit between two given points. A functional-analysis approach to the time-optimal problem is given in the next section. The brachistochron problem of the calculus of variations shows the difficulty of treating time-optimal problems. The other class of problems is concerned with maximizing a function describing returns for effort or with minimizing a cost function subject to constraints, including, of course, the differential equations.

Such a function can often be expressed as an integral, and then variational methods can be used to solve the problem. Sometimes such an integral can be replaced by a first-order differential equation whose left side consists of the time derivative of a new variable x_{n+1} and whose right side consists of the integrand. The problem then becomes one of finding $u(t)$ which maximizes x_{n+1} subject to the system of differential equations.

We first give the Pontryagin principle in the restricted sense and without the use of constraints. We shall then apply it to a simple example with a constraint on u and derive it for the rocket example of the introduction, using a theorem from the calculus of variations. Since in the rocket example we have the constraint $0 \leq u_1(t) \leq 1$, we shall show how to treat this problem in spite of the fact that it is not contained in the statement of the principle.

Finally, we shall present a summary of a characterization approach to the problem, framed in the broadest terminology of the calculus of variations and taking into consideration constraints imposed on both the free and the control variables. Some existence theorems for optimal control may be found in Ref. 64a.

Consider the nonautonomous system

$$\dot{x}_i = f_i(x,u,t) \qquad i = 1, \ldots, n$$
$$x_i(0) \equiv x^0 \tag{5-34}$$

where the f_i are continuous and bounded and possess continuous first-order partial derivatives, and $u_1, \ldots, u_m$ are piecewise-continuous. The vector u is referred to as an *admissible control vector*.

Define the hamiltonian

$$H[x(t),\lambda(t),u(t)] = \sum_{i=1}^{n} \lambda_i(t)f_i(x,u,t) \tag{5-35}$$

and let

$$M(x,\lambda) = \max_u H(x,\lambda,u) \tag{5-36}$$

Consider the hamiltonian system (a set of differential equations associated with H)

$$\dot{x} = \frac{\partial H}{\partial \lambda} \qquad \lambda = -\frac{\partial H}{\partial x} \qquad \frac{\partial H}{\partial u} = 0 \tag{5-37}$$

Here $\partial H/\partial u = 0$ gives the necessary condition for H to be maximum with respect to u, provided that the critical u is interior to the set of admissible u's.

If $\bar{x}(t)$ is a solution of (5-34) corresponding to $\bar{u}(t)$, a choice of $u(t)$ which maximizes

$$\sum_{i=1}^{n} c_i x_i(T)$$

then Pontryagin's maximum principle states that there is a $\bar{\lambda}(t)$ which, together with $\bar{x}(t)$, solves the hamiltonian system, and

$$H(x,\lambda,\bar{u}) \geq H(x,\lambda,u)$$

In the autonomous case,

$$H[\bar{x}(t),\bar{\lambda}(t),\bar{u}(t)]$$

is a constant, as can be verified by differentiating H with respect to t. Our purpose is to derive this principle through variational methods. Now we apply the principle to an example.

Example 1:   Assume that an optimization problem has been resolved (by introducing an additional variable) into a first-order linear system of the form

$$\dot{x}_1 = x_2 \qquad \dot{x}_2 = -ax_2 + u(t)$$

subject to $|u(t)| \leq 1$.   Then

$$H = \lambda_1 x_2 + \lambda_2(-ax_2 + u) \qquad \lambda_1 = 0 \qquad \lambda_2 = -\lambda_1 + a\lambda_2$$

from which $\qquad\qquad \lambda_1 = b \qquad \lambda_2 = ce^{at} + \dfrac{b}{a}$

where b and c are determined by the condition $x_i(T)$, $i = 1, 2$, which is given.   Clearly the value of u which maximizes H is either $+1$ or -1, depending on whether the sign of λ_2 is positive or negative.   Since u is a boundary point $\partial H/\partial u = 0$ is not true.   Since $ce^{at} + b$ changes sign at most once, the values of t are divided into two intervals.   Thus the optimum control $\bar{u}(t)$ assumes the value $+1$ in one time interval and -1 in the other, an illustration of the "bang-bang" principle.

Exercise 5-4   Verify for this autonomous system that H is constant and that

$$H(x,\lambda,\bar{u}) \geq H(x,\lambda,u)$$

Example 2:   Before we study the rocket example, we shall derive the hamiltonian system for an integral subject to constraints—of which our problem is a special case.   Note that one may write

$$X = \sum_{i=1}^{n} c_i x_i(T) = \int_0^T \sum_{i=1}^{n} c_i \dot{x}_i(t) \, dt + \sum_{i=1}^{n} c_i x_i(0) \qquad (5\text{-}38)$$

Our problem then is to find $u(t)$ such that on substitution in the differential equations, the $\dot{x}_i = f_i$ yield a set of $x_i(t)$ for which the integral in (5-38) is maximum.   The conditions necessary for a stationary value of the integral, subject to the constraints, are obtained from the general version of the following well-known theorem from the calculus of variations and from a remark on the behavior of the variation at T.

Theorem   If $\bar{x}(t)$, $\bar{y}(t)$, $\bar{z}(t)$ is a relative minimum of

$$\int_{t_0}^{t_1} f(t,x,y,z,\dot{x},\dot{y},\dot{z}) \, dt$$

subject to $g(t,x,y,z) = 0$

and if $\dfrac{\partial}{\partial t} g(t,\bar{x},\bar{y},\bar{z}) \neq 0$ $t_0 \leq t \leq t_1$

then there exists a continuous function $\lambda(t)$ in $t_0 \leq t \leq t_1$ such that, if $F = f + \lambda(t)g$, then

$$F_x - \frac{d}{dt} F_{\dot{x}} = 0 \qquad F_y - \frac{d}{dt} F_{\dot{y}} = 0 \qquad F_z - \frac{d}{dt} F_{\dot{z}} = 0$$

where, after the operations indicated in the last set of equations are performed, the argument of F is $(t,\bar{x},\bar{y},\bar{z},\dot{\bar{x}},\dot{\bar{y}},\dot{\bar{z}})$ [Ref. 86].

If we multiply the constraints $\dot{x}_i - f_i$ which are equal to zero by

$$\lambda_i(t) \qquad i = 1, \ldots, n$$

and add to (5-38), we obtain, using $x_i(0) \equiv x_i{}^0$,

$$X = \int_0^T \left[\sum_{i=1}^{n} c_i \dot{x}_i + \sum_{i=1}^{n} \lambda_i(t)(\dot{x}_i - f_i) \right] dt + \sum_{i=1}^{n} c_i x_i{}^0 \qquad (5\text{-}39)$$

If we denote the quantity in brackets by F, then our independent variable is t, and the unknown functions are the x_j and the u_k.   Note that the above theorem assumes that the functions x, y, z are fixed at the end points.   However, if they are arbitrary at one end point (as here), then there are additional conditions to those given above.

Using the above theorem, the necessary conditions for a stationary point give

$$\frac{d}{dt} \frac{\partial F}{\partial \dot{x}_j} - \frac{\partial F}{\partial x_j} = 0 \qquad \frac{d}{dt} \frac{\partial F}{\partial \dot{u}_k} - \frac{\partial F}{\partial u_k} = 0 \qquad (5\text{-}40)$$

However, $\dot{u}_k \equiv du_k/dt$ does not appear in F, and $\dot{x}_i$ does not occur in f_i. Now

$$\frac{d}{dt}\frac{\partial F}{\partial \dot{x}_j} = \frac{d}{dt}[c_j + \lambda_j(t)] = \dot{\lambda}_j$$

and

$$\frac{\partial F}{\partial x_j} = -\sum_{i=1}^{n} \lambda_i(t)\frac{\partial f_i}{\partial x_j}$$

Therefore the first set of equations in (5-40) gives

$$\dot{\lambda}_j = -\sum_{i=1}^{n} \lambda_i(t)\frac{\partial f_i}{\partial x_j} \qquad j = 1, \ldots, n \tag{5-41}$$

This system of differential equations requires a set of initial (or final) conditions to yield a unique solution. Similarly the second set of equations in (5-40) gives

$$\sum_{i=1}^{n} \lambda_i(t)\frac{\partial f_i}{\partial u_k} = 0 \qquad k = 1, \ldots, m \tag{5-42}$$

Exercise 5-5 Show the equivalence of the necessary conditions (5-41) and (5-42) and the hamiltonian system.

The reader will recall that in deriving Euler's equation in Chap. 4 we integrated by parts and equated to zero terms of the form $(\partial F/\partial \dot{x})\eta(t)\Big|_0^T$ because of the end-point assumption on the variation $\delta x \equiv \eta$, that is, $\eta(0) = \eta(T) = 0$. In our problem $\eta(0) = 0$, since the x_i are prescribed at the origin, but $\eta(T)$ is arbitrary, since the problem is to determine $x_i(T)$. Thus there is a factor of the form

$$\sum_{i=1}^{n} [c_i + \lambda_i(t)]\,\delta x_i(t)\Big|_0^T \tag{5-43}$$

In order for this expression to vanish [we have $\delta x_i(0) = 0$, with $\delta x_i(T)$ arbitrary], we must have

$$\lambda_i(T) = -c_i \tag{5-44}$$

which is our third set of conditions for a stationary value.

If we consider the hamiltonian function

$$H(x,\lambda,u) = \sum_{i=1}^{5} \lambda_i f_i(x,u) \tag{5-45}$$

then $\dot{x} = \partial H / \partial \lambda$ is the original system of differential equations, i.e., (5-1) to (5-4); $\lambda = -\partial H / \partial x$ gives

$$\dot{\lambda}_1 = -\lambda_4 \qquad \dot{\lambda}_2 = -\lambda_5 \qquad \dot{\lambda}_3 = \frac{A u_1}{x_3{}^2} (\lambda_1 \cos u_2 + \lambda_2 \sin u_2)$$

$$\dot{\lambda}_4 = 0 \qquad \dot{\lambda}_5 = 0 \qquad (5\text{-}46)$$

and, since f_4 and f_5 contain neither u_1 nor u_2, for $\partial H / \partial u = 0$ it suffices to consider, instead of H,

$$H^*(x,\lambda,u) = u_1 \left[\frac{(\lambda_1{}^2 + \lambda_2{}^2)^{\frac{1}{2}}}{x_3} \sin (u_2 + \varphi) - \lambda_3 \alpha \right] \qquad (5\text{-}47)$$

where
$$\sin \varphi = \frac{\lambda_1}{(\lambda_1{}^2 + \lambda_2{}^2)^{\frac{1}{2}}} \qquad \cos \varphi = \frac{\lambda_2}{(\lambda_1{}^2 + \lambda_2{}^2)^{\frac{1}{2}}}$$

To minimize this expression, recall that $0 \leq u_1 \leq 1$, and hence first minimize with respect to u_2, substitute its value, and study the minimum for u_1. Thus the absolute minimum is attained at

$$u_1 = \text{sign} \left[\frac{(\lambda_1{}^2 + \lambda_2{}^2)^{\frac{1}{2}}}{x_3} + \alpha \lambda_3 \right] \equiv \text{sign } \theta(\lambda,x)$$

$$u_2 = -\frac{\pi}{2} - \arctan \frac{\lambda_1}{\lambda_2} \qquad (5\text{-}48)$$

where
$$\text{sign } \theta(\lambda,x) = \begin{cases} 1 & \text{if } \theta > 0 \\ 0 & \text{if } \theta < 0 \end{cases}$$

With this choice of optimal values of $u_1(t)$ and $u_2(t)$, we must solve the nonlinear system of differential equations

$$\dot{x}_1 = -\frac{A \text{ sign } \theta}{x_3} \frac{\lambda_2}{(\lambda_1{}^2 + \lambda_2{}^2)^{\frac{1}{2}}}$$

$$\dot{x}_2 = -\frac{A \text{ sign } \theta}{x_3} \frac{\lambda_2}{(\lambda_1{}^2 + \lambda_2{}^2)^{\frac{1}{2}}} - g$$

$$\dot{x}_3 = -\alpha \text{ sign } \theta \qquad (5\text{-}49)$$

$$\dot{x}_4 = x_1$$

$$\dot{x}_5 = x_2$$

$$\dot{\lambda}_1 = -\lambda_4 \qquad \dot{\lambda}_2 = -\lambda_5 \qquad \dot{\lambda}_3 = -\frac{A \text{ sign } \theta}{\lambda_2{}^3} (\lambda_1{}^2 + \lambda_2{}^2)^{\frac{1}{2}}$$

$$\dot{\lambda}_4 = 0 \qquad \dot{\lambda}_5 = 0$$

We are now where we were in the previous chapter; i.e., it is not easy to solve a system of nonlinear differential equations, and we will not pursue the solution of this problem. The additional conditions (5-44) must also be included. Later in this section, we shall characterize the solution of the original system, Eqs. (5-1) to (5-4).

Exercise 5-6 For the case of spatial (rather than planar) motion, show that if $y_1(t)$, $y_2(t)$, $y_3(t)$ are the projections of the velocity on the three coordinate axes $x = y_5$, $y = y_6$, $z = y_7$, if $u_2(t)$ and $u_3(t)$ are the projections of the thrust direction along the tangent and along the trajectory, respectively, and if

$$y_4(t) = \frac{m(t)}{m(0)} \qquad u_1(t) = \frac{P(t)}{P_{\max}}$$

then, as in the planar case, the equations of motion are

$$\dot{y}_1 = \frac{Au_1 \cos u_2 \cos u_3}{y_4} \qquad \dot{y}_2 = \frac{Au_1 \cos u_2 \sin u_2}{y_4}$$

$$\dot{y}_3 = \frac{Au_1 \sin u_2}{y_4} - g \qquad \dot{y}_4 = -\alpha u_1 \qquad \dot{y}_5 = y_1 \qquad \dot{y}_6 = y_2 \qquad \dot{y}_7 = y_3 \tag{5-50}$$

Apply the maximum principle to this problem.   Obtain the two-dimensional problem as a special case.

The General Formulation of Berkovitz[10,11]

The following formulation was first given by Hestenes.[44] A number of optimum control problems can be put in the following form:

Find vectors

$$x(t) = [x_1(t), \ \dots \ , x_n(t)] \qquad y(t) = [y_1(t), \ \dots \ , y_m(t)]$$

[where in a control problem $\dot{y}(t) = u(t)$, $y(t_0) = 0$] which minimize the integral

$$I = F_1(s) + \int_{t_0}^{t_1} F_2(t, x, \dot{y}) \, dt \tag{5-51}$$

where $s = (s_1, \ \dots \ , s_p)$ is a parametric vector describing the set of admissible end points on which both t and x depend (see the discussion of the transversality conditions given below), and which satisfy:

1. The differential-equations constraints

$$\dot{x}_j - f_j(t, x, \dot{y}) = 0 \qquad j = 1, \ \dots \ , n \tag{5-52}$$

or simply $\dot{x} - f(t, x, \dot{y}) = 0$

2. The differential inequalities with r components

$$g_i(t, x, \dot{y}) \geq 0 \qquad i = 1, \ \dots \ , r \qquad \text{or} \qquad G(t, x, \dot{y}) \geq 0† \tag{5-53}$$

3. The constraints $\theta(t, x) \geq 0$ in which no component of the control vector appears

4. The end conditions

$$x(t_0) = x^0 \qquad y(t_0) \equiv y^0 = 0$$
$$\text{and} \qquad t = t_1(s) \qquad x^1 \equiv x(t_1) = x^1(s) \qquad y^1 \equiv y(t_1) \tag{5-54}$$

† If $r > m$, then at any point (t, x, u) at most m components of G are allowed to vanish, and the matrix $(\partial g_i / \partial u_j)$ for these vanishing inequalities must have maximum rank.

Other end conditions will be added after 2 and 3 are reduced to equivalent equality systems.

The inequality constraints 2 are reduced to equations by writing

$$g_i(t,x,\dot{y}) - (\dot{z}_i)^2 = 0 \qquad i = 1, \ldots, r \qquad (5\text{-}55)$$

with $\qquad z_i(t_0) \equiv z^0 = 0 \qquad z_i(t_1) \equiv z_i{}^1$

Constraints 3 cannot be simply adjoined to $G \geq 0$, for the resulting set of inequalities does not satisfy the maximum-rank condition imposed on the matrix $(\partial g_i / \partial u_j)$ for the g_i which vanish at a point (t,x,u) at which $\theta(t,x) = 0$.

If we define

$$\gamma(t,x,\eta) = \begin{cases} \eta^4 - \theta(t,x) & \text{for } \eta \geq 0 \\ -\theta(t,x) & \text{for } \eta < 0 \end{cases} \qquad (5\text{-}56)$$

then γ is twice continuously differentiable in (t,x,η). The constraints $\theta \geq 0$ are reduced to the equations

$$\theta_t + \theta_x \dot{x} - \gamma_\eta \dot{\eta} = 0 \qquad (5\text{-}57)$$

and adjoined to the above equations with the additional end conditions

$$\gamma(t_0, x^0, \eta^0) = 0 \qquad \gamma(t_1, x^1, \eta^1) = 0$$

where $\qquad \eta(t_0) \equiv \eta^0 \qquad \eta(t_1) = \eta^1$

Without loss of generality we assume that $F_1(s) \equiv 0$ and form the following function, analogous to the lagrangian function of Chap. 3:

$$F[t,x,y,z,\dot{x},\dot{y},\dot{z},\lambda_0,\lambda(t),\mu(t),\nu(t)] = \lambda_0 F_2 + \lambda(t)(f - \dot{x})$$
$$+ \mu(t)(G - \dot{z}^2) + \nu(t)(\theta_t + \theta_t \dot{x} - \gamma_\eta \dot{\eta}) \qquad (5\text{-}58)$$

where $\lambda_0 \geq 0$ is a constant, $\lambda(t)$ is an n vector, $\mu(t)$ is an r vector, and $\nu(t)$ is a q vector defined and continuous on $t_0 < t < t_1$, and where

$$[\lambda_0, \lambda(t), \mu(t), \nu(t)]$$

is never zero except possibly at values of t corresponding to corners of the optimal solution. Note that F_2 in (5-51) does not contain $\dot{x}$, since $\dot{x}$ may be replaced by f from (5-52), which itself contains no $\dot{x}$. The usual procedure for deriving necessary conditions for solving the equality problem is to obtain for F:

1. Euler's equation (as in the last chapter). This gives

$$\frac{d}{dt} F_{\dot{x}} = F_x \qquad \frac{d}{dt} F_{\dot{y}} = F_y \qquad \frac{d}{dt} F_{\dot{z}} = F_z \qquad \frac{d}{dt} F_{\dot{\eta}} = F_\eta \qquad (5\text{-}59)$$

Note that in control problems $\dot{y}$ is the control vector, and hence y does not occur in F; thus $F_y = 0$, from which $F_{\dot{y}} = \text{const.}$ Similarly $F_{\dot{z}} = \text{const.}$ We shall see from the transversality condition and from the corner conditions to be given next that, for example, $F_{\dot{y}}$, $F_{\dot{z}}$ are continuous and equal to zero at the terminal point t_1, and hence $F_{\dot{y}} = 0$ and $F_{\dot{z}} = 0$ everywhere. Every piecewise continuously differentiable solution of the Euler equation is called an *extremal curve* of the variational problem. Note that we require only piecewise smoothness. This prevents us from simplifying Euler's equation to the form in which the derivatives occur, because they may not exist. Thus the Euler condition requires that every arc of the extremal curve on which the first derivatives have no discontinuities be a solution of Euler's equation. The corner condition to be given later answers the question of what happens at possible points of discontinuity of some of the derivatives of the variables. The corner condition, by ensuring continuity, forces $F_{\dot{z}}$ to zero everywhere because it is zero at the terminal point.

2. The transversality conditions. These conditions arise when the initial and terminal points are themselves variable. In simple variational problems, each point may be anywhere on a prescribed curve, and hence we must search for the optimum from among all curves joining the two end points, each of which assumes an arbitrary position on its curve. Each of the two end-point curves may be described in parametric form. In a higher-dimensional problem such as we have in control theory, instead of a curve describing a point, we have a manifold on which the end point can vary. A number of conditions on the variables may be prescribed at the end point; then the number of parameters $(s_1, \ldots, s_p)$ describing the manifold is equal to the number of variables minus the number of independent conditions. In a number of control problems the initial point is completely fixed; however there is a manifold associated with the terminal point. If the initial point has any arbitrary components, then conditions similar to those we now give for the terminal point must be included. Now we distinguish between (1) the parametric vector s describing the manifold and associated with the terminal values of t and x and (2) the parametric vectors v associated with y and w associated with z. Naturally all three vectors s, v, and w taken together describe the manifold. However we shall soon see that v and w will pose no problem, as the conditions under which they arise simplify considerably in the control problem. If we define $\nabla_s = (\partial/\partial s_1, \ldots, \partial/\partial s_p)$, then $\nabla_s t_1$ is obtained by applying ∇_s to t and evaluating the result at the terminal point t_1; similarly for $\nabla_s x^1$ and $\nabla_s \eta^1$. The matrices $\nabla_v y$ and $\nabla_w z$ are similarly defined. We associate s with η because η occurs in constraints involving x and t only.

The transversality conditions are

$$(F - \dot{x}F_{\dot{x}} - \dot{y}F_{\dot{y}} - \dot{z}F_{\dot{z}})\nabla_s t_1 + F_{\dot{x}}\nabla_s x^1 + F_{\dot{\eta}}\nabla_s \eta' = 0$$
$$F_{\dot{y}}\nabla_v y^1 = 0 \qquad F_{\dot{z}}\nabla_w z^1 = 0 \tag{5-60}$$

We have not specified the number of components in v and w, as we shall not make explicit use of them. Now the vector y is the integral of the control vector $\dot{y}$. Since no condition on y on the terminal manifold is given in the control problems that we are considering, $F_{\dot{y}}\nabla_v y^1 = 0$ implies that $\nabla_v y^1$ is the identity matrix, and hence that $F_{\dot{y}} = 0$ (see Refs. 10 and 11). By a similar argument we have $F_{\dot{z}} = 0$ from the third condition.

Since F_2 can be expressed as a function of t, x, and $\dot{y}$ only, and since $\dot{x} = f$, the first condition in (5-60) becomes

$$(\lambda_0 F_2 + \lambda f)\nabla_s t_1 - \lambda\nabla_s x^1 + \nu(\theta_t\nabla_s t_1 + \theta_x\nabla_s x^1 - \gamma_\eta\nabla_s\eta^1) = 0 \quad (5\text{-}60a)$$

But γ is zero at the terminal point, and hence this condition simplifies to

$$(\lambda_0 F_2 + \lambda f)\nabla_s t_1 - \lambda\nabla_s x^1 = 0$$

Exercise 5-7 Obtain the transversality condition of the variational integral

$$\int_{\alpha_0}^{\alpha_1} f(x,y,y')\,dx \qquad \text{where} \qquad y' = \frac{dy}{dx}$$

with α_0 fixed and α_1 variable on a curve $y = \varphi(x)$ as a special case of condition (5-60a). Show that it is given by

$$f[\alpha_1,y(\alpha_1),y'(\alpha_1)] + [\varphi'(\alpha_1) - y'(\alpha_1)]f_{y'}[\alpha_1,y(\alpha_1),y'(\alpha_1)] = 0$$

3. The **Weierstrass-Erdmann corner conditions**, which give the requirement for a solution at points of discontinuity (the corners of the extremal curve). Because the first derivatives of F, for example, $F_{\dot{x}}$ and $F_{\dot{y}}$, are required to be continuous, the conditions for an extremal curve require that

$$F_{\dot{x}}, \qquad F_{\dot{y}}, \qquad F_{\dot{z}}, \qquad \text{and} \qquad (F - \dot{x}F_{\dot{x}} - \dot{y}F_{\dot{y}} - \dot{z}F_{\dot{z}} - \dot{\eta}F_{\dot{\eta}}) \quad (5\text{-}61)$$

have well-defined one-sided limits that are equal. Since F_2 can be expressed as a function of t, x, and $\dot{y}$, (5-60a) simplifies to

$$\lambda_0 F_2 + \lambda f + \nu\theta_t \qquad (5\text{-}61a)$$

because $F_{\dot{y}} = F_{\dot{z}} = 0$ [use the equality constraints (5-52), (5-55), and (5-57)]. Note that along an admissible extremal curve (x,y,z), the functions F_x, F_y, F_z are piecewise-continuous, and since $F_{\dot{x}}$, $F_{\dot{y}}$, $F_{\dot{z}}$ are integrals of these piecewise functions, they are themselves continuous.

Exercise 5-8 Obtain as a special case, the following condition for the function $f(x,y,y')$ at a point of discontinuity:

$$f_{y'}[\xi, y(\xi), y'(\xi - 0)] = f_{y'}[\xi, y(\xi), y'(\xi + 0)]$$

4. The Weierstrass necessary condition for a minimum, i.e.,

$$E \equiv \{F(t,x,y,z,\dot{X},\dot{Y},\dot{Z}) - F(t,x,y,z,\dot{x},\dot{y},\dot{z}) - (\dot{X} - \dot{x})F_{\dot{x}}$$
$$- (\dot{Y} - \dot{y})F_{\dot{y}} - (\dot{Z} - \dot{z})F_{\dot{z}}\} \geq 0 \quad (5\text{-}62)$$

where $F_{\dot{x}}$, $F_{\dot{y}}$, and $F_{\dot{z}}$ are evaluated at the solution (x,y,z), and $\dot{X}$, $\dot{Y}$, and $\dot{Z}$ are admissible variables, i.e., they satisfy the equality constraints. From the previous discussion, this condition, with F_2 as a function of t, x, and $\dot{y}$ only simplifies to

$$\lambda_0[F_2(t,x,\dot{Y}) - F_2(t,x,\dot{y})] + \lambda(\dot{X} - \dot{x}) \geq 0 \quad (5\text{-}62a)$$

There is another interesting condition due to Clebsch[†] which is a consequence of the Weierstrass condition.

We now apply the foregoing theory to prove that for the rocket problem optimal control is obtained by applying minimum or maximum thrust. This type of control, referred to as *bang-bang* control, is discussed on several occasions in this chapter.

Consider the following specialization of the rocket problem: Let it be desired to deliver a payload $x_3(T) \equiv x_3^1$ at time $t = T$ (where T is not prescribed) to an altitude $x_5(T) \equiv x_5^1$ such that the velocity vector forms a prescribed angle with the horizontal [that is, $x_2(T)/x_1(T) = k \equiv \tan \beta$, $k \neq 0$] and the range $x_4(T)$ is maximum. Thus we know that $c_4 = 1$. We shall see that $\lambda_4(T) = -1$. Since $\lambda_4 = 0$, it follows that $\lambda_4(t) = -1$, and hence $\lambda_1(t) = 1$. Since $x_5(T)$ is given, we know that $x_5(T)$ is not being maximized at T; $c_5 = 0$, from which $\lambda_5(T) = 0$; hence $\lambda_5(t) = 0$, from which it follows that $\lambda_2(t)$ is constant. First we reduce the constraint $0 \leq u_1(t) \leq 1$ to an equality of the form

$$u_1(1 - u_1) - \dot{z}^2 = 0 \quad (5\text{-}63a)$$

In general a condition of the form $a \leq u_1(t) \leq b$ reduces to

$$(b - u_1)(u_1 - a) - \dot{z}^2 = 0$$

In (5-58) we put $\lambda_0 = 1$ and $F = \dot{x}_4(t) = f_4$, and we have, applying the conditions for a minimum,

$$F = -\dot{x}_4(t) + \lambda(t)(f - \dot{x}) + \mu_1(t)[u_1(1 - u_1) - \dot{z}^2]$$

[†] The necessary condition of Clebsch states that along the arc in the (t,x,y,z,η) space corresponding to an optimal control vector, $\mu \leq 0$ and

$$e[(\lambda_0 F_2 + \lambda f + \mu G)_{\dot{y}\dot{y}}]e \geq 0$$

for all m-dimensional vectors e which are interior to $\theta(t,x) \geq 0$ and satisfy $\tilde{G}_{\dot{y}}e = 0$. On the boundary $\theta(t_1,x) = 0$, the condition must hold for all e satisfying $\tilde{G}_{\dot{y}}e = 0$, $\theta_x f_{\dot{x}}e = 0$. Here $\tilde{G}$ corresponds to those components of G which vanish at the point.

The Euler conditions give, in addition to (5-46), two equations of the form $F_{\dot y} = $ const, where $\dot y = (u_1, u_2)$, and from the transversality condition we showed that $F_{\dot y} = 0$. Similarly we have one condition from $F_{\dot z} = 0$. These three conditions are (with $\alpha = 1$):

$$\frac{A}{x_3} (\lambda_1 \cos u_2 + \lambda_2 \sin u_2) - \lambda_3 - \mu_1(1 - 2u_1) = 0$$

$$\frac{A u_1}{x_3} (\lambda_1 \sin u_2 - \lambda_2 \cos u_2) = 0 \qquad (5\text{-}63b)$$

$$\mu_1 \dot z = 0$$

Note from the second equation that $\tan u_2 = \lambda_2/\lambda_1$ (continuous since λ_1 and λ_2 are continuous), which admits two principal solutions $u_2 = \bar u_2$ and $u_2 = \bar u_2 + \pi$, so that discontinuity in u_2 may occur. To decide on which principal value to use, we need the subsequent conditions. We now apply the transversality condition (5-60). First we have, for the parameters of our terminal manifold,

$$x_1^1 = s_1 \quad x_2^1 = ks_1 \quad x_3^1 = \text{const} \quad x_4^1 = s_2 \quad x_5^1 = \text{const} \quad t_1 = T = s_3$$

$$\nabla_s t_1 = \left(\frac{\partial}{\partial s_1}, \frac{\partial}{\partial s_2}, \frac{\partial}{\partial s_3}\right) (s_3) = (0,0,1) \qquad (5\text{-}63c)$$

$$F_{\dot x} = -(\lambda_1, \ldots, \lambda_5) - (0,0,0,1,0)$$

Using (5-60), we obtain

$$\lambda \dot x (0,0,1) + (\lambda_1, \lambda_2, \lambda_3, \lambda_4, \lambda_5) \begin{pmatrix} 1 & 0 & 0 \\ k & 0 & 0 \\ 0 & 0 & 0 \\ 0 & 1 & 0 \\ 0 & 0 & 0 \end{pmatrix} = 0 \qquad (5\text{-}63d)$$

where the last matrix is $\nabla_s x^1$, and all the quantities are evaluated on the manifold. This simplifies to

$$\lambda_1(T) + k\lambda_2(T) = 0$$

$$\lambda_4(T) = -1 \qquad (5\text{-}63e)$$

$$\lambda_1 \dot x_1 + \lambda_2 \dot x_2 + \lambda_3 \dot x_3 + \lambda_4 \dot x_4 + \lambda_5 \dot x_5 = 0$$

evaluated at T.

Exercise 5-9 Verify the above expressions.

The corner condition simply tells us that certain functions previously described should be continuous. It is, essentially, that $\lambda_i, i = 1, \ldots, 5$ and the left side of the last equation in (5-63e) have equal two-sided limits.

To apply the Weierstrass condition, we note that, since

$$F_2 = -\dot{x}_4 = -x_1$$

Eq. (5-62a) reduces to $\lambda(\dot{X} - \dot{x}) \geq 0$. Since $\dot{x}_4 = x_1$ and $\dot{x}_5 = x_2$, we have $\dot{X}_4 - \dot{x}_4 = 0$, and $\dot{X}_5 - \dot{x}_5 = 0$. However, Eqs. (5-1) to (5-3) involve $\dot{y}$, and hence the expressions for $\dot{X}$ involve $\dot{Y}$. The expressions $\lambda_2 g$ cancel, and we have

$$\frac{\lambda_1 A}{x_3} \dot{Y}_1 \cos \dot{Y}_2 + \frac{\lambda_2 A \dot{Y}_1}{x_3} \sin \dot{Y}_2 - \lambda_3 \dot{Y}_1$$

$$\geq \frac{\lambda_1 A}{x_3} \dot{y}_1 \cos \dot{y}_1 + \frac{\lambda_2 A}{x_3} \dot{y}_1 \sin \dot{y}_2 - \lambda_3 \dot{y}_1$$

which we write in the form

$$\dot{Y}_1 \kappa^* - \dot{y}_1 \kappa \geq 0 \qquad\qquad (5\text{-}63f)$$

We now follow Leitmann's proof[66] of the bang-bang condition for our problem. Inequality (5-63f) must be satisfied for all admissible variations not all zero. If we put $\dot{y}_1 = \dot{Y}_1$ and $\dot{y}_2 \neq \dot{Y}_2$, then the condition becomes $\kappa^* - \kappa \geq 0$, which is satisfied by choosing $\dot{y}_2 = u_2$ to minimize κ. This is the principal value for which $0 \geq \kappa$. On the other hand, if $\dot{y}_1 \neq \dot{Y}_1$ and $\dot{y}_2 = \dot{Y}_2$, then (5-63f) becomes $(\dot{Y}_1 - \dot{y}_1)\kappa \geq 0$. Now when $\dot{y}_1 = u_1 = 0$, we have $\dot{Y}_1 \kappa \geq 0$, and when $\dot{y}_1 = u_1 = 1$, we have $(\dot{Y}_1 - 1)\kappa \geq 0$; since $0 < \dot{Y} < 1$, we have

$$\kappa \geq 0 \qquad \text{when} \qquad u_1 = 0$$
$$\kappa \leq 0 \qquad \text{when} \qquad u_1 = 1$$

Remark: Note from the Euler condition $\mu_1 \dot{z} = 0$ that $\mu_1 = 0$ and $\dot{z} \neq 0$ (that is, $0 < u_1 < 1$) or $\mu_1 \neq 0$, $\dot{z} = 0$ (that is, $v_1 = 0$ or $u_1 = 1$) or $\mu_1 = 0$, $\dot{z} = 0$ (that is, $u_1 = 0$ or $u_1 = 1$), and hence the extremal arc consists in subarcs of minimum, maximum, and intermediate thrust.

We now show that regardless of the end-point specification either the solution is nonunique or, if it is unique, then it is also bang-bang; i.e., the thrust satisfies $u_1 = 0$ or $u_1 = 1$, but never $0 < u_1 < 1$.

Let us first differentiate κ and use Eq. (5-46) to obtain

$$\dot{\kappa} = -\frac{A}{x_3} (\lambda_4 \cos u_2 + \lambda_5 \sin u_2)$$

Note from the Euler equations that $\lambda_4 = \text{const}$, $\lambda_5 = \text{const}$. Suppose that λ_4 and λ_5 are not both zero. If $\dot{\kappa} = 0$, then $\cos u_2 = -\lambda_4/\lambda_5$. But since we also have $\cos u_2 = \lambda_2/\lambda_1$, where λ_1 and λ_2 are functions of t, it follows that there is at most one instant of time when $\kappa = 0$. (If both λ_1 and λ_2 are zero, and if $\kappa = 0$, then all λ's and μ_1 are zero.) From this it follows that $\kappa \neq$ const, and hence $\kappa \neq 0$. From the first of Eqs. (5-63b) we have $\mu_1 \neq 0$ and $1 - 2u_1 \neq 0$. But $\mu \neq 0$ implies that we either have $u_1 = 0$ or $u_1 = 1$, and hence only subarcs of minimum or maximum thrust can occur. If $\dot{\kappa} \neq 0$, the same argument applies. If, now, $\lambda_4 = \lambda_5 = 0$, then from Euler's equation $\lambda_1 = $ const, $\lambda_2 = $ const, and hence u_2 is constant, from which it follows that $\dot{\kappa} = 0$ and $\kappa = $ const. If $\kappa \neq 0$, we have bang-bang as before. If $\kappa = 0$, any of the three arc conditions based on $\mu_1 \dot{z} = 0$ could arise. In that case the third of Eqs. (5-46) [if we use the fact that $\dot{x}_3 = -u_1(t)$, i.e., put $\alpha = 1$] becomes

$$\lambda_3 + \frac{\dot{x}_3}{x_3{}^2} K = 0$$

and the first of Eqs. (5-63b) becomes

$$\lambda_3 x_3 = K$$

where
$$K = A(\lambda_1 \cos u_2 + \lambda_2 \sin u_2) = \text{const}$$

These two equations are not independent. They are satisfied by an arbitrary $x_3(t)$, and hence $u_1(t)$ must also be arbitrary [except for the fact that $0 \leq u_1(t) \leq 1$]. Thus for $\lambda_4 = \lambda_5 = 0$, the solution is nonunique. According to Ref. 66, one comes to the following conclusions regarding the optimum trajectory of a single-stage rocket in constant-gravity flight in vacuum: It consists of maximum and minimum (coasting) thrust arcs, of which there are at most three. Maximum thrust occurs when $\kappa > 0$, and minimum thrust occurs when $\kappa < 0$, with transition between areas occurring at $\kappa = 0$. Burnout, that is, $x_3 = x_3(T)$, occurs at the junction between a thrust and final coasting arc, or during a final thrust arc.†

Exercise 5-10 Study the nature of the solution for the case of maximum horizontal velocity and hence minimize $F_2 = -x_1(T)$, with $x_3(T)$ given. Do the same for maximum vertical velocity, that is, $F_2 = -x_2(T)$, with $x_3(T)$ given. Finally, for maximum payload take $F_2 = -x_3(T)$, with $x_4(T)$ and $x_5(T)$ given, and repeat.

We derive the Pontryagin principle for the case in which $\theta(x,t) \equiv 0$. If we apply the variational conditions (using this assumption on θ) to the control problem with hamiltonian (recall that F_2 does not involve $\dot{x}$ since

---

† R. Isaacs has achieved similar results within the framework of game theory in a series of Rand Corporation reports: RM-1391 (1954), RM-1399 (1954), RM-1411 (1955), and RM-1486 (1955). These are contained also in his forthcoming book, "Differential Games."

this can be replaced by f)

$$H(t,x,\dot{y},\lambda_0,\lambda) = \lambda_0 F_2 + \lambda f$$

we have $F = H - \lambda \dot{x} + \mu(G - \dot{z}^2)$ $F_x = H_x + \mu G_x$

 $F_{\dot{z}} = -\lambda$ $F_{\dot{y}} = H_{\dot{y}} + \mu G_{\dot{y}}$

Since $F_{\dot{y}} = 0$, we have

$$H_{\dot{y}} + \mu G_{\dot{y}} = 0$$

Also the second and third conditions give

$$\dot{\lambda} = -(H_x + \mu G_x)$$

From $F_{\dot{y}} = 0$, $F_{\dot{z}} = 0$, the first and third conditions, and the fact that $G = \dot{z}^2$, we have

$$F - \dot{x} F_{\dot{x}} - \dot{y} F_{\dot{y}} - \dot{z} F_{\dot{z}} = H$$

which, when combined with the Weierstrass condition, gives the Pontryagin minimum condition:

$$H(t,x,\dot{Y},\lambda_0,\lambda) \geq H(t,x,\dot{y},\lambda_0,\lambda)$$

The transversality condition is

$$H \nabla_s t_1 - \lambda \nabla_s x^1 = 0$$

Berkovitz shows how to reduce the above Pontryagin condition to the optimality principle of Bellman.

Exercise 5-11   Derive the last two conditions.

Exercise 5-12   Apply the Pontryagin principle to the rocket problem to obtain the condition

$$\dot{y}_{1\kappa}{}^* - \dot{y}_{1\kappa} \geq 0$$

Remark: If the constraints are of the form

$$B_i(t,x) \leq u_i \leq A_i(t,x) \qquad i = 1, \ldots, n$$

where both A_i and B_i are continuous and twice-differentiable, then

$$H_{u_i} \begin{cases} \geq 0 & \bar{u}_i = B_i \\ = 0 & B_i < \bar{u}_i < A_i \\ \leq 0 & \bar{u}_i = A_i \end{cases}$$

where $\bar{u}_i$ is the value of the optimal control.

Remark: If we have integral constraints of the form

$$\int_{t_0}^{t_1} g_i(t,x,u) \, dt \leq C_i \qquad i = 1, \ldots, p$$

then they can be transformed by means of new variables $x_{n+1}, \ldots, x_{n+p}$ as follows:

$$\dot{x}_{n+i} = g_i(t,x,u) \qquad x_{n+i}(t_0) = 0 \qquad x_{n+i}(t_1) \leq C_i \qquad i = 1, \ldots, p$$

Berkovitz also derives sufficiency conditions for an absolute minimum by generalizing from known results in the calculus of variations.

Exercise 5-13 For a system linear in the control variables of the form

$$\dot{x}_i = f_i(x,t) + Au \qquad x_i(0) = x_i^0 \qquad i = 1, \ldots, n$$

where $A = (a_{ij})$ is an $n \times m$ matrix, and $u = (u_1, \ldots, u_m)$, with the constraints

$$|u_k| \leq a_k$$

where $a_k > 0$ is constant, show that a necessary condition for a stationary solution is that

$$u_k = a_k \operatorname{sign} \sum_{i=1}^{n} \lambda_i a_{ik} \qquad k = 1, \ldots, m$$

unless the quantity

$$\sum_{i=1}^{n} \lambda_i a_{ik}$$

is zero, in which case (as yet unsolved) the solution of the hamiltonian system is not sensitive to the choice of u.

The foregoing exercise shows that for a system with linear control the Pontryagin principle asserts that a necessary condition for optimality is the use, at each instant, of the maximum available force and then a change of its algebraic sign at the next instant of application. This property characterizes bang-bang or on-off control systems. This theoretical instantaneous switching of the force can be approximately realized in practice by means of relays with intermittent action. The effect is a reduction in the size of the error between the actual and desired trajectories and in its sustained periodic oscillations about the zero value. The problem here is to minimize the amplitude of these oscillations.

LaSalle[59] has examined the linear control system

$$\dot{x} = A(t)x + B(t)u + f(t)$$

where $A(t)$ is $n \times n$, $B(t)$ is $n \times m$, and $f(t)$ is an n vector, under the constraint

$$|u_i(t)| \leq 1$$

which will be treated in a general context in the next section. Let $z(t)$, $t \geq 0$, be the continuous trajectory of a particle moving in phase space. Let $x(0) = x_0$, and suppose it is desired to have $x(T) = z(T)$, $T > 0$, in minimum time. If $x(t)$ is the solution of $\dot{x} = A(t)x$, $x(0) = I$, then

$$x(T) = X(T)x_0 + X(T) \int_0^T X^{-1}(t)B(t)u(t) \, dt + X(T) \int_0^T X^{-1}(t)f(t) \, dt$$

Let

$$w(T) = X^{-1}(T)x(T) - x_0 - \int_0^T X^{-1}(t)f(t) \, dt$$

and

$$Y(t) = X^{-1}(t)B(t)$$

Then

$$w(T) = \int_0^T Y(t)u(t) \, dt$$

We wish to minimize $w(T)$ with respect to $u(t)$ subject to the constraints. The set of all $u(t)$ with $|u_i| = 1$, $i = 1, \ldots, m$, is the set of bang-bang steering functions.

Exercise 5-14   Show that the optimum $u(t)$ is given by the bang-bang solution

$$u(t) = \text{sign} \, [\lambda Y(t)]$$

where λ is a nonzero n vector

This solution for $u(t)$ is sufficient for a minimum time if we have $\lambda Y(t) \neq 0$ on any interval in $[0,T]$ and

$$\lambda[w(t) - w(T)] \geq 0 \qquad \text{for } t < T$$

Systems of the form $\lambda Y(t) \equiv 0$ are called *proper* systems. Every proper system is completely controllable; that is, $u(t)$ can be found for a trajectory between any two given points. In a normal system no component of $\lambda Y(t)$ is identically zero for $\lambda \neq 0$. In this case $u(t) = \text{sign} \, [\lambda Y(t)]$ is unique.

In the bang-bang solution the quantity $\lambda Y(t)$ may be zero for a certain time interval, and hence $u(t)$ is not completely determined by this equation. Thus the optimal steering is not necessarily bang-bang. However, it can be shown that bang-bang steering can be used without loss of optimality.[59]

The most significant general result for a nonlinear system is given by Pontryagin's principle.

5-4 FUNCTIONAL ANALYSIS AND OPTIMUM CONTROL

In this section we study a class of optimum control problems using methods of functional analysis. This procedure provides a path which is in some ways simpler and more direct than the calculus-of-variations approach (see Kulikowski[56]).

Let us suppose that for each interval $0 \leq t \leq T$, and for each set of m control functions $u_1(t), \ldots, u_m(t)$, the "state variables" (functions) $x_1(t), \ldots, x_n(t)$ are uniquely determined. For example, we may be given a system of n differential equations

$$\dot{x}_i = f_i(t,x_1, \ldots, x_n; u_1, \ldots, u_m) \qquad (5\text{-}64)$$

in which the initial values of x_i, and the functions $u_j(t)$, are prescribed. Suppose also that, at time $t = T$, the x_i must coincide with prescribed values z_i which may depend on T:

$$x_i(T) = z_i(T) \qquad i = 1, \ldots, n$$

This is exemplified by the situation in which $x_i(T)$ and $z_i(T)$ denote, respectively, the coordinates of a missile and its target. Finally, suppose that the functions $u_j(t)$ must satisfy

$$a_j \leq u_j(t) \leq b_j \qquad j = 1, 2, \ldots, m \qquad (5\text{-}65)$$

The problem here is to choose the control functions $u_j(t)$ so that T is a minimum. If we write

$$u_j(t) = c_j + d_j v_j(t)$$

with appropriate constants c_j, d_j, the constraints (5-65) become

$$-1 \leq v_j(t) \leq 1 \qquad j = 1, \ldots, m$$

Therefore we may suppose from the outset that our constraints are already in the form

$$|u_j(t)| \leq 1 \qquad j = 1, \ldots, m \qquad (5\text{-}66)$$

To take advantage of the function-space viewpoint, let E denote the real Banach space of m-tuples $(g_1(t), \ldots, g_m(t))$ whose elements belong to $L_1(0,T)$ (see Sec. 1-9). The end point $T > 0$ is fixed here. The norm of a vector $\{g_j(t)\} = g$ in E is

$$\|g\| = \sum_{j=1}^{m} \int_0^T |g_j(t)| \, dt$$

The dual space E^* consists of all m-tuples $v = (v_1(t), \ldots, v_m(t))$ with elements in $L_\infty(0,T)$. We have

$$\|v\| = \max_{1 \leq j \leq m} \max_{0 \leq t \leq T} |v_j(t)| \qquad (5\text{-}67)$$

for every v in E^*. For every g in E and v in E^*, we have

$$(g,v) = \sum_{j=1}^{m} \int_0^T g_j(t) v_j(t) \, dt \qquad (5\text{-}68)$$

Now we can consider the control functions $u_1(t), \ldots, u_m(t)$ as elements of a vector $u = \{u_j(t)\}$ in E^*. The constraints (5-66) can be written

$$\|u\| \leq 1$$

a *The Linear Case*

Let us return to our original problem now, and suppose first that the functions $x_i(t)$ arise from a linear system in the u's; for example, suppose the x_i are the solutions of the system (5-64), which is now assumed to be linear:

$$\dot{x} = A(t)x + B(t)u + h(t)$$

where x, u, h, A, and B are matrices of dimensions, respectively, $n \times 1$, $m \times 1$, $n \times 1$, $n \times n$, and $n \times m$. We may then write

$$x(t) = y(t) + \int_0^t k(t,s)u(s)\,ds \tag{5-69}$$

where the matrices x, y, k, and u are, respectively, of dimensions $n \times 1$, $n \times 1$, $n \times m$, and $m \times 1$. This can be written more explicitly as

$$x_i(t) = y_i(t) + \int_0^t \sum_{j=1}^m k_{ij}(t,s)u_j(s)\,ds \qquad i = 1, \ldots, n \tag{5-70}$$

Exercise 5-15   Let

$$\begin{pmatrix} \dot{x}_1 \\ \dot{x}_2 \end{pmatrix} = \begin{pmatrix} 4 & 3 \\ 3 & -4 \end{pmatrix} \begin{pmatrix} x_1 \\ x_2 \end{pmatrix} + \begin{pmatrix} 0 & 1 & 1 \\ -1 & 0 & 2 \end{pmatrix} \begin{pmatrix} u_1 \\ u_2 \\ u_3 \end{pmatrix} + \begin{pmatrix} 1 \\ t \end{pmatrix}$$

in which the $u_j(t)$ are given. Solve for $x_1(t)$ and $x_2(t)$, assuming $x_1(0) = -1$ and $x_2(0) = 3$, and show that the solution is of the form of (5-69).

From (5-70), we have

$$x_i(T) = y_i(T) + (g_i, u) \tag{5-71}$$

where $g_i(t) = (k_{i1}(T,t), \ldots, k_{im}(T,t))$, and (g_i, u) is defined in (5-68). Since it is required that $x_i(T) = z_i(T)$, we must have

$$(g_i, u) = z_i(T) - y_i(T) \qquad i = 1, \ldots, n \tag{5-72}$$

We are now ready to solve our linear problem. For fixed $T > 0$, we find the u in E^* with minimum norm [see Eq. (5-67)] for which (5-72) holds. We shall discuss this facet of the problem in the next paragraph. If, for each $T > 0$, the smallest norm $\|u\|$ exceeds unity, our problem has no solution. Otherwise, let T_0 be the smallest T for which $\|u\| \leq 1$. Either $T_0 = 0$, in which case $z_i(0) = y_i(0)$ and the control functions do not enter the picture, or else $T_0 > 0$. In the latter case, we must have $\|u\| > 1$ for $T < T_0$, so that $\|u\| = 1$ at T_0. Clearly T_0 is the solution to our original problem.

It only remains, in the linear situation, to find, for each fixed $T > 0$, that u in E^* of minimum norm which satisfies (5-72). For this purpose, set $c_i = z_i(T) - y_i(T)$, and let F be the n-dimensional subspace of E consisting of all linear combinations of $g_1, \ldots, g_n$ (we suppose that the g_i are linearly independent). Let u_0 denote the linear functional on

F that is uniquely determined by the equations

$$(g_i, u_0) = c_i \qquad i = 1, \ldots, n$$

Let M denote the norm of u_0. Then

$$M = \|u_0\| = \max_{\varphi \text{ in } F} \frac{(\varphi, u_0)}{\|\varphi\|} = \max_{\lambda} \frac{\left(\sum_1^n \lambda_i g_i, u_0 \right)}{\left\| \sum_1^n \lambda_i g_i \right\|}$$

or

$$M = \max_{\lambda} \frac{\sum_1^n \lambda_i c_i}{\left\| \sum_1^n \lambda_i g_i \right\|} \tag{5-73}$$

By the Hahn-Banach theorem (see Sec. 1-9), the functional u_0 can be extended to all E; the resulting linear functional u will have the same norm M as u_0. The functional u in E^* provides our solution; any other functional v in E^* satisfying (5-72) must coincide with u on F, and therefore the norm of v cannot be less than the value M given in (5-73).

To describe u more explicitly, let us observe that if $\lambda_1^*, \ldots, \lambda_n^*$ provides the maximum in (5-73), then

$$\begin{aligned}
M \left\| \sum_i \lambda_i^* g_i \right\| &= \sum_i \lambda_i^* c_i \\
&= \sum_j \int_0^T \sum_i \lambda_i^* k_{ij}(T, t) u_j(t) \, dt \\
&\leq M \sum_j \int_0^T \left| \sum_i \lambda_i^* k_{ij}(T, t) \right| dt \\
&= \left\| \sum_i \lambda_i^* g_i \right\| M
\end{aligned}$$

so that we must have equality between the third and fourth quantities. But equality can hold there only if

$$u_j(t) = M \operatorname{sgn} \sum_{i=1}^n \lambda_i^* k_{ij}(T, t) \qquad j = 1, \ldots, m \tag{5-74}$$

at least for those values of t for which the right side of (5-74) is nonzero.

We shall suppose that for each $j = 1, \ldots, m$, the n functions $k_{ij}(T, t)$, $i = 1, \ldots, n$, have the property that, for $\lambda_1, \ldots, \lambda_n$ not all zero, the function

$$\sum_{i=1}^n \lambda_i k_{ij}(T, t)$$

vanishes for only a finite number of t's.   In this case, the $u_j(t)$ are completely determined by (5-74).

We summarize the linear situation thus: For each $T > 0$, let $c = z(T) - y(T)$. Find the $\lambda_1^*, \ldots, \lambda_n^*$ that provide the maximum M in (5-73), in which

$$\left\| \sum_1^n \lambda_i g_i \right\| = \sum_{j=1}^m \int_0^T \left| \sum_{i=1}^n \lambda_i k_{ij}(T,t) \right| dt \qquad (5\text{-}75)$$

This is a problem in the n-dimensional λ space.

Remark: Note that if the c_i are not all zero (if $c_i = 0$ for every i, we need only take $u_j = 0$), then (5-73) is equivalent to

$$M^{-1} = \min_{\Sigma \lambda_i c_i = 1} \left\| \sum_1^n \lambda_i g_i \right\|$$

i.e., we are to minimize (5-75) subject to the constraint $\Sigma \lambda_i c_i = 1$, and we have

$$\frac{\partial}{\partial \lambda_i} \left\| \sum \lambda_i g_i \right\| = \sum_{j=1}^m \int_0^T \left[\operatorname{sgn} \sum_{r=1}^n \lambda_r k_{rj}(t) \right] k_{ij}(t) \, dt$$

The $u_j(t)$ are given by (5-74).   If T_0 is the smallest T for which $\|u\| \leq 1$, then $\|u\| = 1$ [unless $T_0 = 0$, $y(0) = z(0)$], and T_0 is the solution to our minimal-time problem.

Our assumption that $\Sigma \lambda_i k_{ij}(T,t)$ vanishes only for a finite number of t's implies that each $u_j(t)$ given by (5-74) assumes its maximum or its minimum value for all other t's; the optimal u_j are bang-bang functions.

Exercise 5-16   Let

$$\dot{x}_1 = -x_1 + u_1 \qquad \dot{x}_2 = -2x_2 + u_1 + u_2$$

where $x_1(0) = x_2(0) = 0$.   Suppose we require

$$x_1(5) = 2 \qquad x_2(5) = -1 \qquad (5\text{-}76)$$

Find the smallest M for which

$$\max_{j=1,2} \ \max_{0 \leq t \leq 5} |u_j(t)| \leq M$$

and for which (5-76) holds, and find the corresponding $u_j(t)$.   *Hint*: According to the results in the last section, we must find λ_1, λ_2 such that

$$2\lambda_1 - \lambda_2 = 1$$

and such that

$$J = \int_0^5 |\lambda_1 e^{-(5-t)} + \lambda_2 e^{-2(5-t)}| \, dt + \int_0^5 |\lambda_2 e^{-2(5-t)}| \, dt$$

is a minimum.   Then

$$u_1(t) = M \text{ sgn } [\lambda_1 e^{-(5-t)} + \lambda_2 e^{-2(5-t)}]$$
$$u_2(t) = M \text{ sgn } [\lambda_2 e^{-2(5-t)}]$$

(5-77)

To minimize J analytically, even in this simple problem, is not easy.   However, we see that $u_2(t)$ is a constant; either $u_2(t) = M$ or $u_2(t) = -M$ for every t.   We require

$$x_1(5) = \int_0^5 e^{-(5-t)} u_1(t) \, dt = 2$$
$$x_2(5) = \int_0^5 e^{-2(5-t)} [u_1(t) + u_2(t)] \, dt = -1$$

(5-78)

From the first of Eqs. (5-77), we see that if $0 < a < 5$, we can choose λ_1 (and $\lambda_2 = 2\lambda_1 - 1$) such that

$$\lambda_1 e^{-(5-a)} + (2\lambda_1 - 1)e^{-2(5-a)} = 0$$

Complete the discussion, and show that

$$\begin{aligned} u_1(t) &= +M & \text{on } (0,a) \\ &= -M & \text{on } (a,5) \\ u_2(t) &= -M & \text{on } (0,5) \end{aligned}$$

Calculate a and M using (5-78).

Exercise 5-17   Discuss the case

$$\ddot{x} + x = u \qquad |u(t)| \leq 1$$

(5-79)

where $x(0) = \dot{x}(0) = 0$.   Here we require

$$x(T) = 8$$

(5-80)

Find the smallest T for which (5-79) and (5-80) hold, and find the corresponding $u(t)$.

b   *The Nonlinear Case*

The previous results could have been given for u_j in $L_p(0,T)$ (see Sec. 1-9), but the case $p = \infty$ is perhaps the most common application for the present problem.   In the nonlinear case, we shall use a Newton-Raphson procedure, for which we shall require that the u_j belong to $L_p(0,T)$, $1 < p < \infty$.

Here we suppose that

$$x_i(t) = F_i(u,t) \qquad i = 1, \ldots, n$$

(5-81)

(which may arise as the solution of a system of nonlinear equations) with $F_i(u,t)$ not necessarily linear in $u = (u_1, \ldots, u_m)$.   Note that in (5-70) the x_i were linear in u.   Again, we have $(z_1(t), \ldots, z_n(t))$, and we seek the smallest T for which $x_i(T) = z_i(T)$, $i = 1, \ldots, n$, subject to the constraint

$$\Big[\sum_{j=1}^{m} \int_0^T |u_j(t)|^p \, dt \Big]^{1/p} \leq 1 \qquad (5\text{-}82)$$

The discussion given in the linear case allows us to rephrase the problem as follows: For fixed $T > 0$, find $u_1(t), \ldots, u_m(t)$ such that

$$F_i(u,T) - z_i(T) = 0 \qquad i = 1, \ldots, n \qquad (5\text{-}83)$$

and such that

$$M = \Big[\sum_{j=1}^{m} \int_0^T |u_j(t)|^p \, dt \Big]^{1/p}$$

is smallest. (Then, if for some $T > 0$ there is an M for which $M \leq 1$, let T_0 be the smallest such T, and the problem is solved.)

We shall first define the appropriate Banach space for this problem. For fixed $T > 0$, let E denote the vector space of m-tuples

$$(g_1(t), \ldots ,g_m(t)) = g(t)$$

of functions $g_j(t)$ defined on $0 \leq t \leq T$ for which the norm

$$\|g\| = \Big[\sum_{j=1}^{m} \int_0^T |g_j(t)|^q \, dt \Big]^{1/q} \qquad (5\text{-}84)$$

is finite; the exponent q is the "conjugate" of p; that is, $q^{-1} + p^{-1} = 1$. Then E^*, the dual space of E, consists of m-tuples

$$v(t) = (v_1(t), \ldots ,v_m(t))$$

defined on $0 \leq t \leq T$ for which the norm is given by

$$\|v\| = \Big[\sum_{j=1}^{m} \int_0^T |v_j(t)|^p \, dt \Big]^{1/p} \qquad (5\text{-}85)$$

Our problem now is to find the $u(t)$ in E^*, satisfying (5-83), for which $M = \|u\|$ is smallest. If we take

$$G(u) = \frac{1}{p} \|u\|^p = \frac{1}{p} \sum_{j=1}^{m} \int_0^T |u_j(t)|^p \, dt \qquad (5\text{-}86)$$

then our problem is equivalent to that of minimizing $G(u)$ subject to the constraints

$$F_i(u,T) - z_i(T) = 0 \qquad i = 1, \ldots , n$$

Assuming that the F_i are smooth functions of u, we have "gradients"

$\nabla F_i = f_i(u,t)$, which are vectors in E such that

$$F_i(u + \Delta u, T) = F_i(u,T) + (f_i,\Delta u) + 0(\|\Delta u\|^2)$$

as $\|\Delta u\| \to 0$ in E^*, in which $f_i(u,t)$ is an m-tuple $\{f_{ij}(u,t)\}$, and

$$(f_i,\Delta u) = \sum_{j=1}^{m} \int_0^T f_{ij}(u,t)\, \Delta u_j(t)\, dt \qquad (5\text{-}87)$$

For example, if

$$F_i(u,T) = \int_0^T k_i(u,t,T)\, dt$$

then

$$\nabla F_i = f_i(u,t) = \{f_{ij}(u,t)\}$$

with

$$f_{ij}(u,t) = \frac{\partial}{\partial u_j} k_i(u,t,T)$$

It is easily seen from (5-86) that

$$G(u + \Delta u) = G(u) + \sum_{j=1}^{m} \int_0^T |u_j(t)|^{p-1} \operatorname{sgn} u_j(t)\, \Delta u_j(t)\, dt + \cdots$$

so that

$$\nabla G(u) = \{|u_j(t)|^{p-1} \operatorname{sgn} u_j(t)\} \qquad (5\text{-}88)$$

Since $G(u)$ is to be minimized subject to (5-83), a generalization of the finite-dimensional situation allows us to state the following: A necessary condition for $G(u)$ to be a minimum subject to (5-83) is that there exist constants $\lambda_1, \ldots, \lambda_n$ such that

$$\nabla G(u) = \sum_{i=1}^{n} \lambda_i \nabla F_i$$

i.e., from (5-88), for $j = 1, \ldots, m$, we have

$$|u_j(t)|^{p-1} \operatorname{sgn} u_j(t) = \sum_{i=1}^{n} \lambda_i f_{ij}(u,t) \qquad (5\text{-}89)$$

From (5-89), we see that

$$u_j(t) = \Big| \sum_{i=1}^{n} \lambda_i f_{ij}(u,t) \Big|^{1/(p-1)} \operatorname{sgn} \Big[\sum_{i=1}^{n} \lambda_i f_{ij}(u,t) \Big] \qquad j = 1, \ldots, m$$

but we do not know the λ_i.

The problem has now been reduced to one of finding $u_1(t), \ldots, u_m(t)$, $\lambda_1, \ldots, \lambda_n$ satisfying (5-83) and (5-89). This problem can be solved by the Newton-Raphson method in Banach spaces if the $f_{ij}(u,t)$ are smooth functions of u. Suppose, then, that we can write

$$f_{ij}(u + \Delta u, \cdot) = f_{ij}(u,\cdot) + \sum_{r=1}^{m} H_{ijr}(u)\, \Delta u_r + \cdots \qquad (5\text{-}90)$$

in which the dot stands in place of the argument t, and $H_{ijr}(u)$ is, for each u, a linear transformation sending Δu_r into an L_q function.

From (5-88), we have

$$\nabla G(u + \Delta u) = \{|u_j(t)|^{p-1} \operatorname{sgn} u_j(t)\}$$
$$+ \{(p-1)|u_j(t)|^{p-2} \Delta u_j(t)\} + \cdots \quad (5\text{-}91)$$

The operator acting on Δu_j to produce the increment in ∇G is simply multiplication by the function $(p-1)|u_j(t)|^{p-2}$.

Let $u, \lambda_1, \ldots, \lambda_n$ be an approximation to the solution of the system consisting in (5-83) and (5-89). Just as in the finite-dimensional case, we shall find the correction $\Delta u, \Delta\lambda_1, \ldots, \Delta\lambda_n$ that satisfies the "linearized" form of our system, evaluated at $u, \lambda_1, \ldots, \lambda_n$. From Eqs. (5-89) to (5-91), we obtain the equations

$$|u_j(t)|^{p-1} \operatorname{sgn} u_j(t) + (p-1)|u_j(t)|^{p-2} \Delta u_j = \sum_{i=1}^{n} \lambda_i f_{ij}(u,t)$$
$$+ \sum_{i=1}^{n} \Delta\lambda_i f_{ij}(u,t) + \sum_{i=1}^{n} \lambda_i \sum_{r=1}^{m} H_{ijr}(u)\,\Delta u_r$$

or $\quad (p-1)|u_j(t)|^{p-2}\,\Delta u_j - \sum_{r=1}^{m} \sum_{i=1}^{n} \lambda_i H_{ijr}(u)\,\Delta u_r$

$$= \sum_{i=1}^{n} \Delta\lambda_i f_{ij}(u,t) + \Big\{ \sum_{i=1}^{n} \lambda_i f_{ij}(u,t) - |u_j(t)|^{p-1} \operatorname{sgn} u_j(t)\Big\} \quad (5\text{-}92)$$

This is a linear system in $\Delta u_1, \ldots, \Delta u_m$, with $\Delta\lambda_1, \ldots, \Delta\lambda_n$ on the right unknown. The solution of (5-92) will yield $\Delta u_1, \ldots, \Delta u_m$ containing $\Delta\lambda_1, \ldots, \Delta\lambda_n$ linearly, say

$$\Delta u_j = \Delta\lambda_1 v_{1j} + \cdots + \Delta\lambda_n v_{nj} + w_j \quad (5\text{-}93)$$

Then, from (5-83), we obtain (dropping the constant T)

$$F_i(u) + \sum_{j=1}^{m} (f_{ij}, \Delta u_j) - z_i = 0$$

or
$$\sum_{j=1}^{m} (f_{ij}, \Delta u_j) = z_i - F_i(u)$$

We apply this to (5-93), which then yields

$$\sum_{k=1}^{n} \Delta\lambda_k \sum_{j=1}^{m} (f_{ij}, v_{kj}) = -\sum_{j=1}^{m} (f_{ij}, w_j) + z_i - F_i(u)$$

$$i = 1, 2, \ldots, n \quad (5\text{-}94)$$

a system of n linear equations in $\Delta\lambda_1, \ldots, \Delta\lambda_n$. When (5-94) has been solved for the $\Delta\lambda_i$, and these values have been substituted in (5-93), we obtain $\Delta u_1(t), \ldots, \Delta u_m(t)$. We then use $u_j + \Delta u_j$ and $\lambda_i + \Delta\lambda_i$ for our next approximation. Conditions ensuring the convergence of the process can be obtained from Theorem 2-1, suitably interpreted. One of the main requirements, of course, is the ability to solve the system (5-92) for Δu_j.

If the constraints were originally given in the form (5-65) or (5-66), so that the transformed problem is to find the u satisfying (5-83) and having the smallest L_∞ norm

$$\|u\|_\infty = \max_{1 \leq j \leq m} \max_{0 \leq t \leq T} |u_j(t)|$$

as in the linear case, we can still use the above procedure in L_p, with p large, to obtain an approximate solution. The formula following (5-89) shows that for large p (and thus large λ_i), the solution $\{u_j(t)\}$ will approximate a bang-bang function, and in the limit as $p \to \infty$, the u_j's will indeed be of the bang-bang type:

$$u_j(t) = M \operatorname{sgn}\left[\sum_{i=1}^n \lambda_i f_{ij}(u,t)\right]$$

We assume as before that for no $\lambda_1, \ldots, \lambda_n$ and for no j does the function $\sum_1^n \lambda_i f_{ij}(u,t)$ vanish in an entire t interval.

REFERENCES

1. Alex, F. R.: A Study of Russian Feed-back Control Theory, part II, vol. 1, A Survey of Russian Control Literature—Linear Methods of Analysis, *Convair*, a Division of General Dynamics Corp., San Diego, Calif., 1961, ASTIA AD 270874.

2. ———: A Study of Russian Feed-back Control Theory, part II, vol. 2, A Survey of Russian Control Literature—Lyapunov's Theory of Stability, *Convair Rept.* ZU 044, 1961, ASTIA AD 270872.

3. ———: A Study of Russian Feed-back Control Theory, part II, vol. 3, A Survey of Russian Control Literature—Krylov and Bogoliubov's Method of Harmonic Linearization, *Convair Rept.* ZU 0004, 1961, ASTIA AD 270873.

4. Aoki, Masanao: On Successive Approximations to Lyapunov Functions in Control Optimization Problems, *University of California at Los Angeles Rept.* 62-50, October, 1962.

5. Bass, R. W.: Reactor Dynamics and Nonlinear Control System Stability Theory, Aeronca Manufacturing Corp., Aerospace Division, *Tech. Rept.* 60-4, Baltimore, Maryland, 1960.

6. Bellman R., I. Glicksberg, and O. Gross: On the "Bang-Bang" Control Problem, *Quart. Appl. Math.*, vol. 14, pp. 11–18, 1956.

7. ———: "Dynamic Programming," Princeton University Press, Princeton, N.J., 1957.

8. Bergen, A. R., and I. J. Williams: Verification of Aizerman's Conjecture for a Class of 3rd Order Systems, *University of California, Berkeley, Electronics Research Laboratory Series* no. 60, issue no. 409, 1961, AF 19(604)-5466.

9. Berkovitz, Leonard D.: An Optimum Thrust Control Problem, *J. Math. Analysis and Applications*, vol. 3, pp. 122–132, 1961.

10. ———: Variational Methods in Problems of Control and Programming, *J. Math. Analysis and Applications*, vol. 3, pp. 145–169, 1961.

11. ———: On Control Problems with Bounded State Variables, Rand Corp. Memo no. RM 3207-PR, 1962, ASTIA AD 278262.

11a. ———: A Survey of Certain Aspects of the Mathematical Theory of Control, Rand Corp. Memo. no. RM3309-PR, 1962.

12. Bliss, G. A.: "Lectures on the Calculus of Variations," The University of Chicago Press, Chicago, 1946.

13. Bodewadt, U. T.: Die Drehstromung uber Festem Grunde, *ZAMM*, vol. 20, pp. 241–253, 1940.

14. Bogner, I., and L. F. Kazda: An Investigation of the Switching Criteria for Higher Order Contractor Servomechanisms, *Trans. Am. Inst. Elec. Engrs.*, vol. 73, part II, pp. 118–127, 1954.

15. Boltianskii, V. G., R. V. Gamkrelidze, and L. S. Pontryagin: On the Theory of Optimal Processes, *Dokl. Akad. Nauk SSSR*, vol. 110, no. 1, pp. 7–10, 1956.

16. Boltianskii, V. G.: The Maximum Principle in the Theory of Optimal Processes, *Dokl. Akad. Nauk SSSR*, vol. 119, no. 6, pp. 1070–1073, 1958.

17. ———: Optimal Processes with Parameters, *Dokl. Akad. Nauk Uz SSR*, vol. 10, pp. 9–13, 1959.

18. ———, R. V. Gamkrelidze, and L. S. Pontryagin: Theory of Optimal Processes, *Izv. Akad. Nauk SSSR, Ser. Mat.*, vol. 24, no. 1, 1960; English translation in *Am. Math. Soc. Translation, ser.* 2, vol. 18, pp. 341–382, 1961.

19. Breakwell, J. V.: The Optimization of Trajectories, *J. Soc. Ind. Appl. Math.*, vol. 7, pp. 215–247, 1959.

20. Bushaw, D. W.: Optimal Discontinuous Forcing Terms, Ph.D. thesis, Department of Mathematics, Princeton University, 1952, in "Contributions to the Theory of Nonlinear Oscillations," vol. 4, Princeton University Press, Princeton, N.J., 1958.

21. Butkovsky, A. G.: L. S. Pontryagin's Maximum Principle in Optimal Control Systems with Linear Controlling Function (translated by Office of Technical Services, Washington 25, D.C.), *Avtomaticheskoye Upravleniye (Automatic Control)*, Moscow Publishing House of the USSR Academy of Sciences, pp. 9–16, 1960.

22. Carathedory, C.: Vorlesungen uber Reelle Funktionnen, *Leipzig*, 1927.

23. ———: Sur une Méthode Directe du Calcul des Variations, *Rend. Circ. Mat. Palermo*, vol. 25, pp. 36–49, 1908.

24. ———: Uber die Einteilung der Variationsprobleme von Lagrange nach Klassen, *Comm. Math. Helv.*, vol. 5, pp. 1–19, 1933.

24a. Chang, S. S. L.: "Synthesis of Optimum Control Systems," McGraw-Hill Book Company, Inc., New York, 1961.

25. Desoer, C. A.: The Bang-Bang Servo Problem Treated by Variational Techniques, *Inform. and Control*, vol. 2, pp. 333–348, 1959.

26. ——— and J. Wing: An Optimal Strategy for a Saturating Sampled-data System, *IRE Trans. on Automatic Control*, vol. AC-6, no. 1, February, 1961.

27. ——— and———: The Minimal Time Regulator Problem for Linear Sampled-data Systems: General Theory, *J. Franklin Inst.*, vol. 272, no. 3, September, 1961.

28. Dreyfus, S. E.: Dynamic Programming and the Calculus of Variations, *J. Math. Anal. and Appls.*, vol. 1, pp. 228–239, 1960.

29. Feldbaum, A. A.: On the Synthesis of Optimal Systems Using Phase Space, *Avtomat. i. Telemeh.*, vol. 16, no. 2, pp. 129–149, 1955.

30. Forsyth, A. R.: "Calculus of Variations," Dover Publications, Inc., New York, 1926.

31. Fried, B. D.: On the Powered Flight Trajectory of an Earth Satellite, *Jet Propulsion*, vol. 27, no. 6, 1957.

32. ———— and J. Richardson: Optimal Rocket Trajectories, *J. Appl. Phys.*, vol. 27, no. 8, 1956.

33. Gaffney, M. P.: Symposium on Optimal Control and Non-linear Systems, *European Scientific Notes*, Office of Naval Research, London, July, 1962.

34. Gamkrelidze, R. V.: On the Theory of the Optimal Processes in Linear Systems, *Dokl. Akad. Nauk, SSSR*, vol. 116, no. 1, pp. 9–11, 1957.

35. ————: The Theory of Time-optimal Processes for Linear Systems, *Izv. Akad. Nauk SSSR. Ser. Mat.*, vol. 22, pp. 449–474, 1958.

36. ————: Time Optimal Processes with Bounded Phase Coordinates, *Dokl. Akad. Nauk SSSR*, vol. 125, no. 3, pp. 475–478, 1959.

37. ————: On the General Theory of Optimal Processes, *Dokl. Akad. Nauk SSSR*, vol. 123, no. 2, pp. 223–226, 1958; English translation in *Automation Express*, vol. 1, no. 7, pp. 37–39, 1959.

38. ————: Optimal Control Processes with Bounded Phase Coordinates, *Izv. Akad. Nauk SSSR. Ser. Mat.*, vol. 25, no. 3, pp. 315–356, 1960.

39. Geis, T.: Ahnlichen Grenzschichten an Rotationskorpern, 50 *Jahre Grenzschichtforschung*, pp. 294–303, 1955.

40. Goldstein, S.: A Note on the Boundary-layer Equation, *Proc. Cambridge Philos. Soc.*, vol. 34, p. 338, 1939.

41. Gorelov, Yu. A.: On Two Classes of Planar Extremal Motions of a Rocket in Space, *Prikl. Mat. Meh.*, vol. 24, no. 2, 1960.

42. Hagin, E. J., and G. H. Fett: Phase Space Metrization for Relay Control of Two Time Constant Servomechanism, *Proc. National. Electronics Conf.*, vol. 13, pp. 537–548, 1957.

43. Halmos, P. R.: The Range of a Vector Measure, *Bull. Am. Math. Soc.*, vol. 54, pp. 416–421, 1948.

43a. Harvey, C. A.: Determining the Switching Criterion for Time-optimal control, *J. Math. Anal. and Applications*, vol. 5, no. 2, pp. 245–257, 1962.

43b. ———— and E. B. Lee: On the Uniqueness of Time-optimal Control for Linear Processes, *J. Math. Anal. and Application*, vol. 5, no. 2, pp. 258–268, 1962.

44. Hestenes, M. R.: A General Problem in the Calculus of Variations with Applications to Paths of Least Time, Rand Corp. *Res. Memo* RM-100, February, 1949.

45. Hsi-yung, Chiang: On The Theory of Optimal Regulation, *Prikl. Mat. Meh.*, vol. 25, no. 3, pp. 413–449, May–June, 1961, translated by Office of Technical Services, Washington 25, D.C.

46. Hughes, W. L.: "Nonlinear Electrical Networks," The Ronald Press Company, New York, 1960.

47. Isayev, V. K.: Pontryagin's Maximality Principle and Optimal Rocket Thrust Programming, translated by Office of Technical Services, Washington, 1961.

48. Kalman, R. E.: A New Approach to Linear Filtering and Prediction Problems, *J. Basic Eng.*, pp. 35–45, March, 1960.

49. ————and J. E. Bertram: Control System Analysis and Design via the "Second Method" of Lyapunov, I: Continuous-time Systems, *J. Basic Eng.*, pp. 371–393, June, 1960.

50. ————: Contributions to the Theory of Optimal Control, *Bol. Soc. Mat. Mexicana*, 1961.

51. ————: The Theory of Optimal Control and the Calculus of Variations, RIAS, Baltimore, Maryland, *Tech. Rept.* 61-3, 1961.

52. Kashmar, C. M., E. L. Peterson, and F. X. Remond: Optimum System Synthesis, a General Approach to the Numerical Solution of Multi-dimensional, Nonlinear, Boundary-valued Variational Problems, R60TMP-27, April, 1960.

53. Kopp, Richard E., "Pontryagin Maximum Principle," Academic Press, Inc., 1961.

54. Krasovskii, N. N.: On the Theory of Optimal Control, *Avtomat. i. Telemeh.*, vol. 18, pp. 960–970, 1957.

55. ————: On a Problem of Optimum Control of Nonlinear Systems, *Prikl. Mat. Meh.*, vol. 23, pp. 209–229, 1959.

56. Kulikowski, Roman: Optimum Systems of Automatic Control, First International Congress IFAC (International Federation of Automatic Control), translated by U.S. Joint Publications Research Service from "Archiwum Automatyki i Tele-mechaniki," vol. 6, no. 2-3, Warsaw, 1961.

57. Landyshev, A. N.: Optimum System Synthesis, *Recent Soviet Progress in Adaptive and Optimal Control*, SP 121, March, 1961.

58. LaSalle, J. P.: Study of the Basic Principle Underlying the Bang-Bang Servo, Goodyear Aircraft Corporation Rept. GER-5518, July, 1953, Abstract 247t, *Bull. Am. Math. Soc.*, vol. 60, 1954.

59. ————: The Bang-Bang Principle, RIAS, Baltimore, Maryland, *Tech. Rept.* 59-5, 1959 (reprinted from *Automatic and Remote Control*).

60. ————: Time Optimal Control Systems, *Proc. National Acad. Sci. U.S.*, vol. 45, pp. 573–577, 1959.

61. ————: The Time Optimal Control Problem, in "Contributions to the Theory of Non-linear Oscillations," vol. 5, Annals of Mathematics Studies, Princeton University Press, Princeton, N.J., 1960.

62. ————: Stability and Control, *J. SIAM Control*, ser. A, vol. 1, no. 1, 1962.

63. Lawden, D.: Dynamic Problems of Interplanetary Flight, *Aeronautical Quart.*, vol. 6, p. 3, 1955.

64. Lee, E. B.: Design of Optimum Multi-variable Control Systems, *J. Basic Eng. (Trans. ASME)*, vol. 83, pp. 85–90, 1961.

64a. ———— and L. Markus: Optimal Control for Nonlinear Processes, *Arch. for Rat. Mech. and Analysis*, vol. 8, no. 1, pp. 36–58, 1961.

64b. ———— and ————: Synthesis of Optimal Control for Nonlinear Processes with One Degree of Freedom, *J. Inst. Math. Acad. Sci. Ukrainian SSR*, 1961.

65. Lefschetz, Solomon: Some Mathematical Considerations on Nonlinear Automatic Controls, RIAS *Tech. Rept.* 62-10, April, 1962, AFOSR 2501.

66. Leitmann, G.: On a Class of Variational Problems in Rocket Flight, *J. Aerospace Sci.*, vol. 27, no. 9, 1959.

67. ————: "Optimization Techniques with Applications to Aerospace Systems," Academic Press Inc., New York, 1962.

68. Letov, A. M.: "Stability in Nonlinear Control Systems," Princeton University Press, Princeton, N.J., 1961.

69. Lyapunov, A.: Sur les Fonctions-vecteurs Complètement Additives, *Bull. Acad. Sci., U.R.S.S. Ser. Math.*, vol. 4, pp. 465–478, 1940.

70. Lur'e, A. I.: "Some Non-linear Problems in the Theory of Automatic Control Translated from Russian," Her Majesty's Stationery Office, London, 1957.

71. ————: Remarks Concerning the Paper "On Some Rocket Trajectories" by G. Leitmann *PMM, J. Appl. Math. and Mech.*, vol. 26, no. 2, p. 391, 1962.

72. Malkin, I. G.: "Theory of Stability of Motion," translated from a publication of the State Publishing House of Technical-Theoretical Literature, Moscow, 1952, by U.S. Atomic Energy Commission, Office of Technical Information.

73. Markus, Lawrence: Optimal Automatic Control for Nonlinear Processes, *University of Minnesota Tech. Rept.* 1, July, 1961 (prepared under Office of Naval Research contract NR 041-255).

74. McRuer, D., and D. Graham: "Analysis of Nonlinear Control Systems," John Wiley & Sons, Inc., New York, 1961.

75. McShane, E. J.: On Multipliers for Lagrange Problems, *Am. J. Math.*, vol. 61, pp. 809–819, 1939.

76. ———: Necessary Conditions in Generalized-curve Problems of the Calculus of Variations, *Duke Math. J.*, vol. 7, pp. 1–27, 1940.

77. Miele, A.: An Extension of the Theory of the Optimum Burning Program for the Level Flight of a Rocket Powered Aircraft, *J. Aeronaut. Sci.*, vol. 24, no. 12, 1957.

78. ——— and J. Capellari: Topics in Dynamic Programming for Rockets, *Z. Flugwissenschaften*, vol. 1, no. 1, 1959.

79. Millsaps, K., and K. Pohlhausen: Heat Transfer by Laminar Flow from a Rotating Plate, *J. Aeronaut. Sci.*, vol. 19, no. 2, pp. 120–126, February, 1952.

80. Mitropol'sky, Yu. A.: "Nonstationary Processes in Nonlinear Oscillatory Systems" ("Nestatsionarnyye Protsessy V Nelineynykh Kolebatel'nykh Sistemake"), Kiev, 1955, chaps. I and II, pp. 7–108, Air Technical Intelligence translation, Wright Patterson Air Force Base, Ohio, ASTIA AD162846.

81. Neustadt, Lucien W.: Synthesizing Time Optimal Control Systems, *J. Math. Anal. and Appls.*, vol. 1, pp. 484–493, 1960.

82. ———: Time Optimal Control Systems with Position and Integral Limits, *J. Math. Anal. and Appls.*, vol. 3, pp. 406–427, 1961.

83. Okhotsmsky, D. Ye, and T. M. Eneyev: Some Variational Problems in the Launching of Artificial Earth Satellites, *Uspehi Fiz. Nauk*, vol. 63, no. 1a, 1957.

83a. Pennisi, Louis L.: An Indirect Sufficiency Proof for the Problem of Lagrange with Differential Inequalities as Added Side Conditions, *Trans. Am. Math. Soc.*, vol. 74, pp. 177–198, 1953.

84. Pontryagin, L. S.: Optimal Control Processes, *Uspehi Mat. Nauk*, vol. 14, no. 1, pp. 3–20, 1959; English translation in *Automation Express*, vol. 1, no. 10, pp. 15–18, 1959, and vol. 2, no.1, pp. 26–30, 1959.

85. ———, V. G. Boltyanskii, R. V. Gamkrelidze, and E. F. Mishchenko: "The Mathematical Theory of Optimal Processes," John Wiley & Sons, Inc., New York, 1962.

86. Rheinboldt, W. C.: Lectures on the Calculus of Variations, American University, Washington, D.C., 1957–1958.

87. Ross, S.: Composite Trajectories Yielding Maximum Coasting Apogee Velocity, *ARS J.*, vol. 26, no. 11, 1959.

88. Roxin, Emilio: A Geometric Interpretation of Pontryagin's Maximum Principle, RIAS, Baltimore, Maryland, *Tech. Rept.* 15, December, 1961.

89. ———: The Existence of Optimal Controls, *Michigan Math. J.*, vol. 9, pp. 109–119, 1962.

90. Rozonoer, L. I.: L. S. Pontryagin's Maximum Principle in the Theory of Optimum Systems, parts I, II, and III, *Avtomat. i. Telemeh.*, vol. 20, pp. 1320–1334, 1441–1458, 1561–1578, 1959 (translated in *Automation and Remote Control*, vol. 20, pp. 1288–1302, 1405–1421, 1517–1532, 1960).

91. ———: On Sufficient Conditions for Optimality, *Dokl. Akad. Nauk SSSR*, vol. 127, no. 3, pp. 520–523, 1959 (English translation in *Automation Express*, vol. 2, no. 7, pp. 5–7, 1960).

92. Sarachik, Philip E.: Cross-coupled Multi-dimensional Feedback Control Systems, *Columbia University Tech. Rept.* T-30/B, May, 1958, ASTIA AD 158374.

93.   Schuh, H.: Uber Ahnlichen Losungen der Instationdren Laminaren Grenz-schichtgleichungen in Incompressiblen Stromungen, 50 *Jahre Grenzschichtforschung*, pp. 147–152, 1955.

94.   Solodovnikov, V. V.: "Introduction to the Statistical Dynamics of Automatic Control Systems" (translation of the Russian edition of 1952, edited by J. B. Thomas and L. A. Zadeh), Dover Publications, Inc., New York, 1960.

95.   Stewart, J. L.: "Circuit Theory and Design," John Wiley & Sons, Inc., New York, 1956.

96.   Troitskii, V. A.: Variational Problems of Optimization of Control Processes for Equations with Discontinuous Right-hand Sides, *J. Appl. Math. and Mech.*, vol. 26, no. 2, pp. 233–246, 1962.

97.   Valentine, F. A.: The Problem of Lagrange with Differential Inequalities as Added Side Conditions, in "Contributions to the Calculus of Variations," 1933–1937, pp. 407–448, University of Chicago Press, Chicago, 1937.

98.   Wing, Jack: Minimal Time Control Problem for Linear Discrete Systems, Ph.D. thesis, University of California, 1962.

SIX

LINEAR AND NONLINEAR PREDICTION THEORY

6-1 INTRODUCTION

In Chap. 5 we encountered examples suggesting the necessity of a theory whose aim is the prediction of values of state variables in the presence of random disturbances. There we saw that a rocket could not be adequately controlled on the basis of differential equations alone. In the communications field, random noise, both internally generated and from external sources, precludes the possibility of obtaining useful signal levels from arbitrarily weak input signals simply by using sufficiently many amplification stages. Consequently the problem of adequately filtering the noise must be faced. These problems of random disturbances in control and communications have been treated on the basis of work done in the field of random processes.

The earliest and some of the most significant results in this area are due to Kolmogorov, Khintchine, Wiener, and Doob. The treatise by Doob[7] is the definitive work on the general mathematical theory of random processes. Subsequent advances in prediction theory for multidimensional random processes have been made by Masani and Wiener,[19] Rozanov,[22] and Helson and Lowdenslager.[10]

We are particularly indebted to Prof. A. V. Balakrishnan for constructive comments on an early version of this chapter, and to Prof. L. A. Zadeh for additional suggestions which broadened its scope. The subject matter in this chapter has been materially influenced by the works of Grenander and Szego,[9] Grenander and Rosenblatt,[8] Balakrishnan,[1] and Zadeh and Ragazzini.[26]

Suppose that we have a record or graph of a variable x versus time t, for $-T < t \leq 0$. Suppose also that $x(t)$ is a random variable, determined by many small "random" or unrecognized causes. Examples are furnished by hourly temperatures, daily commodity prices, electrical voltages containing noise (from the shot effect in tubes, or "static" or lightning discharge), and by the history of the errors of a gun camera tracking an evading target.

Each such t function is called a *sample* from a random process. A random process is a collection of random variables; in our case the collection is the set of $x(t)$'s, for example, $x(t_1)$, $x(t_2)$, . . . where $t_1, t_2,$ . . . are any fixed values of t.

The problem that will concern us in this chapter is that of determining the "best" guess at the value of a given random variable w when we have observed $x(t)$ for $-T \leq t \leq 0$ or $-\infty < t \leq 0$. The criterion by which we recognize the best guess, or estimate, will be the least-squares criterion, about which we shall say more below. We may call this general problem the problem of *optimal estimation*. A special case arises when the desired random variable w is $w = x(h)$, with $h > 0$. In this case, the problem can be called the *prediction* or *extrapolation* problem. (However, the term prediction has also been used for the more general problem which we are now calling estimation, and we even call this chapter "prediction theory." The term "estimation," which is perhaps more appropriate, is not entirely satisfactory either, inasmuch as in statistics the word is used to mean estimating a parameter, e.g., the mean, by means of a function of a sample. There, a random variable is used to estimate the value of a parameter which is not a random variable. We shall be using random variables to estimate other random variables.)

The construction of optimal estimates, then, is the object of prediction theory. By the "optimal" estimate we shall mean, for the case of linear estimation, that random variable $\hat{w}$ which is a limit of linear combinations of $x(t)$, $t \leq 0$, having the smallest deviation from w in the sense of least squares, i.e.,

$$E((w - \hat{w})^2) < E((w - y)^2) \tag{6-1}$$

for all admissible y's; in the linear case the admissible y's are all linear combinations of the $x(t)$'s. We shall suppose throughout that all our random variables have finite second moments.

In the general (nonlinear) case, we demand that $\hat{w}$ be a limit of functions of $x(t)$, not necessarily linear functions, and that (6-1) again be valid, but this time for all y's that are functions of the $x(t)$'s. In the final section, we depart from the least-squares criterion and use the maximum-likelihood estimate.

Since every least-squares problem can be formulated as a Hilbert-space problem (see Sec. 1-6), we see that if we take the inner product of u, v to be $E(u\bar{v})$, and $\mathfrak{M}$ to be the subspace generated by the class of admissible y's, then $\hat{w}$ is the projection of w onto $\mathfrak{M}$; that is, $\hat{w}$ is that vector in $\mathfrak{M}$ such that $w - \hat{w}$ is orthogonal to $\mathfrak{M}$,

$$(w,y) = (\hat{w},y)$$

for every y in $\mathfrak{M}$. If our process is stationary, as defined below, and if we are dealing with linear estimates, we can say much more by using power series and Fourier transforms to get explicit solutions for $\hat{w}$. However we have not assumed a knowledge of Hilbert space for the study of this chapter, and therefore we have stated, where they are needed, certain facts and definitions that already appear in Chap. 1.

We shall devote the major part of the chapter to linear prediction for the simple reason that the nonlinear theory is still quite new and is far less rich in results than the linear theory; furthermore, even the linear theory is still not so widely known that it may be called standard material. We have placed particular emphasis on providing a lucid picture of the linear stationary case in this technically difficult area, using somewhat stronger hypotheses than are needed (but which are almost always satisfied in practice) in order to simplify the proofs. The nonlinear version is then given in as painless a manner as is available. As indicated above, the main difference between the linear stationary case and the general case is the subspace of admissible candidates for the optimal estimate. Although the characterization of the best estimate, in the least-squares problem, is the same in all cases, the construction of this estimate differs considerably in the various cases.

We shall often use the term *gaussian process*, which is defined as follows: If, for every $t_1, \ldots, t_n$, the variables $x(t_1), \ldots, x(t_n)$ have a joint distribution that is gaussian [i.e., the distribution is defined by a density of the form

$$f(u_1, \ldots, u_n) = \frac{|A|^{1/2}}{(2\pi)^{n/2}} e^{-u'Au/2}$$

in which u is the column vector $(u_1, \ldots, u_n)$ and u' is its transpose, and

A is positive-definite and depends on $t_1, \ldots, t_n$], then the process is called a gaussian process. If the process is gaussian, and if when w is added to the collection $x(t)$ the process is still gaussian, then, as we shall see later, the nonlinear problem has for its solution the same random variable that is the solution in the linear case. Bars will indicate complex conjugates.

6-2 THE DISCRETE STATIONARY CASE

We shall start with the problem of linear estimation and make several simplifying assumptions. In the first place, instead of treating random functions $x(t)$ defined over a t interval, we shall consider sequences

$$\{x_n\} \qquad n = 0, \pm 1, \pm 2, \ldots$$

of random variables. Examples can be obtained, among others, by considering the results of sampling (measuring) random functions $x(t)$, of the sort described above, at discrete times $t_n = n\,\Delta t$, and setting $x_n = x(t_n)$.

In addition, we shall suppose that our *discrete process* $\{x_n\}$ is stationary in the wide sense, that is, $E(x_n^2) < \infty$ for every n (E denotes the expectation), and $E(x_n x_m) = \rho_{m-n}$ depends only on the difference between the indices. It is customary and convenient to suppose also that $E(x_n) = 0$ for each n, and we now make that assumption. So $E(x_n x_m) = \rho_{m-n}$ are the covariances.

Example: Suppose $\{u_n\}, n = 0, \pm 1, \pm 2, \ldots$ is a collection of independent (or simply uncorrelated) random variables from the same population, all with mean zero and variance σ^2. Let $\{\gamma_n\}$ be a real sequence such that

$$\sum_{n=0}^{\infty} \gamma_n^2 < \infty$$

and put

$$x_n = \sum_{j=0}^{\infty} \gamma_j u_{n-j} \qquad n = 0, \pm 1, \pm 2, \ldots \qquad (6\text{-}2)$$

Then $E(x_n) = 0$ for each n, and

$$E(x_n x_m) = \sigma^2 \sum_{j=0}^{\infty} \gamma_j \gamma_{m-n+j} \equiv \rho_{m-n}$$

that is, $E(x_n x_m)$ depends only on $m - n$ (γ_{m-n+j} is zero if $m - n + j <$ Thus the process $\{x_n\}$ is stationary. In a process such as (6-2), x_n called a *moving average of uncorrelated random variables*. We shall see i. Eq. (6-26) that conversely, under fairly general conditions on $\{x_n\}$, there is an uncorrelated process $\{u_n\}$ such that x_n is a moving average of the u_m's. A representation of $\{x_n\}$ as such a moving average is often useful in certain problems in statistics.

The main purpose of this section is to show that, under conditions that are nearly always satisfied in practice, there is associated with a stationary process $\{x_n\}$, a power density $f(\theta)$ on $-\pi \leq \theta \leq \pi$, defined below, and that there is a function $g(\theta) = \sum\limits_{n=0}^{\infty} \gamma_n e^{in\theta}$ with $\sum\limits_{n=0}^{\infty} |\gamma_n|^2 < \infty$, such that $f(\theta) = |g(\theta)|^2$. This will be used to solve the linear discrete prediction problem.

We consider now a general wide-sense discrete stationary process $\{x_n\}$, with $E(x_n) = 0$ and with covariances

$$\rho_{m-n} = E(x_n x_m) = \rho_{n-m}$$

We shall suppose that the power series $\sum\limits_{n=0}^{\infty} \rho_n \lambda^n$ converges for $|\lambda| < R$ with $R > 1$. It follows that the function

$$f(\theta) = \sum_{-\infty}^{\infty} \rho_n e^{in\theta} \tag{6-3}$$

is continuous, since the series converges uniformly in θ (note that $\rho_{-n} = \rho_n$). The function $f(\theta)$ is nonnegative. Indeed, let

$$s_N(\theta) = \sum_{-N}^{N} x_n e^{in\theta}$$

Then
$$E(|s_n(\theta)|^2) \geq 0$$

or
$$\sum_{-N}^{N} \sum_{-N}^{N} e^{i(n-m)\theta} \rho_{n-m} \geq 0$$

This is the same as

$$N \sum_{-N}^{N} \left(1 - \frac{|\nu|}{N}\right) \rho_\nu e^{i\nu\theta} \geq 0 \tag{6-4}$$

Since

$$\sum_{-N}^{N} \left(1 - \frac{|\nu|}{N}\right) \rho_\nu e^{i\nu\theta}$$

converges to $f(\theta)$ as $N \to \infty$, we have $f(\theta) \geq 0$. The function $f(\theta)$ is called the *power density* of the process $\{x_n\}$. We have, of course,

$$\rho_n = \frac{1}{2\pi} \int_{-\pi}^{\pi} f(\theta) e^{-in\theta} \, d\theta \tag{6-5}$$

for all n.

We shall suppose that $f(\theta) \geq \epsilon > 0$ for all θ. Then for a sufficiently narrow annulus containing the circle $|z| = 1$ in the z plane, the values of

$$w = F(z) = \sum_{-\infty}^{\infty} \rho_n z^n$$

fill out a neighborhood of a bounded closed interval of the positive real axis in the w plane, and the negative real axis is bounded away from this neighborhood. It follows that $\log w = \log F(z)$ is a single-valued analytic function of z in an annulus containing $|z| = 1$. Therefore, $\log F(z)$ has a Laurent series expansion converging in the annulus, and we can write $\log F(z) = H_1(z) + H_2(z)$, in which $H_1(z)$ is a power series converging for $|z| < R'$ for some $R' > 1$, and $H_2(z)$ is a power series in z^{-1} with no constant term converging for $|z| > r'$ for some $r' < 1$. We note that

$$H_1(0) = \frac{1}{2\pi i} \int_{|z|=1} \frac{\log F(z)}{z}\, dz$$

and since $\log F(z) = \log f(\theta)$ is real for $|z| = 1$, we see that $H_1(0)$ is real. Again, since $\log F(z)$ is real for $|z| = 1$, it follows that the analytic function $\overline{\log F(1/\bar{z})}$ coincides with $\log F(z)$, and we find that

$$H_1(z) + H_2(z) = \overline{H_1\left(\frac{1}{\bar{z}}\right)} + \overline{H_2\left(\frac{1}{\bar{z}}\right)} \tag{6-6}$$

This equality of two Laurent series is possible only if

$$H_1(z) = \overline{H_2\left(\frac{1}{\bar{z}}\right)} + H_1(0) \qquad H_2(z) = \overline{H_1\left(\frac{1}{\bar{z}}\right)} - H_1(0)$$

Exercise 6-1   Verify the last statement.

The result is that

$$\log F(z) = H_1(z) + \overline{H_1\left(\frac{1}{\bar{z}}\right)} - H_1(0)$$

Note that if $|z| = 1$, $\bar{z} = 1/z$, so that

$$\log f(\theta) = H_1(z) + \overline{H_1(z)} - H_1(0) \qquad |z| = 1$$

Now let $\qquad G(z) \equiv \exp\left[H_1(z) - \tfrac{1}{2}H_1(0)\right]$

Then $G(z)$ is analytic for $|z| < R'$, $R' > 1$, and $f(\theta) = |G(z)|^2$ for $|z| = 1$. Furthermore, $G(z) \neq 0$ for $|z| < R'$, since the exponential function never vanishes.

We conclude that

$$f(\theta) = |g(\theta)|^2 \tag{6-7}$$

where $g(\theta) = G(e^{i\theta})$ is the boundary value of an analytic function $G(\lambda)$; moreover $G(\lambda) \neq 0$ for $|\lambda| < R'$, $R' > 1$. [Our hypotheses made the conclusion easy to prove. The best theorem of this sort is as follows: Let $f(\theta)$ be nonnegative in $|\theta| \leq \pi$, and let

$$\int_{-\pi}^{\pi} f(\theta)\, d\theta < \infty$$

Suppose $\qquad \displaystyle\int_{-\pi}^{\pi} \log f(\theta)\, d\theta > -\infty$

Then and only then there is a function $G(\lambda)$ analytic in $|\lambda| < 1$; $G(\lambda) \neq 0$ if $|\lambda| < 1$,

$$g(\theta) = \lim_{r \to 1} G(re^{i\theta})$$

exists for almost every θ, and $f(\theta) = |g(\theta)|^2$ almost everywhere. The proof requires the theory of the Lebesgue integral and is much more delicate than what we have presented above.]

6-3 CONSTRUCTION OF THE DISCRETE - CASE ESTIMATE

Now suppose we have a random variable w which we want to express as a (possibly infinite) linear combination of $x_0, x_{-1}, \ldots$, plus an (hopefully small) error term r:

$$w = \sum_{\nu=0}^{\infty} \alpha_\nu x_{-\nu} + r \tag{6-8}$$

This procedure could perhaps be applied in a situation in which we have observed the variable x_n for $n \leq 0$, and, on the basis of these observations, would like to "predict" w using a linear function of the x's.

The problem is to determine the coefficients α_ν. We shall take, as our criterion of "best fit," the *least-squares criterion*: that set $\{\alpha_\nu\}$ will be best which makes $E(r^2)$ a minimum. We have

$$r^2 = w^2 - 2w \sum_0^{\infty} \alpha_\nu x_{-\nu} + \sum_0^{\infty} \sum_0^{\infty} \alpha_\nu \alpha_k x_{-\nu} x_{-k} \tag{6-9}$$

$$E(r^2) = E(w^2) - 2 \sum_0^{\infty} \alpha_\nu E(w x_{-\nu}) + \sum_0^{\infty} \sum_0^{\infty} \alpha_\nu \alpha_k \rho_{k-\nu} \tag{6-10}$$

(With our assumptions on the ρ_n's, the series will converge if $\sum_0^{\infty} |\alpha_\nu|^2 < \infty$.) If we now formally differentiate (6-10) with respect to α_k and set the derivative equal to zero, we get

$$\sum_{\nu=0}^{\infty} \rho_{k-\nu} \alpha_\nu = E(w x_{-k}) \qquad k = 0, 1, 2, \ldots \tag{6-11}$$

We shall restrict ourselves to those cases in which the random variable w satisfies the following condition: Let $c_n = E(w x_{-n})$, $n = 0, 1, 2, \ldots$. Then $\sum_{n=0}^{\infty} c_n \lambda^n$ is required to have a radius of convergence larger than 1. This says roughly that the correlation of w with x_{-n} tends to zero quite rapidly as $n \to \infty$. In such a case the system (6-11) will have a unique solution, which we shall exhibit shortly, and the set $\{\alpha_\nu\}$ will in fact yield the minimum value of $E(r^2)$. The α_ν's will also tend to zero rapidly, that is, $\sum_{\nu=0}^{\infty} \alpha_\nu \lambda^\nu$ will converge if $|\lambda| < R$, for some $R > 1$, and the random variable

$$\hat{w} = \sum_{n=0}^{\infty} \alpha_n x_{-n} \tag{6-11a}$$

that best approximates w is the limit as $N \to \infty$ of the sequence $\{\hat{w}_N\}$

where $$\hat{w}_N = \sum_{n=0}^{N} \alpha_n x_{-n}$$

in the following sense:

$$E((\hat{w} - \hat{w}_N)^2) \to 0 \text{ as } N \to \infty$$

Let us now make use of the fact that $f(\theta) = |g(\theta)|^2$, where $g(\theta) = G(e^{i\theta})$, and $G(\lambda)$ has a series expansion $G(\lambda) = \sum_{n=0}^{\infty} \gamma_n \lambda^n$. The $\{\gamma_n\}$ are known, at least theoretically. They arise from the knowledge of $\{\rho_n\}$, the construction of $f(\theta)$, and $g(\theta)$, as indicated in the last section. Then

$$f(\theta) = \sum_{n=0}^{\infty} \gamma_n e^{in\theta} \sum_{m=0}^{\infty} \bar{\gamma}_m e^{-im\theta}$$

$$= \sum_{m=-\infty}^{\infty} \sum_{n=0}^{\infty} \gamma_n \bar{\gamma}_{n-m} e^{im\theta}$$

so that $$\rho_m = \sum_{n=0}^{\infty} \gamma_n \bar{\gamma}_{n-m} \tag{6-12}$$

for every m.

Now, with every sequence $\{\xi_n\}$, $n = 0, 1, 2, \ldots$, we shall associate the power series $\sum_{n=0}^{\infty} \xi_n \lambda^n$; if the ξ_n's tend to zero geometrically (i.e., like r^n with $0 < r < 1$), then the power series has a radius of convergence larger than 1.

Consider the unknown α's and the associated power series,

$$\sum_{n=0}^{\infty} \alpha_n \lambda^n \equiv A(\lambda)$$

We have $$G(\lambda)A(\lambda) = \sum_{n=0}^{\infty} \sum_{m=0}^{\infty} \gamma_n \alpha_m \lambda^n \lambda^m = \sum_{n=0}^{\infty} \sum_{m=0}^{n} \gamma_{n-m} \alpha_m \lambda^n \tag{6-13}$$

For every power series

$$B(\lambda) = \sum_{n=0}^{\infty} \beta_n \lambda^n$$

let us define

$$B^*(\lambda) = \sum_{n=0}^{\infty} \bar{\beta}_n \lambda^n$$

and calculate

$$G^*(\lambda^{-1})B(\lambda)$$

We obtain

$$G^*(\lambda^{-1})B(\lambda) = \sum_{n=0}^{\infty} \bar{\gamma}_n \lambda^{-n} \sum_{m=0}^{\infty} \beta_m \lambda^m$$

$$= \sum_{n=-\infty}^{\infty} \sum_{m=0}^{\infty} \bar{\gamma}_{m-n} \beta_m \lambda^n \qquad (6\text{-}14)$$

For every Laurent series $H(\lambda) = \sum_{-\infty}^{\infty} \eta_n \lambda^n$, let us define $PH(\lambda)$, called the *inner part*, by the equation

$$PH(\lambda) = \sum_{n=0}^{\infty} \eta_n \lambda^n \qquad (6\text{-}15)$$

Finally, let us calculate $PG^*(\lambda^{-1})G(\lambda)A(\lambda)$, using (6-12) to (6-15). We find, after some manipulation, that

$$PG^*(\lambda^{-1})G(\lambda)A(\lambda) = \sum_{n=0}^{\infty} \sum_{j=0}^{\infty} \rho_{n-j}\alpha_j \lambda^n \qquad (6\text{-}16)$$

By (6-11), we have

$$PG^*(\lambda^{-1})G(\lambda)A(\lambda) = C(\lambda) = \sum_{n=0}^{\infty} c_n \lambda^n \qquad (6\text{-}17)$$

where

$$c_n = E(wx_{-n}) \qquad n = 0, 1, 2, \ldots$$

Conversely, if we start with (6-17), i.e., if we can find $A(\lambda)$ satisfying (6-17), then its coefficients α_n will satisfy (6-11), and our problem is solved. We need the following fact.

Lemma 1 If $F(\lambda)$, $G(\lambda)$, and $H(\lambda)$ are power series, then

$$PF(\lambda^{-1})PG(\lambda^{-1})H(\lambda) = PF(\lambda^{-1})G(\lambda^{-1})H(\lambda) \qquad (6\text{-}18)$$

Proof: Let

$$G(\lambda^{-1})H(\lambda) = K_1(\lambda) + K_2(\lambda^{-1})$$

where K_2 is a power series with no constant term. Then

$$PG(\lambda^{-1})H(\lambda) = K_1(\lambda)$$

and the left side of (6-18) is

$$PF(\lambda^{-1})K_1(\lambda)$$

The right side of (6-18) is

$$PF(\lambda^{-1})(K_1(\lambda) + K_2(\lambda^{-1})) = PF(\lambda^{-1})K_1(\lambda) + PF(\lambda^{-1})K_2(\lambda^{-1})$$

But
$$PF(\lambda^{-1})K_2(\lambda^{-1}) = 0$$
since
$$F(\lambda^{-1})K_2(\lambda^{-1})$$

has no constant term.   The proof is complete.

From the results of Sec. 6-2 we know that $G(\lambda) \neq 0$ for $|\lambda| \leq 1$, and since

$$f(\theta) = |G(e^{i\theta})|^2 \geq \epsilon > 0$$

we can conclude that

$$|G(\lambda)| \geq \epsilon^{\frac{1}{2}}$$

for all λ such that $|\lambda| \leq 1$.   Then if $H(\lambda) = 1/G(\lambda)$, $H(\lambda)$ is analytic in a disk with radius larger than 1.   The power series for $H(\lambda)$ multiplied by the power series $G(\lambda)$ yields 1.

We can now solve (6-17).

***Theorem* 6.1**   Under the hypotheses stated [regarding ρ_n and $f(\theta)$], the equation (6-17) has the unique solution

$$A(\lambda) = H(\lambda)PH^*(\lambda^{-1})C(\lambda) \tag{6-19}$$

and the radius of convergence of $A(\lambda)$ exceeds 1.

Proof: We have from (6-19)

$$
\begin{aligned}
PG^*(\lambda^{-1})G(\lambda)A(\lambda) &= PG^*(\lambda^{-1})G(\lambda)H(\lambda)PH^*(\lambda^{-1})C(\lambda) \\
&= PG^*(\lambda^{-1})PH^*(\lambda^{-1})C(\lambda) && \text{(since } GH = 1\text{)} \\
&= PG^*(\lambda^{-1})H^*(\lambda^{-1})C(\lambda) && \text{(by Lemma 1)} \\
&= PC(\lambda) = C(\lambda)
\end{aligned}
$$

If there were a second solution $A_1(\lambda)$ satisfying (6-17), then, by subtraction, we would have

$$PG^*(\lambda^{-1})G(\lambda)[A(\lambda) - A_1(\lambda)] = 0$$
This means that
$$G^*(\lambda^{-1})G(\lambda)[A(\lambda) - A_1(\lambda)] = K(\lambda^{-1})$$

for some power series in λ^{-1} with no constant term.   Multiplying both sides by $H^*(\lambda^{-1})$, we get

$$G(\lambda)[A(\lambda) - A_1(\lambda)] = H^*(\lambda^{-1})K(\lambda^{-1})$$

in which the right side is a power series in λ^{-1} with no constant term, and the left side is a power series in λ.   Therefore, both sides vanish identically,

$$G(\lambda)[A(\lambda) - A_1(\lambda)] = 0$$

and we now multiply by $H(\lambda)$ and find that $A(\lambda) = A_1(\lambda)$.

Finally, let us note in (6-19) that $C(\lambda)$ is analytic if $|\lambda| < R$, for some $R > 1$. $H^*(\lambda^{-1})$ is analytic if $|\lambda| > r$, with $r < 1$, so $H^*(\lambda^{-1})C(\lambda)$ is analytic in an annulus containing $|\lambda| = 1$. Therefore, $PH^*(\lambda^{-1})C(\lambda)$ is a power series with a radius of convergence exceeding 1. So is $H(\lambda)$; therefore, so is $A(\lambda)$.

The discrete stationary problem is now solved. If we are given $\{x_n\}$, the corresponding $\{\rho_n\}$, and the random variable w, we get $\hat{w}$ from (6-11a). We use $\{\rho_n\}$ to obtain $f(\theta)$, $g(\theta) = G(e^{i\theta}) = G(\lambda) = \sum_{n=1}^{\infty} \gamma_n \lambda^n$,

as in Sec. 6-2. Then $H(\lambda) = 1/G(\lambda)$, $C(\lambda) = \sum_{n=0}^{\infty} c_n \lambda^n$ with $c_n = E(wx_{-n})$.

Finally, the α_n required in (6-11a) are the coefficients in the series expansion of $A(\lambda)$, which is given in Theorem 6-1. We shall see in Sec. 6-6 how the steps can be easily performed when $G(\lambda)$ is a rational function of λ.

6-4 THE DISCRETE PREDICTION PROBLEM

As an example of the preceding, suppose that $w = x_1$. This is the problem of predicting x one unit ahead. We have

$$E(x_1 x_{-n}) = c_n \qquad \text{or} \qquad c_n = \rho_{n+1} \qquad n = 0, 1, 2, \ldots$$

Then
$$C(\lambda) = \sum_{n=0}^{\infty} \rho_{n+1} \lambda^n$$

or, since
$$G^*(\lambda^{-1})G(\lambda) = \sum_{-\infty}^{\infty} \rho_n \lambda^n$$

we have
$$C(\lambda) = P\lambda^{-1}G^*(\lambda^{-1})G(\lambda)$$

and from (6-19) and Lemma 1,

$$A(\lambda) = H(\lambda)PH^*(\lambda^{-1})\lambda^{-1}G^*(\lambda^{-1})G(\lambda) = H(\lambda)P\lambda^{-1}G(\lambda)$$
$$= \frac{1}{G(\lambda)} \frac{G(\lambda) - G(0)}{\lambda}$$

or
$$A(\lambda) = \frac{1 - G(0)H(\lambda)}{\lambda} \qquad (6\text{-}20)$$

Let us again consider the more general situation in which w is a random variable for which $E(wx_{-n}), n = 0, 1, 2, \ldots$, tends to zero rapidly, and let us formally state the following important observation:

Theorem 6-2 The random variable $\hat{w}$ is the best linear prediction of w, given $x_0, x_{-1}, \ldots, x_{-n}, \ldots$, if and only if $\hat{w}$ is a linear combination of the x_{-n}, $n \geq 0$, and $E[(w - \hat{w})x_{-n}] = 0$, $n \geq 0$.

Proof: This follows directly from (6-11).

We shall call two random variables u and v *orthogonal* if $E(u\bar{v}) = 0$. (If v is real, the complex conjugate bar is superfluous.)   Then Theorem 6-2 says that $\hat{w}$ is the best linear prediction of w if and only if $\hat{w}$ is a linear combination of the x_{-n}, $n \geq 0$, and $w - \hat{w}$ is orthogonal to every x_{-n}, $n \geq 0$.

Now let us return to the case in which $w = x_1$.   Let $A(\lambda)$, determined by (6-20), have the power-series expansion

$$A(\lambda) = \sum_0^\infty \alpha_n \lambda^n$$

Then
$$\hat{w} = \hat{x}_1 = \sum_{n=0}^\infty \alpha_n x_{-n}$$

and if
$$v = w - \hat{w} = x_1 - \hat{x}_1$$

or
$$v = x_1 - \sum_{n=0}^\infty \alpha_n x_{-n} \tag{6-21}$$

then v is orthogonal to $x_0, x_{-1}, \ldots, x_{-n}, \ldots$ Let

$$v_m = x_m - \sum_{n=0}^\infty \alpha_n x_{-n+m-1} \tag{6-22}$$

Then v_m is obtainable from v by augmenting all the indices on the x's by $m - 1$, and since $\{x_n\}$ is stationary, it follows that v_m is orthogonal to $x_{m-1}, x_{m-2}, x_{m-3}, \ldots$ If $k > m$, then v_k is orthogonal to $x_{k-1}, x_{k-2}, \ldots$, and v_m is a linear combination of the latter x's, so v_k is orthogonal to v_m.   The conclusion is that the random variables v_m are pairwise-orthogonal.

Each v_m is a linear combination of $x_m, x_{m-1}, \ldots$ with coefficients, from (6-22), given by $1, -\alpha_0, -\alpha_1, \ldots$ The power series formed with these coefficients is, from (6-20),

$$1 - \sum_{n=1}^\infty \alpha_{n-1}\lambda^n = 1 - \lambda \sum_{n=0}^\infty \alpha_n \lambda^n = 1 - \lambda \frac{1 - G(0)H(\lambda)}{\lambda}$$

that is,
$$1 - \sum_{n=1}^\infty \alpha_{n-1}\lambda^n = G(0)H(\lambda) \tag{6-23}$$

If we let U denote the operation of transforming each x_n into x_{n-1}, so that $Ux_n = x_{n-1}$ for every n, and for every $\{c_n\}$,

$$U \sum_n c_n x_n = \sum_n c_n x_{n-1} \qquad (\text{also } U^k x_n = x_{n-k})$$

then also $Uv_m = v_{m-1}$, and (6-22) can be written

$$v_m = G(0)H(U)x_m \tag{6-24}$$

which we can invert to obtain

$$x_m = H(0)G(U)v_m \tag{6-25}$$

or
$$x_m = \frac{1}{\gamma_0}(\gamma_0 v_m + \gamma_1 v_{m-1} + \cdots)$$

or
$$x_m = \frac{1}{\gamma_0}\sum_{n=0}^{\infty}\gamma_n v_{m-n} \tag{6-26}$$

Equation (6-26) exhibits the x_m's as a one-sided moving average of orthogonal random variables. This is one of the more important results in the theory. It is true in more general situations than we are considering. In fact, whenever we can write

$$\rho_n = \frac{1}{2\pi}\int_{-\pi}^{\pi}f(\theta)e^{-in\theta}\,d\theta$$

for all n, and $\int_{-\pi}^{\pi}\log f(\theta)\,d\theta > -\infty$, then (6-26) is valid. (If the process $\{x_n\}$ is gaussian, i.e., the joint distributions of $x_m, x_{m-1}, \ldots, x_{m-k}$ are gaussian, then so is $\{v_n\}$. But orthogonal gaussian random variables are independent. So if the process $\{x_m\}$ is gaussian and our hypotheses hold, then x_m is a one-sided moving average of independent variables.)

We now have a second way, which is of theoretical importance, to solve the more general problem, with w a given random variable. We want w to be a linear combination of $x_0, x_{-1}, \ldots$ such that $w - \hat{w}$ is orthogonal to $x_0, x_{-1}, \ldots$. But by (6-22) and (6-26), this is the same as saying that w is a linear combination of $v_0, v_{-1}, v_{-2}, \ldots$, and $w - \hat{w}$ is orthogonal to $v_0, v_{-1}, v_{-2}, \ldots$. We easily find, since the v_m's are pairwise-orthogonal, that

$$\hat{w} = \sum_{n=0}^{\infty}\frac{E(wv_{-n})}{E(v^2)}\,v_{-n} \tag{6-27}$$

A simple and classical example of the above is provided by considering the case in which

$$x_n = \alpha x_{n-1} + u_n \tag{6-28}$$

where $-1 < \alpha < 1$, and the $\{u_n\}$ are independent variables, all with the same distribution. We have, automatically,

$$u_n = x_n - \alpha x_{n-1} \tag{6-29}$$

which is already in the form of (6-22).   From (6-24), we see that

$$G(0)H(\lambda) = 1 - \alpha\lambda \qquad H(0)G(\lambda) = 1 + \alpha\lambda + \alpha^2\lambda^2 + \cdots$$

$$G(\lambda) = G(0)(1 + \alpha\lambda + \alpha^2\lambda^2 + \cdots) = \frac{G(0)}{1 - \alpha\lambda}$$

$$f(\theta) = G^*(\lambda^{-1})G(\lambda) \qquad \text{for } |\lambda| = 1$$

$$= |G(0)|^2 \frac{1}{|1 - \alpha\lambda|^2} \qquad \lambda = e^{i\theta}$$

Since $\rho_0 = E(x_0{}^2) = (1/2\pi) \int_{-\pi}^{\pi} f(\theta) \, d\theta$, we have

$$|G(0)|^2 = (1 - \alpha^2)\sigma^2 \qquad \text{where} \qquad \sigma^2 = E(x_0{}^2)$$

and we may take $G(0) = \sqrt{1 - \alpha^2}\,\sigma$.   [Of course, $G(\lambda)$ is determined only up to a multiplicative constant of unit modulus.]   In this case, if $w = x_1$, we use (6-20) and find that $A(\lambda) = \alpha$, or $\hat{x}_1 = \alpha x_0$.   The best linear predictor one step ahead, in this case, is α times the last observation.   [This is also obvious from (6-29) and Theorem 6-2.]

6-5   THE PREDICTION ERROR

The error of prediction e is given by

$$e^2 = E(|w - \hat{w}|^2)$$

Since $\hat{w}$ is orthogonal to $w - \hat{w}$, this is the same as $e^2 = E(|w|^2) - E(|\hat{w}|^2)$.

In the important special case in which $w = x_1$, we have, from (6-21),

$$E(|x_1 - \hat{x}_1|^2) = E(|v|^2)$$

or

$$E(|x_1 - \hat{x}_1|^2) = E\left(x_1{}^2 - 2\sum \alpha_n x_1 x_{-n} + \sum\sum \alpha_n \alpha_m x_{-n} x_{-m}\right)$$

$$= \rho_0 - 2\sum \alpha_n \rho_{n+1} + \sum\sum \alpha_n \alpha_m \rho_{n-m}$$

$$= \rho_0 - \sum_0^\infty \alpha_n \rho_{n+1} \qquad \text{using (6-11)}$$

$$= \frac{1}{2\pi} \int_{-\pi}^{\pi} \left(1 - \sum_0^\infty \alpha_n e^{i(n+1)\theta}\right) f(\theta) \, d\theta$$

$$= \frac{1}{2\pi} \int_{-\pi}^{\pi} G(0)H(e^{i\theta})f(\theta) \, d\theta \qquad \text{using (6-23)}$$

$$= \frac{1}{2\pi} \int_{-\pi}^{\pi} G(0)H(e^{i\theta})G(e^{i\theta})\overline{G(e^{i\theta})} \, d\theta$$

$$= \frac{1}{2\pi} \int_{-\pi}^{\pi} G(0)\overline{G(e^{i\theta})} \, d\theta$$

$$= \sum_{n=0}^\infty \frac{1}{2\pi} \int_{-\pi}^{\pi} G(0)\bar{\gamma}_n e^{-in\theta} \, d\theta$$

$$= G(0)\bar{\gamma}_0 = |G(0)|^2$$

Now

$$G(\lambda) \neq 0 \qquad \text{for} \qquad |\lambda| < R, R > 1$$

so log $G(\lambda)^2$ is analytic for $|\lambda| < R$, and Re $[\log G(\lambda)^2] = \log |G(\lambda)|^2$ is harmonic. It follows, since the value of a harmonic function is the mean of its values around a circle with center at the point in question, that

$$\log |G(0)|^2 = \frac{1}{2\pi} \int_{-\pi}^{\pi} \log |G(e^{i\theta})|^2 \, d\theta$$

or

$$|G(0)|^2 = \exp \left[\frac{1}{2\pi} \int_{-\pi}^{\pi} \log f(\theta) \, d\theta \right]$$

and, finally,

$$E(|x_1 - \hat{x}_1|^2) = \exp \left[\frac{1}{2\pi} \int_{-\pi}^{\pi} \log f(\theta) \, d\theta \right] \qquad (6\text{-}30)$$

This result is true even if the integral diverges to $-\infty$, in which case $E(|x_1 - \hat{x}_1|^2) = 0$. This says that $\hat{x}_1$ is the limit of linear combinations of x_n for $n \leq 0$. A process $\{x_n\}$ for which this is true is called *deterministic*. For such a process, *every* x_k, $k > 0$ is a limit of linear combinations of x_n for $n \leq 0$, so that a knowledge of the past x_n's theoretically furnishes exact knowledge of all future x's. But simple formulas like (6-11a) for the $x_k = \hat{x}_k$ will not ordinarily exist.

6-6 THE SPECIAL CASE OF RATIONAL DENSITIES

We saw in Sec. 6-2, under our hypotheses on ρ_n, that $f(\theta) = |G(\lambda)|^2$ where $\lambda = e^{i\theta}$, and $G(\lambda) \neq 0$ for $|\lambda| \leq 1$. It may happen in some problems that we can write, by inspection, $f(\theta) = |g(\lambda)|^2$ with $g(\lambda)$ analytic for $|\lambda| \leq 1$, but $g(\lambda)$ may have zeros in $|\lambda| < 1$. In this case, if there are only a finite number of zeros, we may find $G(\lambda)$ as follows: Let the zeros of $g(\lambda)$ in $|\lambda| < 1$ be $\lambda_1, \lambda_2, \ldots, \lambda_k$ (not necessarily distinct). Then set

$$G(\lambda) = g(\lambda) \frac{(1 - \lambda \bar{\lambda}_1) \cdots (1 - \lambda \bar{\lambda}_k)}{(\lambda - \lambda_1) \cdots (\lambda - \lambda_k)}$$

The resulting analytic function $G(\lambda)$ has no zeros in $|\lambda| < 1$, and $|G(\lambda)| = |g(\lambda)|$ for $|\lambda| = 1$, because each factor $(1 - \lambda \bar{\lambda}_j)/(\lambda - \lambda_j)$ has unit absolute value for $|\lambda| = 1$, as is easily seen by multiplying the numerator by $\bar{\lambda}$, which has modulus 1 when $|\lambda| = 1$. (If there are infinitely many zeros, a similar procedure is possible, but we shall restrict ourselves to situations in which there are only a finite number.)

In the following, we shall assume that $G(\lambda)$ is a rational function of λ. Its zeros and poles will be located in $|\lambda| > 1$. Furthermore, we shall assume that the random variable w has the property that $\sum_{n=0}^{\infty} E(wx_{-n})\lambda^n$ also represents a rational function. We again set $E(wx_{-n}) = c_n$. Recall Eq. (6-19):

$$A(\lambda) = H(\lambda)PH^*(\lambda^{-1})C(\lambda)$$

where $H(\lambda) = 1/G(\lambda)$.

In our present situation, $H^*(\lambda^{-1})$ and $C(\lambda)$ are rational, so their product is also rational. If we write

$$H^*(\lambda^{-1})C(\lambda) = Q(\lambda) + \sum_j \frac{k_j}{(\lambda - \lambda_j)^{p_j}}$$

in which $Q(\lambda)$ is a polynomial, and the λ_j denote the distinct poles of $H^*(\lambda^{-1})C(\lambda)$ with multiplicity p_j, then

$$PH^*(\lambda^{-1})C(\lambda) = Q(\lambda) + \sum_{|\lambda_j|>1} \frac{k_j}{(\lambda - \lambda_j)^{p_j}}$$

Example (Filtering): Suppose $x_n = y_n + z_n$, in which z_n is a stationary process which we may call noise. We observe only the sum x_n, for $n \le 0$, and we want to obtain $\hat{y}_0$; that is, our random variable here is $w = y_0$. We suppose that y_n and z_n are uncorrelated, $E(y_n z_m) = 0$ for all m, n; we also suppose that we know $\rho_{y,n} = E(y_0 y_n)$ and $\rho_{z,n} = E(z_0 z_n)$. Then

$$\rho_{x,n} = \rho_{y,n} + \rho_{z,n}$$

and similarly for the power densities:

$$f_x(\theta) = f_y(\theta) + f_z(\theta)$$

We suppose $f_z(\theta) = 1$ for all θ,

$$f_y(\theta) = \frac{\lambda}{(\lambda - 2)(1 - 2\lambda)} \qquad \lambda = e^{i\theta}$$

Then
$$f_z(\theta) = 1 + \frac{\lambda}{(\lambda - 2)(1 - 2\lambda)}$$
$$= \frac{-2\lambda^2 + 6\lambda - 2}{(\lambda - 2)(1 - 2\lambda)}$$
$$= \frac{(\lambda - r)(\lambda - 1/r)}{(\lambda - 2)(\lambda - \frac{1}{2})}$$

in which
$$r = \tfrac{3}{2} + \tfrac{1}{2}\sqrt{5}$$

So
$$f_z(\theta) = \frac{(\lambda - r)}{(\lambda - 2)} \frac{2}{r} \frac{(1 - r\lambda)}{(1 - 2\lambda)} \frac{\bar{\lambda}}{\bar{\lambda}}$$

if $\lambda = e^{i\theta}$; finally,

$$f_z(\theta) = \frac{2}{r} \frac{(\lambda - r)}{(\lambda - 2)} \frac{(\bar{\lambda} - r)}{(\bar{\lambda} - 2)}$$

$$G(\lambda) = \sqrt{\frac{2}{r}} \frac{\lambda - r}{\lambda - 2}$$

We have also $E(wx_{-n}) = E(y_0 x_{-n}) = E(y_0 y_{-n}) = E(y_0 y_n) = \rho_{y,n}$,

so that
$$C(\lambda) = \sum_0^\infty E(wx_{-n})\lambda^n = P \sum_{n=-\infty}^\infty E(wx_{-n})\lambda^n$$

$$= P \sum_{n=-\infty}^\infty \rho_{y,n}\lambda^n = P \frac{\lambda}{(\lambda - 2)(1 - 2\lambda)}$$

Since $H(\lambda) = 1/G(\lambda) = \sqrt{r/2}\,(\lambda - 2)/(\lambda - r)$, we have

$$A(\lambda) = \frac{r}{2}\frac{\lambda - 2}{\lambda - r}\, P\, \frac{1 - 2\lambda}{1 - r\lambda}\, \frac{\lambda}{(\lambda - 2)(1 - 2\lambda)}$$

$$= \frac{r}{2}\frac{\lambda - 2}{\lambda - r}\, P\, \frac{\lambda}{(\lambda - 2)(1 - r\lambda)}$$

Now
$$\frac{\lambda}{(\lambda - 2)(1 - r\lambda)} = \frac{2/(1 - 2r)}{\lambda - 2} + \frac{K}{1 - r\lambda}$$

in which K need not be calculated, since

$$P\, \frac{\lambda}{(\lambda - 2)(1 - r\lambda)} = \frac{2}{(1 - 2r)(\lambda - 2)}$$

Finally,
$$A(\lambda) = \frac{r}{2}\frac{\lambda - 2}{\lambda - r}\, \frac{2}{(1 - 2r)(\lambda - 2)}$$

or
$$A(\lambda) = \frac{r}{1 - 2r}\frac{1}{\lambda - r} = \frac{-1}{1 - 2r}\frac{1}{1 - \lambda/r}$$

$$= \frac{-1}{1 - 2r}\left(1 + \frac{\lambda}{r} + \frac{\lambda^2}{r^2} + \cdots\right)$$

and
$$\hat{y}_0 = \frac{1}{2r - 1}\left(x_0 + \frac{1}{r}\, x_{-1} + \frac{1}{r^2}\, x_{-2} + \cdots\right)$$

6-7 THE CONTINUOUS STATIONARY CASE

Here we are given the stationary process $x(t)$, $-\infty < t < \infty$,

$$E[x(t)x(s)] = \rho(t - s) = \rho(s - t)$$

in which we suppose $\rho(t)$ is continuous.   Corresponding to what we assumed in the discrete case, we suppose here that the integral $\int_0^\infty \rho(t)e^{-pt}\, dt$ is absolutely convergent for all complex p satisfying Re $p > -\epsilon$, for some $\epsilon > 0$.   In this case, the Fourier transform $f(\omega) = \int_{-\infty}^\infty \rho(t)e^{-i\omega t}\, dt$ is analytic in a strip containing the real ω axis; $f(\omega)$, for real ω, is called the *power density* or *spectral density* of the process $x(t)$. We have $f(\omega) \geq 0$ for all real ω.   Also, $f(\omega) = f(-\omega)$, since $\rho(t) = \rho(-t)$. [That $f(\omega) \geq 0$ can be shown in much the same way as in the discrete case.   Let

$$S_T(\omega) = \int_{-T}^T x(t)e^{-i\omega t}\, dt$$

then $(1/2T)E(|S_T(\omega)|^2) \geq 0$ for all ω and all $T > 0$; that is,

$$0 \leq \frac{1}{2T}\iint\limits_{-T}^{T} \rho(t - s)e^{-i\omega(t-s)}\, dt\, ds$$

We now put $t - s = u$; we find that

$$\int_{-2T}^{2T} \left(1 - \frac{|u|}{2T}\right) \rho(u) e^{-i\omega u} \, du \geq 0 \qquad (6\text{-}31)$$

and, as $T \to \infty$, this integral converges to $f(\omega)$.]

We suppose, in addition, that $f(\omega) > 0$ for all finite (real) ω. [Of course, $f(\omega) \to 0$ as $|\omega| \to \infty$.] Finally, we suppose that

$$\int_{-\infty}^{\infty} [|\log f(\omega)|/(1 + \omega^2)] \, d\omega < \infty$$

Then, for all $\omega = u + iv$ such that $\text{Im } \omega < 0$, the integral

$$h(\omega) = h(u + iv) = \frac{1}{2\pi} \int_{-\infty}^{\infty} \log f(\zeta) \frac{-v}{(\zeta - u)^2 + v^2} \, d\zeta \qquad v < 0 \qquad (6\text{-}32)$$

is absolutely convergent and defines a harmonic function in the region $\text{Im } \omega < 0$. So $h(\omega)$ is the real part of a function $h(\omega) + ih_1(\omega)$ analytic for $\text{Im } \omega < 0$. In fact,

$$h_1(u + iv) = \frac{1}{2\pi} \lim_{A \to \infty} \int_{-A}^{A} [\log f(\zeta)] \frac{\zeta - u}{(\zeta - u)^2 + v^2} \, d\zeta$$

Thus we have

$$\lim_{v \to 0} h(\omega) = \frac{1}{2} \log f(u)$$

If we take $g(\omega) = e^{h(\omega) + ih_1(\omega)}$, then $g(\omega)$ is analytic for $\text{Im } \omega < 0$,

$$|g(\omega)| = e^{h(\omega)}$$

and, as $\text{Im } \omega \to 0$, we have $|g(u)| = \sqrt{f(u)}$, or, finally,

$$|g(\omega)|^2 = f(\omega) \qquad (6\text{-}33)$$

for real ω. Furthermore $g(\omega)$ is bounded in the lower half plane. In fact, if $0 \leq f(\zeta) \leq M$ for all real ζ, then $\log f(\zeta) \leq \log M$ and

$$h(\omega) \leq \frac{1}{2}(\log M) \frac{1}{\pi} \int_{-\infty}^{\infty} \frac{|v|}{(\zeta - u)^2 + v^2} \, d\zeta = \frac{1}{2} \log M$$

so that $|g(\omega)| = e^{h(\omega)} \leq M^{\frac{1}{2}}$ for $\text{Im } \omega < 0$. Finally, $g(\omega) \neq 0$ for $\text{Im } \omega < 0$, since g is the exponential of $h + ih_1$.

We state the following without proof: Since

$$\frac{1}{2\pi} \int_{-\infty}^{\infty} |g(\omega)|^2 \, d\omega = \rho_0 < \infty$$

there is a function $\gamma(t)$ such that

$$\int_{-\infty}^{\infty} |\gamma(t)|^2 \, dt = \frac{1}{2\pi} \int_{-\infty}^{\infty} |g(\omega)|^2 \, d\omega$$

and

$$\gamma(t) = \frac{1}{2\pi} \int_{-\infty}^{\infty} g(\omega) e^{i\omega t} \, d\omega$$

where the last equation must be interpreted as

$$\frac{1}{2\pi} \int_{-A}^{A} g(\omega)e^{i\omega t}\, d\omega \rightarrow \gamma(t)$$

for almost every t, as $A \rightarrow \infty$. Since $g(\omega)$ is regular and bounded for $\text{Im}(\omega) < 0$, $\gamma(t) = 0$ for $t < 0$. We have $g(\omega) = \int_0^{\infty} \gamma(t)e^{-i\omega t}\, dt$ for all ω such that $\text{Im}\,\omega < 0$. This corresponds to the one-sided Fourier series for $g(\theta)$ in the discrete case (Sec. 6-2). If $\varphi(t)$ and $\psi(t)$ are both square-integrable with Fourier transforms $\check{\varphi}(\omega)$ and $\check{\psi}(\omega)$, respectively, then

$$\int_{-\infty}^{\infty} \varphi(t)\overline{\psi(t)}\, dt = \frac{1}{2\pi} \int_{-\infty}^{\infty} \check{\varphi}(\omega)\overline{\check{\psi}(\omega)}\, d\omega \qquad \text{(Parseval's equality)}$$

We put $\varphi(t) = \gamma(t + s)$ and $\psi(t) = \gamma(t)$, so that $\check{\varphi}(\omega) = g(\omega)e^{i\omega s}$ and $\check{\psi}(\omega) = g(\omega)$, and we find that

$$\int_{-\infty}^{\infty} \gamma(t + s)\overline{\gamma(t)}\, dt = \frac{1}{2\pi} \int_{-\infty}^{\infty} e^{i\omega s}|g(\omega)|^2\, d\omega = \rho(s) \qquad (6\text{-}34)$$

In the first integral, since $\gamma(t) = 0$ for $t < 0$, the lower limit may be taken to be the larger of $-s$ and 0.

6-8 CONSTRUCTION OF THE CONTINUOUS - CASE ESTIMATE

Now suppose we are given the random variable w, which we want to express as a linear function of the observed values $x(t)$, for $t \leq 0$, plus an error term. Corresponding to the situation in the discrete case, we are tempted to write

$$w = \int_0^{\infty} \alpha(t)x(-t)\, dt + r \qquad (6\text{-}35)$$

where $\alpha(t)$ is to be found so that $E(|r|^2)$ is a minimum. It can be shown that a necessary condition for $E(|r|^2)$ to be a minimum is that $\alpha(t)$ satisfies the Wiener-Hopf equation

$$\int_0^{\infty} \rho(t - s)\alpha(s)\, ds = E(wx(-t)) \qquad \text{for} \qquad t \geq 0 \qquad (6\text{-}36)$$

This corresponds to Eq. (6-11) in the discrete case.

However, in many cases the best $\hat{w}$ cannot be written in the form $\int_0^{\infty} \alpha(t)x(-t)\, dt$. Although solutions in such cases can be obtained formally in which $\alpha(t)$ can be expressed in terms of delta functions and their derivatives when the density function $f(\omega)$ and the Fourier transform of $E(wx(-t))$ are simple enough, we shall use the following more rigorous procedure. (This is not to say that the delta-function approach cannot be made rigorous; it can.)

We shall associate with every random variable $x(t)$ the ω function $e^{i\omega t}g(\omega)$, where $g(\omega)$ satisfies (6-33),

$$x(t) \rightarrow e^{i\omega t}g(\omega) \tag{6-37}$$

and with every linear combination of $x(t)$'s the corresponding linear combination of $e^{i\omega t}g(\omega)$ functions,

$$\sum_\nu \alpha_\nu x(t_\nu) \rightarrow \sum_\nu \alpha_\nu e^{i\omega t_\nu}g(\omega) \tag{6-38}$$

Then if ξ is a linear function of the $x(t)$'s, and if $\xi \rightarrow Z(\omega)$, we shall have

$$E(|\xi|^2) = \frac{1}{2\pi}\int_{-\infty}^{\infty}|Z(\omega)|^2\,d\omega \tag{6-39}$$

Furthermore, if $\eta \rightarrow Y(\omega)$, then

$$E(\xi\bar{\eta}) = \frac{1}{2\pi}\int_{-\infty}^{\infty}Z(\omega)\overline{Y(\omega)}\,d\omega \tag{6-40}$$

We now suppose that w is itself a limit of linear combinations of x_t, $-\infty < t < \infty$. [If this is not true, the following procedure can still be used; of course, it is always assumed that $E(|w|^2) < \infty$.] Let $A(\omega)$ be the ω function corresponding to w. By (6-40), we have

$$E(wx(-t)) = \frac{1}{2\pi}\int_{-\infty}^{\infty}A(\omega)\overline{g(\omega)}e^{i\omega t}\,d\omega \tag{6-41}$$

for every t, and, consequently,

$$A(\omega) = \int_{-\infty}^{\infty}E(wx(-t))e^{-i\omega t}\,dt\,\frac{1}{\overline{g(\omega)}} \tag{6-42}$$

in which the integral may have to be interpreted as a limit, for example

$$\lim_{\delta \rightarrow 0^+}\int_{-\infty}^{\infty}E(wx(-t))e^{-\delta|t|}e^{-i\omega t}\,dt$$

if $\int_{-\infty}^{\infty}|E(wx(-t))|\,dt$ does not exist.

Let $B(\omega)$ be the ω function corresponding to $\hat{w}$. Then, just as in the discrete case, $\hat{w}$ is a limit of linear combinations of $x(t)$'s with $t \leq 0$ and $E(wx(-t)) = E(\hat{w}x(-t))$ for all $t \geq 0$. This is equivalent to saying that $B(\omega)$ is a limit of functions $\sum_\nu \alpha_\nu e^{-i\omega t_\nu}g(\omega)$ with $t_\nu \geq 0$, and, using (6-40),

$$\frac{1}{2\pi}\int_{-\infty}^{\infty}[A(\omega) - B(\omega)]\overline{g(\omega)}e^{i\omega t}\,d\omega = 0 \quad\text{for}\quad t \geq 0 \tag{6-43}$$

Since $e^{-i\omega t_\nu}g(\omega)$ is the Fourier transform of $\gamma(t - t_\nu)$, which vanishes for

$t < 0$ when $t_r \geq 0$, it follows that $B(\omega)$ is the Fourier transform of a t function that vanishes for $t < 0$.

We can now determine $B(\omega)$. We shall write $\mathfrak{F}[a(t)]$ for the Fourier transform of $a(t)$,

$$\mathfrak{F}[a(t)](\omega) = \int_{-\infty}^{\infty} a(t)e^{-i\omega t}\, dt$$

and $\mathfrak{F}^{-1}[\tilde{a}(\omega)]$ for the inverse transform. Set $a(t) = \mathfrak{F}^{-1}[A(\omega)]$, where $A(\omega)$ is obtainable from (6-42). Set

$$B(\omega) = \int_0^{\infty} a(t)e^{-i\omega t}\, dt \tag{6-44}$$

that is,
$$B(\omega) = \mathfrak{F}\{P\mathfrak{F}^{-1}[A(\omega)]\}$$

where, for any t function $a(t)$, $Pa(t)$ is the function equal to $a(t)$ if $t > 0$, and equal to zero if $t < 0$.

This $B(\omega)$ is the ω function corresponding to $\hat{w}$, since (1) it is the Fourier transform of a t function vanishing for $t < 0$, (2) $A(\omega) - B(\omega)$ is the Fourier transform of $a(t) - Pa(t)$, which vanishes for $t > 0$, and (3) $g(\omega)e^{-i\omega s}$ is the Fourier transform of $\gamma(t - s)$, which vanishes for $t < 0$ if $s \geq 0$, so that

$$\frac{1}{2\pi}\int_{-\infty}^{\infty} [A(\omega) - B(\omega)]\overline{g(\omega)}e^{i\omega s}\, d\omega = 0 \qquad \text{for} \qquad s \geq 0$$

which is the required Eq. (6-43).

Let us observe that in solving a particular problem of this type, any function $\hat{A}(\omega)$ of the form

$$\hat{A}(\omega) = \int_{-C}^{\infty} E(wx(-t))e^{-i\omega t}\, dt\, \frac{1}{g(\omega)} \tag{6-45}$$

with $C \geq 0$, may be used in place of $A(\omega)$. [Sometimes it is easier to get such an $\hat{A}(\omega)$ than to evaluate (6-42).] The reason is essentially that only values of $E(wx(-t))$ for $t \geq 0$ enter into the problem.

Let us also note that the basic correspondence of linear combinations of $x(t)$'s and ω functions was of the form (6-38), and for $\hat{w}$ we have $\hat{w} \to B(\omega)$, which, by analogy with (6-38), we can write as

$$\hat{w} \to C(\omega)g(\omega)$$

where
$$C(\omega) = \frac{1}{g(\omega)}\, B(\omega) \tag{6-46}$$

The function $C(\omega)$, analytic for Im $\omega < 0$, is the "transfer function" of the optimal filter in problems in which an electronic "black box" is to be found which performs the desired prediction or filtering. Although

$B(\omega)$ is square-integrable and is the transform of a square-integrable
function, the function $C(\omega)$ need not have this property. In some typical
applications $C(\omega)$ will be a rational function of ω, with poles and zeros
only in the upper half plane and possibly at $\omega = \infty$. In the latter case
we can write

$$C(\omega) = C_1(\omega) + C_2(\omega)$$

in which $C_1(\omega)$ is a polynomial,

$$C_1(\omega) = \sum_{\nu=0}^{N} c_\nu (i\omega)^\nu$$

and $C_2(\omega)$ is a rational function vanishing at $\omega = \infty$. Then $C_2(\omega)$ is
the transform of a square-integrable t function $c(t)$, and we have

$$\hat{w} = \sum_{\nu=0}^{N} c_\nu x^{(\nu)}(0) + \int_0^\infty c(t)x(-t)\, dt$$

where $x^{(\nu)}(0)$ is the νth derivative of $x(t)$ at $t = 0$. It can be shown that
such derivatives exist as limits, in the least-squares sense, of linear combinations of $x(t)$'s.

We can now summarize the solution of the continuous problem. We
are given the process $\{x(t)\}$, its corresponding correlation function $\rho(t)$,
and the random variable w. We then get $f(\omega) = \int_{-\infty}^{\infty} \rho(t)e^{-i\omega t}\, dt$; and $g(\omega)$,
regular and bounded for Im $\omega \leq 0$, can be obtained as in Sec. 6-7 such
that $f(\omega) = |g(\omega)|^2$ for real ω. The inverse Fourier transform $\gamma(t)$ of
$g(\omega)$ vanishes for negative t's, and (6-34) holds. With w given, we get
$A(\omega)$ from (6-42); call its inverse Fourier transform $a(t)$. Then $B(\omega)$ is
given by (6-44), and the transfer function $C(\omega) \equiv B(\omega)/g(\omega)$ gives
the required $\hat{w}$. If $c(t)$ is the inverse transform of $C(\omega)$, we have
$\hat{w} = \int_0^\infty c(t)x(-t)\, dt$. When $g(\omega)$ is a rational function of ω, the steps
given here can be easily performed; see Sec. 6-10.

In engineering applications, the term *physically realizable* is used to
describe a filter that does not violate the requirement that if an input
signal $x(t)$ vanishes for $t < 0$, then the output signal $y(t)$ also vanishes
for $t < 0$, that is, a filter whose transfer function $C(\omega)$ is of the form (6-46),
where $g(\omega)$ and $B(\omega)$ are of the type discussed. The mathematical necessary and sufficient condition for the existence of such a solution is that
the spectral (or power) density $f(\omega)$ satisfy

$$\int_{-\infty}^{\infty} \frac{|\log f(\omega)|}{1 + \omega^2}\, d\omega < \infty$$

In Chap. 1 we pointed out that, in the interval $0 \leq t < \infty$, a convenient complete orthonormal set of functions is the set of Laguerre functions $L_n(t)$, where

$$L_n(t) = \frac{1}{n!} e^{t/2} \frac{d^n}{dt^n} (t^n e^{-t}) \qquad n = 0, 1, 2, \ldots$$

By a change of scale on the t axis, we obtain the functions

$$\sqrt{b} \, L_n(bt) = \frac{\sqrt{b}}{n!} e^{-\frac{1}{2}bt} \frac{d^n}{dt^n} (t^n e^{-bt})$$

where b is any convenient positive constant. The Fourier transforms $l_n(\omega)$ of these functions are given by

$$l_n(\omega) = \sqrt{b} \, \frac{(i\omega - b/2)^n}{(i\omega + b/2)^{n+1}} \qquad n = 0, 1, 2, \ldots$$

Since every square-integrable realizable response function $f(t)$ [i.e., every $f(t)$ for which $f(t) = 0$ for $t < 0$, and for which $\int_0^\infty |f(t)|^2 \, dt < \infty$] can be approximated by a linear combination of Laguerre functions, it follows

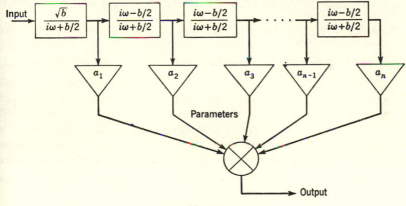

Fig. 6-1

that the transfer function of a realizable filter can be approximated by the type of filter (the Laguerre filter) given in Fig. 6-1.

6-9   THE CONTINUOUS PREDICTION PROBLEM

Let us consider the important special case in which $w = x(h)$, $h > 0$. This is the problem of prediction h units ahead. We have

$$E(wx(-s)) = \rho(h + s)$$

and, from (6-42),

$$A(\omega) = \frac{1}{g(\omega)} e^{i\omega h} f(\omega) = e^{i\omega h} g(\omega)$$

Then $a(t) = \mathcal{F}^{-1}[A(\omega)] = \gamma(t + h)$, and from (6-44) and (6-46),

$$B(\omega) = \int_0^\infty \gamma(t + h) e^{-i\omega t} \, dt$$

$$C(\omega) = \frac{1}{g(\omega)} \int_0^\infty \gamma(t + h) e^{-i\omega t} \, dt$$

Example: Suppose $\rho(t) = e^{-|t|}$. Then $f(\omega) = 2/(1 + \omega^2)$. Here we see that we can take

$$g(\omega) = \frac{\sqrt{2}}{1 + i\omega} \qquad \text{so that} \qquad \gamma(t) = \sqrt{2} \, e^{-t} H(t)$$

Here $H(t)$ is the Heaviside function, given by $H(t) = 0$ for $t < 0$, $H(t) = 1$ for $t \geq 0$. Also,

$$\int_0^\infty \gamma(t + h) e^{-i\omega t} \, dt = \sqrt{2} \int_0^\infty e^{-(t+h)} e^{-i\omega t} \, dt = \sqrt{2} \, e^{-h} \frac{1}{1 + i\omega}$$

and finally $C(\omega) = e^{-h}$. It follows that $\hat{x}(h) = e^{-h} x(0)$ in this case.

Exercise 6-2 Let $w = x(h)$, $h > 0$, and let $f(\omega) = 1/(1 + \omega^4)$. Find $\hat{x}(h)$.

Under the correspondence defined in Eq. (6-38), there is, for each t, a random variable $v(t)$ such that

$$v(t) \to \int_0^t e^{i\omega s} \, ds \qquad (6\text{-}47)$$

[or

$$v(t) \to -\int_t^0 e^{i\omega s} \, ds$$

depending on whether $t \geq 0$ or $t < 0$]. For any $\Delta t > 0$, we have

$$\Delta v(t) = v(t + \Delta t) - v(t) \to \int_t^{t+\Delta t} e^{i\omega s} \, ds$$

and, by (6-39) and Parseval's equality,

$$E(|\Delta v|^2) = \frac{1}{2\pi} \int_{-\infty}^\infty \left| \int_t^{t+\Delta t} e^{i\omega s} \, ds \right|^2 d\omega = \Delta t$$

It follows that $\Delta v / \Delta t$ does not converge to a random variable with finite variance as $\Delta t \to 0$. If $\alpha(t)$ is a smooth function of t vanishing outside a bounded interval, then

$$\sum_\nu \alpha(t_\nu) \, \Delta v(t_\nu) \to \sum_\nu \alpha(t_\nu) \int_{t_\nu}^{t_\nu + \Delta t} e^{i\omega s} \, ds$$

As max $\Delta t_\nu \to 0$, the right-hand side converges to $\int_{-\infty}^{\infty} \alpha(t)e^{i\omega t}\,dt$; therefore so does the left converge, and we have

$$\int_{-\infty}^{\infty} \alpha(t)\,dv(t) \to \int_{-\infty}^{\infty} \alpha(t)e^{i\omega t}\,dt = \tilde{\alpha}(\omega) \qquad (6\text{-}48)$$

in which the left-hand side is to be interpreted as the limit in the least-squares sense of the finite sums of the previous left-hand side. We have then,

$$E|\int \alpha(t)\,dv(t)|^2 = \int |\alpha(t)|^2\,dt \qquad (6\text{-}49)$$

and the validity of (6-48) can now be extended to the class of all square-integrable functions $\alpha(t)$. The left-hand side of (6-49) can be manipulated formally to achieve the result

$$\iint \alpha(t)\overline{\alpha(s)}\,E(dv(t)\,\overline{dv(s)}) = \int |\alpha(t)|^2\,dt \qquad (6\text{-}50)$$

One sometimes sees the statement, as a consequence of (6-50), that $v'(t)$ is a process with the correlation function

$$E(v'(t)\overline{v'(s)}) = \delta(t - s)$$

where δ is the Dirac delta function, and spectral density $f(\omega) = 1$ for all ω. Since there is no such process (the second moments would have to be infinite), such a statement should be interpreted to mean only that (6-49) is valid for all square-integrable $\alpha(t)$.

The process $v(t)$ is called an *orthogonal* process, since $E(\Delta v(t)\,\overline{\Delta v(s)}) = 0$ if Δt and Δs are small enough. If the process is also gaussian, it is called *brownian motion*, or a *Wiener* process. It is not stationary.

The "process" $v'(t)$ is stationary and is called *white noise*, since its power density $f(\omega) = 1$ (containing infinite power!) contains amplitudes of all frequencies with absolute value 1. [The white-noise process may be thought of as a sequence $\{x_\nu(t)\}$, $\nu = 1, 2, \ldots$, of stationary processes wherein $\rho_\nu(t) = \frac{1}{2}\nu e^{-\nu|t|}$ and $f_\nu(\omega) = \dfrac{1}{[1 + (1/\nu^2)\omega^2]}$. As $\nu \to \infty$, the process $x_\nu(t)$ gets closer and closer to what we would like for our white-noise process, but, of course, there is no process in the strict sense to which $x_\nu(t)$ converges.]

Let us return to (6-48) and observe that

$$\int_{-\infty}^{\infty} \alpha(t - r)\,dv(t) \to e^{i\omega r}\int_{-\infty}^{\infty} \alpha(t)e^{i\omega t}\,dt \qquad (6\text{-}51)$$

By (6-37), we have

$$\begin{aligned}
x(r) &\to e^{i\omega r}g(\omega)\\
&= e^{i\omega r}\int_0^{\infty} \gamma(t)e^{-i\omega t}\,dt\\
&= e^{i\omega r}\int_{-\infty}^0 \gamma(-t)e^{i\omega t}\,dt
\end{aligned}$$

When we compare this with (6-51), we see that

$$x(r) = \int_{-\infty}^{\infty} \gamma(-t + r)\, dv(t) \qquad \text{for all } r \tag{6-52}$$

or, equivalently,

$$x(r) = \int_0^{\infty} \gamma(t)\, dv(r - t) \tag{6-53}$$

Thus our process $x(r)$ is a one-sided moving average of an orthogonal process. This can be compared with (6-26) for the discrete case.

It is now not difficult to show that if $h > 0$, then

$$\hat{x}(h) = \int_h^{\infty} \gamma(t)\, dv(h - t)$$

$$x(h) - \hat{x}(h) = \int_0^h \gamma(t)\, dv(h - t) \tag{6-54}$$

and

$$E((x(h) - \hat{x}(h))^2) = \int_0^h |\gamma(t)|^2\, dt$$

6-10 EXAMPLES

We present here a few examples in which $f(\omega)$ is assumed to be rational in ω. $g(\omega)$ is also rational, and if we can exhibit $f(\omega)$ as $f(\omega) = |g(\omega)|^2$ for real ω, by inspection [this will always be true if we can express $f(\omega)$ as a product of factors $(\omega - \omega_j)$ and $(\omega - \kappa_j)^{-1}$], then we can easily modify $g(\omega)$ if necessary to get a new $g(\omega)$ with no poles or zeros for $\text{Im } \omega < 0$. In fact, if $\omega_1, \ldots, \omega_m$ are zeros of $g(\omega)$ with $\text{Im } \omega_j < 0$, and $\kappa_1, \ldots, \kappa_n$ are poles with $\text{Im } \kappa_j < 0$, then we can replace $g(\omega)$ with the function

$$g(\omega)\, \frac{(\omega - \bar{\omega}_1) \cdots (\omega - \bar{\omega}_m)(\omega - \kappa_1) \cdots (\omega - \kappa_n)}{(\omega - \omega_1) \cdots (\omega - \omega_m)(\omega - \bar{\kappa}_1) \cdots (\omega - \bar{\kappa}_n)}$$

Filtering

Suppose $x(t)$ is the sum of two stationary processes $y(t)$ and $z(t)$

$$x(t) = y(t) + z(t)$$

in which $y(t)$ is the desired variable, and $z(t)$ is an unwanted variable superposed on the $y(t)$. In communications problems, $y(t)$ denotes the signal or message, and $z(t)$ the noise. We observe $x(t)$ for $t \leq 0$, and we want $\hat{y}(0)$. Suppose further that $z(t)$ is white noise with spectral density $f_z(\omega) = c^2$, $c > 0$, that the spectral density $f_y(\omega)$ of $y(t)$ is rational, and that the $z(t)$ and $y(t)$ are uncorrelated. Then the corresponding correlation functions are related by $\rho_x(t) = \rho_y(t) + \rho_z(t)$, so that

$$f_x(\omega) = f_y(\omega) + f_z(\omega)$$

We have $E(y(0)x(-t)) = E(y(0)y(-t)) = \rho_y(t)$, and, by (6-45)

$$A(\omega) = f_y(\omega) \frac{1}{g(\omega)}$$

where $|g(\omega)|^2 = f_x(\omega)$. Since $f_y(\omega)$ and $g(\omega)$ are rational, we can write

$$A(\omega) = \frac{f_y(\omega)}{g(\omega)} = \sum_j \frac{\kappa_j}{(\omega - \omega_j)^{\alpha_j}}$$

in which $\{\omega_j\}$ is the set of distinct zeros of $g(\omega)$, and α_j is the multiplicity of ω_j. [Note that $A(\omega)$ can not contain a polynomial term in the last expansion, since $f_y(\omega) \to 0$ as $\omega \to \infty$.]

Now the inverse Fourier transform $a(t)$ of $A(\omega)$ is, for $t > 0$,

$$a(t) = i \sum_{\mathrm{Im}\,\omega_j > 0} \text{residues o} \frac{e^{i\omega t}}{(\omega - \omega_j)^{\alpha_j}} \kappa_j$$

or

$$a(t) = i \sum_{\mathrm{Im}\,\omega_j > 0} \kappa_j e^{it\omega_j} \frac{(it)^{\alpha_j - 1}}{(\alpha_j - 1)!} \qquad t > 0$$

Consequently, from (6-44),

$$B(\omega) = \sum_{\mathrm{Im}\,\omega_j > 0} \frac{\kappa_j}{(\omega - \omega_j)^{\alpha_j}} \tag{6-55}$$

and finally, from (6-46), the transfer function is

$$C(\omega) = \frac{1}{g(\omega)} \sum_{\mathrm{Im}\,\omega_j > 0} \frac{\kappa_j}{(\omega - \omega_j)^{\alpha_j}}$$

Then $\hat{y}(0)$ can be obtained as a linear function of the $x(t)$'s for $t \le 0$ by the discussion following (6-46).

Specifically, suppose $f_y(\omega) = (1 + \omega^2)/(\omega^4 + 6\omega^2 + 25)$; then

$$f_x(\omega) = c^2 + f_y(\omega) = \frac{c^2\omega^4 + (6c^2 + 1)\omega^2 + (25c^2 + 1)}{\omega^4 + 6\omega^2 + 25}$$

The poles of $f_x(\omega)$ are the four numbers $\pm 1 \pm 2i$. The zeros of $f_x(\omega)$ will be denoted ω_1, ω_2, $-\omega_1$, and $-\omega_2$, where $\mathrm{Im}\,\omega_1 > 0$, $\mathrm{Im}\,\omega_2 > 0$. Either ω_1 and ω_2 are purely imaginary, or else $\omega_1 = a + ib$, $\omega_2 = -a + ib$ with $a \ne 0$, $b \ne 0$. We may now take

$$g(\omega) = \frac{c(\omega - \omega_1)(\omega - \omega_2)}{(\omega - 1 - 2i)(\omega + 1 - 2i)}$$

Therefore $A(\omega) = \dfrac{1 + \omega^2}{\omega^4 + 6\omega^2 + 25} \dfrac{(\omega - 1 + 2i)(\omega + 1 + 2i)}{c(\omega - \bar\omega_1)(\omega - \bar\omega_2)}$

$$= \frac{1 + \omega^2}{c(\omega - 1 - 2i)(\omega + 1 - 2i)(\omega - \bar\omega_1)(\omega - \bar\omega_2)}$$

$$= \frac{K_1}{\omega - 1 - 2i} + \frac{K_2}{\omega + 1 - 2i} + \cdots$$

where the dots denote the terms in $1/(\omega_1 - \bar{\omega}_1)$ and $1/(\omega - \bar{\omega}_2)$, which we don't need [see Eq. (6-55)]. The constants K_1 and K_2 are easily obtained by standard methods. Then

$$B(\omega) = \frac{K_1}{\omega - 1 - 2i} + \frac{K_2}{\omega + 1 - 2i}$$

and

$$C(\omega) = \frac{(\omega - 1 - 2i)(\omega + 1 - 2i)}{c(\omega - \omega_1)(\omega - \omega_2)} \left(\frac{K_1}{\omega - 1 - 2i} + \frac{K_2}{\omega + 1 - 2i} \right)$$

$$= \frac{1}{c} \left[\frac{K_1(\omega + 1 - 2i)}{(\omega - \omega_1)(\omega - \omega_2)} + \frac{K_2(\omega - 1 - 2i)}{(\omega - \omega_1)(\omega - \omega_2)} \right]$$

This is the required transfer function, and if we calculate

$$\beta(t) = \frac{1}{2\pi} \int_{-\infty}^{\infty} C(\omega) e^{i\omega t} \, d\omega$$

then $\beta(t) = 0$ for $t < 0$, and

$$\hat{y}(0) = \int_0^{\infty} \beta(t) x(-t) \, dt$$

Filtering and Prediction

Again suppose $x(t) = y(t) + z(t)$, where $x(t)$ is observed for $t \leq 0$; this time we want $\hat{y}(h)$, $h > 0$. We now have

$$E(y(h)x(t)) = E(y(h)y(t)) = \rho_y(t - h)$$

and, from (6-42), we see that

$$A(\omega) = \frac{e^{i\omega h} f_y(\omega)}{g(\omega)} \qquad |g(\omega)|^2 = f_x(\omega) \qquad f_x(\omega) = f_y(\omega) + f_z(\omega)$$

We suppose again that $f_z(\omega) = C^2$, $f_y(\omega)$ rational, so that $g(\omega)$ is rational. This time we can write

$$A(\omega) = \frac{e^{i\omega h}}{g(\omega)} f_y(\omega) = \sum_j \frac{K_j e^{i\omega h}}{(\omega - \omega_j)^{\alpha_j}}$$

and we can now proceed as we did in the last example. However, we shall use a somewhat different, but equivalent, approach. Each term $e^{i\omega h}/(\omega - \omega_j)^{\alpha_j}$, with $\text{Im } \omega_j < 0$, is bounded and regular for $\text{Im } \omega > 0$. Therefore each such term contributes nothing to the t functions for $t > 0$. Each term $e^{i\omega h}/(\omega - \omega_j)^{\alpha_j}$, with $\text{Im } \omega_j > 0$, can be written

$$\frac{e^{i\omega h}}{(\omega - \omega_j)^{\alpha_j}} = \frac{e^{i\omega_j h}}{(\omega - \omega_j)^{\alpha_j}} \sum_{\nu=0}^{\alpha_j - 1} \frac{[i(\omega - \omega_j)h]^\nu}{\nu!} + \frac{e^{i\omega_j h}}{(\omega - \omega_j)^{\alpha_j}} \sum_{\nu=\alpha_j}^{\infty} \frac{[i(\omega - \omega_j)h]^\nu}{\nu!}$$

in which the first term is regular and bounded for Im $\omega < 0$, and the second is regular and bounded for Im $\omega > 0$ and contributes nothing to the t function for $t > 0$.   It follows that

$$B(\omega) = \sum_{\text{Im } \omega_j > 0} K_j \frac{e^{i\omega_j h}}{(\omega - \omega_j)^{\alpha_i}} \sum_{\nu=0}^{\alpha_j - 1} \frac{[i(\omega - \omega_j)h]^\nu}{\nu!}$$

and the solution continues as before.

6-11 CONDITIONAL EXPECTATION

Let $x_1, \ldots, x_N$ be random variables with finite second moments, and suppose their joint probability distribution function is the integral of a density function $f(x_1, \ldots, x_N)$.   Then the expected value of x_1 is

$$E(x_1) = \int \cdots \int_{-\infty}^{\infty} x_1 f(x_1, \ldots, x_N) \, dx_1 \cdots dx_N$$

If we restrict the variables $x_2, \ldots, x_N$ to lie in an interval

$$a_j < x_j < a_j + \Delta a_j \qquad j = 2, 3, \ldots, N$$

then the average of x_1, given that

$$a_j < x_j < a_j + \Delta a_j \qquad \text{for} \qquad 2 \leq j \leq N$$

is defined to be

$$E(x_1 | a_j < x_j < a_j + \Delta a_j; 2 \leq j \leq N)$$
$$= \frac{\int_{a_N}^{a_N + \Delta a_N} \cdots \int_{a_2}^{a_2 + \Delta a_2} \int_{-\infty}^{\infty} x_1 f(x_1, \ldots, x_N) \, dx_1 \cdots dx_N}{\int_{a_N}^{a_N + \Delta a_N} \cdots \int_{a_2}^{a_2 + \Delta a_2} \int_{-\infty}^{\infty} f(x_1, \ldots, x_N) \, dx_1 \cdots dx_N}$$

provided that the denominator does not vanish.   It is then natural to *define* the expected value of x_1, given that $x_2 = a_2, \ldots, x_N = a_N$, as

$$E(x_1 | x_2 = a_2, \ldots, x_N = a_N) = \frac{\int_{-\infty}^{\infty} x_1 f(x_1, a_2, \ldots, a_N) \, dx_1}{\int_{-\infty}^{\infty} f(x_1, a_2, \ldots, a_N) \, dx_1}$$

when the quotient exists.   We shall write, more briefly,

$$E(x_1 | x_2, \ldots, x_N) = \frac{\int_{-\infty}^{\infty} x_1 f(x_1, \ldots, x_N) \, dx_1}{\int_{-\infty}^{\infty} f(x_1, \ldots, x_N) \, dx_1} \qquad (6\text{-}56)$$

Example: Let $x_1, \ldots, x_N$ be gaussian with joint density

$$\frac{|A|^{1/2}}{(2\pi)^{N/2}} e^{-\frac{1}{2}x'Ax}$$

where x is the column vector $(x_1, \ldots, x_N)$, and x' is the transpose of x. The matrix A is symmetric and positive-definite. It is the inverse of the covariance matrix $R = \{r_{ij}\}$, $r_{ij} = E(x_i x_j)$. (The means of the x's are all zero.) Then

$$
\begin{aligned}
E(x_1|x_2, \ldots, x_N) &= \frac{\displaystyle\int_{-\infty}^{\infty} x_1 e^{-\frac{1}{2}x'Ax}\, dx_1}{\displaystyle\int_{-\infty}^{\infty} e^{-\frac{1}{2}x'Ax}\, dx_1} \\[2mm]
&= \frac{-1}{a_{11}} (a_{12}x_2 + a_{13}x_3 + \cdots + a_{1N}x_N)
\end{aligned}
$$

where a_{ij} is the ijth element of A. Note that, in the gaussian case, $E(x_1|x_2, \ldots, x_N)$ is a linear combination of $x_2, \ldots, x_N$.

Now suppose that we have observed $x_2, \ldots, x_N$ and want to estimate x_1 in the "optimum" manner (making use of $x_2, \ldots, x_N$). More precisely, we want to find that function $\varphi(x_2, \ldots, x_N)$ which makes the error $x_1 - \varphi(x_2, \ldots, x_N)$ smallest in the sense of least squares; that is, $E([x_1 - \varphi(x_2, \ldots, x_N)]^2)$ is to be made a minimum with respect to φ. This is the simplest problem in nonlinear prediction; here we do not demand a linear combination of $x_2, \ldots, x_N$. We shall, in general, get a better approximation than we would get if we demanded a linear combination, for the linear combinations are possible candidates for the solution here, along with many other more complicated candidates.

We must minimize

$$J = \int \cdots \int [x_1 - \varphi(x_2, \ldots, x_N)]^2 f(x_1, \ldots, x_N)\, dx_1, \ldots, dx_N$$

with respect to φ. (Of course, we suppose that φ has a finite second moment.) We find, in the usual manner, that, if φ is optimal,

$$\delta J = - \int \cdots \int_{-\infty}^{\infty} 2(x_1 - \varphi)\, \delta\varphi(x_2, \ldots, x_N) f(x_1, \ldots, x_N)$$

$$dx_1 \cdots dx_N = 0$$

for all $\delta\varphi(x_2, \ldots, x_N)$, or

$$\int_{-\infty}^{\infty} x_1 f(x_1, \ldots, x_N)\, dx_1 = \varphi(x_2, \ldots, x_N) \int_{-\infty}^{\infty} f(x_1, \ldots, x_N)\, dx_1$$

which says that

$$\varphi(x_2, \ldots, x_N) = E(x_1|x_2, \ldots, x_N) \tag{6-57}$$

(the conditional expectation of x_1, given $x_2, \ldots, x_N$, is the solution to our estimation problem). We see that if $x_1, \ldots, x_N$ have a gaussian distribution, then the nonlinear estimation or prediction problem has for its solution a linear combination of $x_2, \ldots, x_N$, so that the linear prediction problem gives the same solution as the more general problem. Of course this is not necessarily true if the x_i are not gaussian.

6-12 THE GENERAL ESTIMATION PROBLEM

In this and the next section, we follow Balakrishnan.[1] Suppose we have a random process $x(t)$, $0 \leq t < T$, T finite. We shall assume that the sample functions $x(t)$ are continuous with probability 1, except possibly at a finite number of t's, and such that

$$\int_0^T x(t)^2 \, dt < \infty$$

We are given another random variable w, and, having observed $x(t)$ for $0 \leq t \leq T$, we wish to estimate w. As usual, if $\hat{w}$ is our estimate, we want $E[(w - \hat{w})^2] \leq E[(w - v)^2]$ for all other possible estimates v. The candidates v, and the best v, or $\hat{w}$, are to be functions of $x(t)$, $0 \leq t \leq T$, by which we mean that they depend only on the $x(t)$'s. We are not assuming anything like stationarity here, of course. On the other hand, we do assume that all the required statistics are known. In particular, we suppose that the joint probability distributions of w, $x(t_1)$, $\ldots$, $x(t_N)$ are known for all choices of $t_1, \ldots, t_N$ and N, and, moreover, that these probabilities are expressible in terms of *densities* if the t's are all distinct.

It can be shown that, just as in the case of a finite number of variables, the best estimate $\hat{w}$ is the conditional expectation

$$\hat{w} = E(w|x(t), \, 0 \leq t \leq T)$$

(which we have defined only for a finite number of variables) and that this conditional expectation can be obtained as a limit as $N \to \infty$, in the least-squares sense, of expressions of the type

$$\bar{\varphi} = \underbrace{\int \cdots \int_0^T}_{N} \varphi(t_1, \ldots, t_N; \, x(t_1), \ldots, x(t_N)) \, dt_1 \cdots dt_N \quad (6\text{-}58)$$

So, instead of defining the conditional expectation given infinitely many variables, we shall simply take, as our class of candidates, the variables that can be expressed in the form (6-58) and their limits [which will *not* always be of the form (6-58)]. Furthermore, we shall take N to be fixed; if we can get the optimal $\hat{w}_N$ for each fixed N, then $\hat{w}$ will be $\lim_{N \to \infty} \hat{w}_N$.

In general, we shall not be able to get the $\hat{w}_N$ analytically—this is true of most problems—but after the problem is set up, it will be clear that there exists a numerical procedure which can be carried out on an automatic computer and which will yield as close an approximation to $\hat{w}_N$ as is desired.

We have

$$E((w - \bar{\varphi})^2) = E(w^2) - 2E(w\bar{\varphi}) + E(\bar{\varphi}^2)$$

in which $E(w^2)$ is constant, so that we must minimize, with respect to $\bar{\varphi}$, the expression

$$J = -2E(w\bar{\varphi}) + E(\bar{\varphi}^2)$$

From (6-58), we get

$$J = -2 \int_{T_N} \int_{-\infty}^{\infty} \int_{R_N} w\varphi(t,u)q(t,u,w) \, du \, dw \, dt$$
$$+ \int_{T_N} \int_{T_N} \int_{R_N} \int_{R_N} \varphi(t,u)\varphi(s,v)p(t,s;u,v) \, du \, dv \, dt \, ds \quad (6\text{-}59)$$

in which we have written

$$\int_{T_N} \text{ for } \overbrace{\int_0^T \cdots \int_0^T}^{N} \qquad \int_{R_N} \text{ for } \overbrace{\int_{-\infty}^{\infty} \cdots \int_{-\infty}^{\infty}}^{N}$$

and where $\qquad t = (t_1, \ldots, t_N) \qquad u = (u_1, \ldots, u_N)$

$q(t,u,w)$ is the joint density of $(x(t_1), \ldots, x(t_N), w)$ at the point $(u_1, \ldots, u_n, w)$, and $p(t,s;u,v)$ is the joint density of $(x(t_1), \ldots, x(t_N), x(s_1), \ldots, x(s_N))$ at the point $(u_1, \ldots, u_n, v_1, \ldots, v_n)$.

Let

$$g(t,u) = \frac{\int_{-\infty}^{\infty} wq(t,u,w) \, dw}{\int_{-\infty}^{\infty} q(t,u,w) \, dw}$$

Then $\quad g(t,x(t)) = E(w|x(t)) \qquad t = (t_1, \ldots, t_N)$
$$x(t) = (x(t_1), \ldots, x(t_N))$$

[In some problems, $E(w|x(t))$ can be obtained a priori.]   Since

$$\int_{-\infty}^{\infty} q(t,u,w) \, dw = p(t,u)$$

in which $p(t,x(t))$ is the joint density of $(x(t_1), \ldots, x(t_N))$, we can write (6-59) as

$$J = -2 \int_{T_N} \int_{R_N} \varphi(t,u)g(t,u)p(t,u) \, du \, dt$$
$$+ \int_{T_N} \int_{T_N} \int_{R_N} \int_{R_N} \varphi(t,u)\varphi(s,v)p(t,s;u,v) \, du \, dv \, dt \, ds \quad (6\text{-}60)$$

Finally, let

$$p(s,v|t,u) = \frac{p(t,s;u,v)}{p(t,u)}$$

then $p(s,x(s)|t,x(t))$ is the conditional joint density of $(x(s_1), \ldots, x(s_N))$ given $(x(t_1), \ldots, x(t_N))$, and we can write

$$J = -2 \int_{T_N} \int_{R_N} \varphi(t,u)g(t,u)p(t,u) \, du \, dt$$
$$+ \int_{T_N} \int_{R_N} \int_{T_N} \int_{R_N} \varphi(t,u)\varphi(s,v)p(s,v|t,u)p(t,u) \, dv \, ds \, du \, dt \quad (6\text{-}61)$$

If we proceed in the usual fashion, replacing φ with $\varphi + \delta\varphi$, we find that J will be a minimum if there is a $\varphi(t,u)$ such that

$$\int_{T_N} \int_{R_N} \varphi(s,v)p(s,v|t,u) \, dv \, ds = g(t,u) \quad (6\text{-}62)$$

In general, there will be no such φ. However, there is a sequence $\{\varphi_n\}$ such that

$$\lim_{n \to \infty} J(\varphi_n) = \inf_{\varphi} J(\varphi)$$

and, for all practical purposes, we can be satisfied with such a sequence $\{\varphi_n\}$, provided that we have some procedure for constructing it. In a specific application, then, we would stop the sequence at some suitable point and use the last $\varphi_n(t,u)$ and the corresponding

$$\bar{\varphi}_n = \int_{T_N} \varphi_n(t,x(t)) \, dt$$

as our approximation to $\hat{w}_N$ (for fixed N).

Now it turns out that the steepest-descent method (Chap. 2) for getting the minimum of quadratic functions in finite-dimensional spaces is also valid in infinite-dimensional inner-product spaces, as we shall see. Let U be the set of all real functions $\varphi(t,u)$ for which

$$\int_{T_N} \int_{R_N} \varphi(t,u)^2 p(t,u) \, du \, dt < \infty$$

Define
$$(\varphi,\psi) = \int_{T_N} \int_{R_N} \varphi(t,u)\psi(t,u)p(t,u) \, dt \, du$$

Then U is a real vector space with an inner product.

The linear transformation A defined by

$$\varphi \xrightarrow{A} \psi(t,u) = \int_{T_N} \int_{R_N} \varphi(s,v)p(s,v|t,u) \, dv \, ds \quad (6\text{-}63)$$

satisfies the relations

$$(A\varphi_1,\varphi_2) = (\varphi_1,A\varphi_2) \quad (6\text{-}64)$$

and
$$\|A\varphi\|^2 = (A\varphi,A\varphi) \le |T_N|^2(\varphi,\varphi) \quad (6\text{-}65)$$

in which $|T_N|$ is the volume of the bounded N-dimensional interval

$$0 \le t_i \le T \qquad i = 1, \ldots, N$$

that is, $|T_N| = T^N$. Furthermore, we have

$$(A\varphi, \varphi) \geq 0 \tag{6-66}$$

for every φ in U.

Exercise 6-3   Verify (6-64) through (6-66).   Use the following facts:

1.
$$p(s,v|t,u)p(t,u) = p(t,s;u,v)$$
so that
$$p(t,u|s,v)p(s,v) = p(s,t;v,u)$$

2. From (6-63) and the Schwartz inequality,

$$|\psi(t,u)|^2 \leq \int_{T_N} \int_{R_N} p(s,v|t,u) \, dv \, ds \int_{T_N} \int_{R_N} \varphi^2(s,v) p(s,v|t,u) dv \, ds$$

3.
$$\int_{R_N} p(s,v|t,u) \, dv = 1$$

4.
$$\int_{R_N} p(t,s;u,v) \, du = p(s,v)$$

5.
$$(A\varphi, \varphi) = \int_{T_N} \int_{T_N} E(\varphi(s,x(s))\varphi(t,x(t)) \, ds \, dt$$
$$= E\left\{ \left[\int_{T_N} \varphi(s,x(s)) \, ds \right]^2 \right\}$$

Since $J(\varphi) = (A\varphi, \varphi) - 2(\varphi, g)$, the computational procedure described in Chap. 2 will yield a sequence $\{\varphi_n\}$ for which $J(\varphi_n)$ is monotone-decreasing.   Recall that, in the steepest-descent method, we choose φ_0 arbitrarily and then define, for each n,

$$r_n = A\phi_n - g \qquad \phi_{n+1} = \phi_n - \lambda_n r_n$$
with
$$\lambda_n = \frac{(r_n, r_n)}{(Ar_n, r_n)}$$

[If $(Ar_n, r_n) = 0$ for some n, then we also have $r_n = A\varphi_n - g = 0$, and we have our minimum, $J(\varphi_n)$.   Indeed, since

$$E[(w - \bar{\varphi})^2] \geq 0$$
we have
$$J(\bar{\varphi}) \geq -E(w^2)$$

for all φ in U.   Now if $(Ar_n, r_n) = 0$, then

$$J(\varphi_n + \lambda r_n) = J(\varphi_n) + 2(A\varphi_n - g, r_n)\lambda$$

so that either $(A\varphi_n - g, r_n) \neq 0$ and

$$\inf_\lambda J(\varphi_n + \lambda r_n) = -\infty$$

which is impossible, or else

$$(A\varphi_n - g, r_n) = 0 = (r_n, r_n) \qquad \text{or} \qquad r_n = A\varphi_n - g = 0]$$

Then, just as in Chap. 2, we have

$$J(\varphi_{n+1}) - J(\varphi_n) = - \frac{(r_n, r_n)^2}{(Ar_n, r_n)}$$

so that

$$\sum_{n=0}^{\infty} \frac{(r_n, r_n)^2}{(Ar_n, r_n)} < \infty$$

and, by the Schwartz inequality and (6-65), we have

$$|(A\varphi, \varphi)| \le \sqrt{(A\varphi, A\varphi)} \sqrt{(\varphi, \varphi)} \le |T_N|(\varphi, \varphi)$$

so that

$$\frac{(r_n, r_n)^2}{(Ar_n, r_n)} \ge \frac{(r_n, r_n)}{|T_N|}$$

Consequently, $\displaystyle\sum_0^{\infty} (r_n, r_n) < \infty$ and $r_n = A\varphi_n - g \to 0$ as $n \to \infty$

(However, φ_n does not, in general, converge.) But this is enough to imply that

$$\lim_{n \to \infty} J(\varphi_n) = \inf_{\varphi} J(\varphi) = -E(w^2)$$

Exercise 6-4 Prove the last assertion. [If the assertion is false, then for some y in U and $\epsilon > 0$,

$$J(y) + \epsilon < J(\varphi_n)$$

for all n, so that

$$\epsilon \le 2 \lim_{n \to \infty} (r_n, \varphi_n - y) = 2 \lim_{n \to \infty} (r_n, \varphi_n)$$

But $\dfrac{1}{\lambda_n}[(\varphi_{n+1}, \varphi_{n+1}) - (\varphi_n, \varphi_n)] = \dfrac{1}{\lambda_n}(\varphi_{n+1} - \varphi_n, \varphi_{n+1} + \varphi_n) = -(r_n, 2\varphi_n - \lambda_n r_n)$

$$= -2(r_n, \varphi_n) + \lambda_n(r_n, r_n)$$

and

$$\lambda_n(r_n, r_n) \to 0$$

since

$$\sum_0^{\infty} \lambda_n(r_n, r_n) < \infty$$

as was remarked above. Therefore, the sequence $\{(\varphi_n, \varphi_n)\}$ is monotone-decreasing for n larger than some n_0, so that $\{(\varphi_n, \varphi_n)\}$ is bounded. But we had $\epsilon \le 2 \lim (r_n, \varphi_n)$, and since $r_n \to 0$ and φ_n is bounded, the Schwartz inequality yields $\epsilon \le 0$, a contradiction.]

6-13 POLYNOMIAL ESTIMATION

The preceding discussion was based on the assumption that all the joint densities of $(w, x(t_1), \ldots, x(t_N))$ were known. Balakrishnan[1] describes a different approach in which we assume only the knowledge of all mixed moments

$$E(wx(t_1) \cdots x(t_N)) \quad \text{and} \quad E(x(t_1) \cdots x(t_N))$$

for all N, and all $t_1, \ldots, t_N$. Of course, we suppose that all these moments are finite. In this approach we consider, as possible candidates for approximating $\hat{w}$, all random variables of the form

$$\bar{\varphi}_N = K_0 + \int_0^T K_1(t)x(t)\,dt + \int_0^T \int_0^T K_2(t_1,t_2)x(t_1)x(t_2)\,dt_1\,dt_2 + \cdots$$
$$+ \underbrace{\int_0^T \cdots \int_0^T}_{N} K_N(t_1,t_2,\ldots,t_N)x(t_1)\cdots x(t_N)\,dt_1 \cdots dt_N$$

$$(6\text{-}67)$$

in which we may suppose the $K_j(t_1, \ldots, t_j)$ to be continuous. We require that $E(\bar{\varphi}_N{}^2) < \infty$, which is equivalent to

$$\sum_{i=0}^N \sum_{j=0}^N \underbrace{\int \cdots \int_0}_{i} \underbrace{\int \cdots \int_0}_{j} K_i(t_1,\ldots,t_i)K_j(s_1,\ldots,s_j)$$
$$R_{ij}(t_1,\ldots,t_i,s_1,\ldots,s_j)\,dt_1 \cdots dt_i\,ds_1 \cdots ds_j < \infty \quad (6\text{-}68)$$

in which

$$R_{ij}(t_1,\ldots,t_i,s_1,\ldots,s_j) = E(x(t_1)\cdots x(t_i)x(s_1)\cdots x(s_j)) \quad (6\text{-}69)$$

is supposed to be jointly continuous in all its arguments.

Now in this procedure we consider the class of all random variables $\bar{\varphi}_N$ of the form (6-67) in which the N takes on all integral values 1, 2, . . . , and in which the K's satisfy (6-68), and all limits of sequences of such variables. If the latter class (including its limits) coincides with the class defined (in Sec. 6-12) by elements of the form (6-58), then using random variables of the form (6-67) will be enough to get as good an approximation to

$$\hat{w} = E(w|x(t),\ 0 \le t \le T) \quad (6\text{-}70)$$

as we desire. This is not always the case; however, the former class, defined by elements of the form (6-67), is still quite large, and it is still useful to be able to get a member of this class as close as possible to $\hat{w}$. Call this member w^0; then to summarize: The elements of the form (6-67) and the limits of sequences of such elements form a linear class of random variables; w^0 is that member of the class which is closest to w, in which we use the metric (or measure of closeness) $[E(w - w^0)^2)]^{1/2}$, as usual. No general conditions are known under which we can be sure that $\hat{w} = w^0$, but it is still useful to be able to calculate w^0 or a sequence converging to w^0.

[If the $x(t)$ process is gaussian, we have $\hat{w} = w^0$. If the $x(t)$'s are uniformly bounded, i.e., there is an M such that $|x(t)| \le M$ for all t in $0 \le t \le T$ with probability 1, we have the same conclusion.]

Exercise 6-5 To provide an example in which $\hat{w} \neq w^0$, it is enough to find a one-dimensional distribution $F(\xi)$ for which all moments

$$\int_{-\infty}^{\infty} |\xi|^n \, dF(\xi) < \infty$$

and such that the class of polynomials $P(\xi)$ is not dense in the class of functions $f(\xi)$ for which $\int_{\infty}^{\infty} f(\xi)^2 \, dF(\xi) < \infty$.

(a) Let $\varphi(t)$ be a fixed function of t,

$$0 \leq t \leq T \qquad \int_0^T \phi(t)^2 \, dt = 1$$

Let $x(t) = \xi\varphi(t)$, where ξ is a random variable with distribution function $F(\xi)$. Then $\{x(t)\}$ is a random process (depending on a single random parameter!). Let $w = g(\xi)$ be any function of ξ with a finite second moment. Then

$$\hat{w} = E(w|x(t), 0 \leq t \leq T) = E(w|\xi) = E(g(\xi)|\xi) = g(\xi) = w$$

But w^0 is a limit of sequences of random variables of the form (6-67), which in this case means random variables of the form

$$\tilde{\varphi}_N = a_0 + a_1\xi + a_2\xi^2 + \cdots a_N\xi^N$$

i.e., polynomials in ξ. So if $g(\xi)$ is not a limit of polynomials in ξ (least-squares sense), we have our required example.

(b) Let ξ have the probability density

$$p(\xi) = \tfrac{1}{24}e^{-\xi^{1/4}} \text{ for } \xi \geq 0 \qquad p(\xi) = 0 \text{ for } \xi < 0$$

Let

$$f(\xi) = \sin \xi^{1/4} \qquad \xi > 0$$

Then ξ has finite moments of all orders, and $f(\xi)$ is bounded, so it has a finite second moment, but $f(\xi)$ is not the limit (in least squares) of polynomials in ξ (see Ref. 21, vol. 1, Prob. 153).

We shall call the random variables (6-67) polynomials in $x(t)$. Again, if w^0 is such that

$$E((w - w^0)^2) \leq E((w - v)^2)$$

for every polynomial v, then

$$E(w^0 v) = E(wv) \tag{6-71}$$

for every such v. If w^0 is itself a polynomial,

$$w^0 = \sum_{\nu=0}^N \underbrace{\int \cdots \int_0}_{\nu} M_\nu(t, \ldots, t_\nu) x(t_1) \cdots x(t_\nu) \, dt_1 \cdots dt_\nu \tag{6-72}$$

then take $v = \underbrace{\int \cdots \int_0}_{\lambda} K_\lambda(t_1, \ldots, t_\lambda) x(t_1) \cdots x(t_\lambda) \, dt_1 \cdots dt_\lambda$

to conclude from (6-71), with $G_\lambda(t_1, \ldots, t_\lambda) \equiv E(wx(t_1) \cdots x(t_\lambda))$, that

$$\sum_{\nu=0}^{N} \underbrace{\int_0^T \cdots \int}_{\nu} M_\nu(t_1, \ldots, t_\nu) \underbrace{\int_0^T \cdots \int}_{\lambda} K_\lambda(s_1, \ldots, s_\lambda)$$

$$R_{\nu\lambda}(t_1, \ldots, t_\nu, s_1, \ldots, s_\lambda) \, dt_1 \cdots dt_\nu \, ds_1 \cdots ds_\lambda$$

$$= \underbrace{\int_0^T \cdots \int}_{\lambda} K_\lambda(s_1, \ldots, s_\lambda) G_\lambda(s_1, \ldots, s_\lambda) \, ds_1 \cdots ds_\lambda \quad (6\text{-}73)$$

valid for all λ and all K_λ.   The result is that

$$\sum_{\nu=0}^{N} \underbrace{\int_0^T \cdots \int}_{\nu} M_\nu(t_1, \ldots, t_\nu) R_{\nu\lambda}(t_1, \ldots, t_\nu, s_1, \ldots, s_\lambda) \, dt_1 \cdots dt_\nu$$

$$= G_\lambda(s_1, \ldots, s_\lambda) \quad (6\text{-}74)$$

for $\lambda = 0, 1, \ldots, N$ and for all $(s_1, \ldots, s_\lambda)$. This is a system of integral equations for the unknown functions M_ν; as usual, the system will not have a solution in general, because w^0 will ordinarily not be of the form (6-72), but rather a limit of a sequence of polynomials (6-72); we shall be satisfied with such a system. (We are keeping N fixed, of course.)

Following the general procedure for such problems, we want to form a vector space V with an inner product in which the left side of (6-74) can be regarded as a "nice" linear transformation. We take for V the class of all $(N + 1)$-tuples

$$(K_0, K_1(t_1), K_2(t_1, t_2), \ldots, K_N(t_1, \ldots, t_N))$$

in which the νth element is a function of $(t_1, \ldots, t_\nu)$, such that

$$K_0{}^2 + \sum_{\nu=1}^{N} \underbrace{\int_0^T \cdots \int}_{\nu} K_\nu(t_1, \ldots, t_\nu)^2 \, dt_1 \cdots dt_\nu < \infty$$

Then our inner product of $(K_0, \ldots, K_N)$ and $(L_0, \ldots, L_N)$ will, of course, be the number

$$K_0 L_0 + \sum_{\nu=1}^{N} \underbrace{\int_0^T \cdots \int}_{} K_\nu(t_1, \ldots, t_\nu) L_\nu(t_1, \ldots, t_\nu) \, dt_1 \cdots dt_\nu \quad (6\text{-}75)$$

The transformation A defined by the left side of (6-74) sends $(K_0, K_1, \ldots, K_N)$ into the $(N + 1)$-tuple $(L_0, \ldots, L_N)$ in which

$$L_i(t_1, \ldots, t_i) = \sum_{j=0}^{N} \underbrace{\int \cdots \int}_{j} R_{ij}(t_1 \cdots t_i, s_1, \ldots, s_j)$$

$$K_j(s_1, \ldots, s_j) \, ds_1, \cdots ds_j$$

We shall then have

$$0 \leq (AK, K) = (K, AK) \leq c(K, K)$$

for every K in V, (K here denotes the generic $(N + 1)$-tuple) in which

$$c = \sum_{\nu=0}^{N} \underbrace{\int \cdots \int}_{\nu} R_{\nu\nu}(t_1, \ldots, t_\nu, t_1, \ldots, t_\nu) \, dt_1 \cdots dt_\nu$$

[Note that $R_{ij}(t_1, \ldots, t_i, s_1, \ldots, s_j)$ depends only on the sum $i + j$ and is symmetric in the arguments $t_1, \ldots, t_i, s_1, \ldots, s_j$.]

Then, just as in Sec. 6-12, we can write (6-74) as $AM = G$ and obtain a sequence $\{M_n\}$ of $(N + 1)$-tuples which minimizes the function

$$J(M) = (AM, M) - 2(M, G)$$

by means of the steepest-descent method. When $J(M_n)$ gets close enough to its limit, we stop the sequence and use the random variable

$$\sum_{\nu=0}^{N} \underbrace{\int_0^T \cdots \int}_{\nu} M_{\nu,n}(t_1, \ldots, t_\nu) x(t_1) \cdots x(t_\nu) \, dt_1 \cdots dt_\nu$$

in which $M_{\nu,n}(t_1, \ldots, t_\nu)$ is the νth component of M_n, as our approximation to w^0, for fixed N.

6-14 THE KARHUNEN - LOÉVE EXPANSION

It is of some interest to consider the method of the last section in the situation in which $N = 1$. The system (6-74) now becomes

$$R_{00} M_0 + \int_0^T R_{01}(t) M_1(t) \, dt = G_0$$

$$R_{10}(s) M_0 + \int_0^T R_{11}(s, t) M_1(t) \, dt = G_1(t)$$

where $R_{00} = E(1) = 1$
$R_{01}(t) = E(x(t)) = R_{10}(t)$
$R_{11}(s, t) = E(x(s)s(t))$
$G_0 = E(w)$
$G_1(t) = E(wx(t))$

There is no loss of generality in supposing here that $E(x(t)) = 0$ for all t, that is, $R_{01}(t) = 0$. We get $M_0 = G_0 = E(w)$, and

$$\int_0^T R(s,t)M_1(t)\,dt = G_1(t) = E(wx(t)) \tag{6-76}$$

[We have dropped the subscripts on $R(s,t)$.]

This is just the problem of linear prediction in which we do not assume that the process is stationary and in which we observe the $x(t)$ only for $0 \le t \le T$. We suppose $R(s,t)$ to be continuous in (s,t).

Again, (6-76) may not have a solution; even if it does, it will not in general be expressible in simple terms, and the method of steepest descent can be used to get the required sequence $\{M_n(t)\}$. But we can say a little more here.

We know, from Mercer's theorem in the theory of integral equations, since our kernel is symmetric, continuous, and positive-definite, that there is a complete orthonormal set $\{\varphi_n(t)\}$ of eigenfunctions

$$\int_0^T R(s,t)\varphi_n(t)\,dt = \lambda_n\varphi_n(s)$$

$$n = 1, 2, 3, \ldots \qquad \lambda_1 \ge \lambda_2 \ge \cdots \ge 0 \qquad \sum_{n=1}^\infty \lambda_n{}^2 < \infty$$

and
$$R(s,t) = \sum_{n=1}^\infty \lambda_n\varphi_n(s)\varphi_n(t) \tag{6-77}$$

in which the series converges uniformly.

Now the fact that $R(s,t)$ is continuous, and hence bounded, implies that the sample functions $x(t)$ satisfy $\int_0^T x(t)^2\,dt < \infty$ with probability 1. Therefore, $x(t)$ can be expanded in terms of the eigenfunctions,

$$x(t) = \sum_{n=1}^\infty \xi_n\varphi_n(t) \tag{6-78}$$

in which the *random variables* ξ_n are obtained thus:

$$\xi_n = \int_0^T x(t)\varphi_n(t)\,dt \tag{6-79}$$

It follows from (6-78) that

$$E(x(t)x(s)) = \sum_{n=1}^\infty \sum_{m=1}^\infty \varphi_n(t)\varphi_m(s)E(\xi_n\xi_m)$$

Comparison with (6-77) yields

$$E(\xi_n\xi_m) = \delta_{nm}\lambda_n \qquad \delta_{nm} = \begin{cases} 1 & n = m \\ 0 & n \ne m \end{cases} \tag{6-80}$$

The variables $\{\xi_n\}$ are uncorrelated. If the process is gaussian, the ξ_n are independent. The expansion (6-78) is called the Karhunen-Loéve expansion.

If w is any random variable with finite second moment, then

$$\hat{w} = \lim_{n \to \infty} E(w|\xi_1, \ldots, \xi_n) \qquad (6\text{-}81)$$

where the limit is taken in the least-squares sense.

6-15 DYNAMICAL SYSTEMS WITH CONTROL VARIABLES

Heretofore, we have used the least-squares criterion to determine the best of a given class of possible estimates. In many applications, it is more convenient to use the *maximum-likelihood* estimate, which we now define. Let x be a (real or vector-valued) random variable with a probability density function $f(x;\theta)$ depending on a parameter θ; θ may be real or vector-valued. In many situations, θ is, or may be regarded as, a random variable, and we may have an a priori knowledge of its probability density function $g(\theta)$. Then the joint probability density of x and θ is $f(x;\theta)g(\theta)$. If x is observed, we can calculate the conditional probability of θ given x:

$$h(\theta|x) = \frac{f(x;\theta)g(\theta)}{\int_{-\infty}^{\infty} f(x;\theta')g(\theta')\,d\theta'} \qquad (6\text{-}82)$$

This is called the *a posteriori* probability density of θ, having observed x; Eq. (6-82) is Bayes's theorem. Similar considerations hold in the discrete case.

The maximum-likelihood estimate of θ, given that x is observed, is that θ^* for which $h(\theta|x)$ is largest. Since the denominator in (6-82) does not contain θ, we can say that θ^* is any θ that maximizes $f(x;\theta)g(\theta)$.

If θ is not regarded as a random variable, or if no *a priori* $g(\theta)$ is known, we may still look on θ as a random variable with a uniform distribution in some region. Then the maximum-likelihood estimate of θ is any θ^* that maximizes $f(x;\theta)$.

Example: Let $x_1, x_2, \ldots, x_n$ be independent samples from the one-dimensional distribution

$$\varphi(x) = \frac{1}{\sqrt{2\pi}\,\sigma} e^{-(x-\theta)^2/2\sigma^2}$$

in which σ is known, and θ is unknown. Then $(x_1, \ldots, x_n)$ is a sample from

$$f(x) = \frac{1}{(2\pi)^{n/2}\sigma^n} \exp\left[-\sum_{i=1}^{n} \frac{(x_i - \theta)^2}{2\sigma^2}\right]$$

The maximum-likelihood estimate of θ is therefore

$$\theta^* = \frac{1}{n} \sum_{i=1}^{n} x_i$$

This is the same as the least-squares estimate, and we see that, in the gaussian case, the maximum-likelihood and minimum-variance estimates coincide. The estimation procedure given below can be viewed as the solution to the problem of estimating control variables $u_i(t)$ and state variables $x_j(t)$ of a physical system containing gaussian noise. We suppose that mean values $\overline{x(t_0)}$ and $\overline{u(t)}$ are given, perhaps as the solution of an optimum control problem, in which random disturbances have been neglected. It is also assumed that we are given observations $z(t)$ which contain random errors; here, $z(t)$ is a known function of the state variables $x(t)$. What is required is the optimal estimate of $x(t)$ having observed $z(t)$. The same problem arises in the determination of the optimal filter for a message process if there is a dynamical system, i.e., a system of differential equations, that provides a sufficiently accurate model for the message process. In this context, $u(t)$ is regarded as noise (Ref. 10a).

Now, following Bryson and Frazier,[3a] we suppose that we are given a dynamic system governed by the possibly nonlinear differential equations

$$\dot{x} = f(x,u,t) \tag{6-83}$$

where x, f, and u are matrices of orders $n \times 1$, $n \times 1$, and $m \times 1$. The u_i may be regarded as forcing functions or control functions. It is assumed that the initial conditions $x(t_0)$ come from a gaussian distribution with known mean $\bar{x}_0$ and known covariance matrix

$$P_0 = E\{[x(t_0) - \bar{x}_0][x(t_0) - \bar{x}_0]'\} \tag{6-84}$$

The forcing function $u(t)$ is a vector random process (gaussian) with known mean $\bar{u}(t)$ and known covariance

$$E[(u(t) - \bar{u}(t))(u(\tau) - \bar{u}(\tau))] = Q(t)\, \delta(t - \tau)$$

where $\delta(t - \tau)$ is the Dirac delta function; $u(t)$ is independent of $u(\tau)$ for $t \neq \tau$. The quantities $x(t_0)$ and $u(\tau)$, for $t_0 \leq \tau \leq t_f$, will play the role of the parameter θ.

We suppose also that we are given a theoretical relationship

$$\bar{z}(t) = h(x(t),t) \tag{6-85}$$

in which $\bar{z}$ and h are $p \times 1$ matrices, between the variables $x(t)$ and observable variables $\bar{z}(t)$. Because of the possible errors in measuring $\bar{z}(t)$, we actually observe

$$z(t) = h(x(t),t) + v(t)$$

for $t_0 \leq t \leq t_f$, in which $v(t)$ is gaussian with mean zero and known covariance matrix

$$E[(z(t) - \bar{z}(t))(z(\tau) - \bar{z}(\tau))] = R(t)\, \delta(t - \tau) \qquad (6\text{-}86)$$

Again $z(t)$ and $z(\tau)$ are independent for $\tau \neq t$. The quantities $z(t)$, for $t_0 \leq t \leq t_f$, play the role of our observed sample x described earlier.

We are to find the maximum-likelihood estimates of $x(t_0)$ and $u(t)$, $t_0 \leq t \leq t_f$. This will also yield the maximum-likelihood estimate of $x(t)$, $t_0 \leq t \leq t_f$, since $x(t)$ is a function of $x(t_0)$ and $u(t)$ through (6-83).

If we introduce points

$$t_0 < t_1 < \cdots < t_N = t_f$$

with $t_i - t_{i-1} = \Delta t$, and set

$$\zeta_i = \int_{t_{i-1}}^{t_i} (z(t) - \bar{z}(t))\, dt$$

$$\eta_i = \int_{t_{i-1}}^{t_i} (u(t) - \bar{u}(t))\, dt \qquad i = 1\ \ldots, n$$

then ζ_i and η_i are gaussian, with mean zero. We then have

$$E(\zeta_i \zeta_j') = \delta_{ij} \int_{t_{i-1}}^{t_i} R(t)\, dt \approx R(t_i)\, \Delta t \delta_{ij}$$

$$E(\eta_i \eta_j') = \delta_{ij} \int_{t_{i-1}}^{t_i} Q(t)\, dt \approx Q(t_i)\, \Delta t \delta_{ij}$$

If we set

$$\xi = x(t_0) - \bar{x}_0$$

then the joint probability density of $\zeta_1, \ldots, \zeta_N, \xi, \eta_1, \ldots, \eta_n$ is

$$k \exp \left\{ -\frac{1}{2} \left[\sum_{i=1}^{N} \zeta_i' R(t_i)^{-1}\, \Delta t^{-1} \zeta_i + \xi' P_0^{-1} \xi + \sum_{i=1}^{N} \eta_i' Q(t_i)^{-1}\, \Delta t^{-1} \eta_i \right] \right\}$$

where k is the appropriate constant. This quantity will be a maximum when the expression in brackets is a minimum. If we now replace ζ_i formally with $(z(t_i)) - \bar{z}(t_i))\, \Delta t$, and η_i with $(u(t_i) - \bar{u}(t_i))\, \Delta t$, and let $\Delta t \to 0$, we find that the expression to be minimized, with respect to $x(t_0)$ and $u(t)$, $t_0 \leq t \leq t_f$, is

$$J = (x(t_0) - \bar{x}_0)' P_0^{-1} (x(t_0) - \bar{x}_0)$$
$$+ \int_{t_0}^{t_f} [(z(t) - \bar{z}(t))' R(t)^{-1}(z(t) - \bar{z}(t))$$
$$+ (u(t) - \bar{u}(t))' Q(t)^{-1}(u(t) - \bar{u}(t))]\, dt \qquad (6\text{-}87)$$

in which $z(t)$ is known (observed), and

$$\bar{z}(t) = h(x(t),t) \qquad \text{and} \qquad \dot{x}(t) = f(x,u,t)$$

[Of course, $\bar{x}_0$ and $\bar{u}(t)$ are also known.] The last two relations are constraints involving $\bar{z}(t), x(t)$, and $u(t)$. We have here a problem in the calculus of variations of the type already treated in Sec. 5-3.

We introduce lagrangian multipliers $\lambda(t)$ and $\mu(t)$, which are matrices of order $n \times 1$ and $p \times 1$, add the terms

$$2 \int_{t_0}^{t_f} \mu'(t)[\bar{z}(t) - h(x(t),t)] \, dt + 2 \int_{t_0}^{t_f} \lambda'(t)[\dot{x}(t) - f(x,u,t)] \, dt$$

to the expression for J in (6-87), and then regard $x(t_0)$, $u(t)$, $x(t)$, and $\bar{z}(t)$ as independent variables. We obtain the following necessary conditions for J to be a minimum:

$$\begin{aligned}
x(t_0) - \bar{x}_0 &= P_0\lambda(t_0) \\
u(t) - \bar{u}(t) &= Q(t)f_u'\lambda(t) \\
\lambda(t_f) &= 0 \\
\dot{\lambda} &= -f_x'\lambda(t) - h_x'\mu(t) \\
z(t) - \bar{z}(t) &= R(t)\mu(t)
\end{aligned} \tag{6-88}$$

together with

$$\dot{x} = f(x,u,t) \qquad \text{and} \qquad \bar{z}(t) = h(x(t),t)$$

The solution to the last two-point boundary system of equations in λ, μ, x, and $\bar{z}$, in which $z(t)$ is known, provides the desired estimates.

If f is linear in x and u, and h is linear in x, the system can be solved as follows: Suppose we have

$$\dot{x} = F(t)x + G(t)u \qquad \bar{z} = H(t)x$$

in which F, G, and H are of orders $n \times n$, $n \times m$, and $p \times n$. Equations (6-88) become

$$\frac{d}{dt}\begin{pmatrix} x \\ \lambda \end{pmatrix} = \begin{pmatrix} F & GQG' \\ H'R^{-1}H & -F' \end{pmatrix}\begin{pmatrix} x \\ \lambda \end{pmatrix} + \begin{pmatrix} G\bar{u} \\ -H'R^{-1}z \end{pmatrix} \tag{6-89}$$

with boundary conditions

$$\lambda(t_f) = 0 \qquad x(t_0) = \bar{x}_0 + P_0\lambda(t_0) \tag{6-90}$$

Since these equations are linear, we can proceed as follows: Let $x_p(t)$, $\lambda_p(t)$ be the particular solution of (6-89) satisfying the initial conditions

$$x_p(t_0) = \bar{x}(t_0) \qquad \lambda_p(t_0) = 0 \tag{6-91}$$

Let $M_x(t)$ and $M_\lambda(t)$ be $n \times n$ matrices satisfying

$$\frac{d}{dt}\begin{pmatrix} M_x \\ M_\lambda \end{pmatrix} = \begin{pmatrix} F & GQG' \\ H'R^{-1}H & -F' \end{pmatrix}\begin{pmatrix} M_x \\ M_\lambda \end{pmatrix} \tag{6-92}$$

with

$$M_x(t_0) = P_0 \qquad M_\lambda(t_0) = I$$

where I is the identity matrix.   Then

$$x(t) = x_p(t) + M_x(t)\lambda(t_0)$$
$$\lambda(t) = \lambda_p(t) + M_\lambda(t)\lambda(t_0) \tag{6-93}$$

and (6-90) will be satisfied if and only if

$$\lambda(t_0) = -M_\lambda(t_f)^{-1}\lambda_p(t_f) \tag{6-94}$$

Therefore, (6-93) yields

$$x(t) = x_p(t) - M_x(t)M_\lambda(t_f)^{-1}\lambda_p(t_f) \tag{6-95}$$

and $u(t)$ and $\bar{z}(t)$ can now be evaluated.   The formula (6-95) is the solution of the *smoothing* problem; i.e., given the observables $z(t)$ for $t_0 \leq t \leq t_f$, containing "noise" $u(t)$ and $v(t)$, find the optimal estimates for $x(t)$, for $t_0 \leq t \leq t_f$.

The *filtering* problem requires only the estimate of $x(t_f)$.   In this problem, an alternative procedure is possible which requires only the solution of an initial-value problem, but it is nonlinear.   In fact, we may set

$$y(t) = x_p(t) - M_x(t)M_\lambda(t)^{-1}\lambda_p(t) \tag{6-96}$$

Then $$y(t_f) = x(t_f)$$

by (6-95).   Now we set

$$S(t) = M_x(t)M_\lambda(t)^{-1} \tag{6-97}$$

to find that

$$\dot{S} = \dot{M}_x M_\lambda^{-1} - M_x M_\lambda^{-1}\dot{M}_\lambda M_\lambda^{-1} \tag{6-98}$$

Now we use (6-92) and (6-97) to get

$$\dot{S} = FS + SF' - SH'R^{-1}HS + GQG' \tag{6-99}$$
and $$S(t_0) = P_0 \tag{6-100}$$

Equations (6-99) and (6-100) provide an initial-value problem for the determination of $S(t)$.   From (6-96),

$$y(t) = x_p(t) - S(t)\lambda_p(t) \tag{6-101}$$
and, therefore, $$\dot{y} = \dot{x}_p - \dot{S}\lambda_p - S\dot{\lambda}_p$$

Now use (6-89), (6-98), and (6-101) to obtain

$$\dot{y} = (F - SH'R^{-1}H)y + G\bar{u} + SH'R^{-1}z \tag{6-102}$$
with $$y(t_0) = \bar{x}(t_0) \tag{6-103}$$

Equations (6-102) and (6-103) define an initial-value problem for $y(t)$ in which $S(t)$ comes from (6-99) and (6-100).   We then have the estimate $x(t_f) = y(t_f)$.   Of course, we also see that, for any t, $y(t)$ is the maximum-likelihood estimate of $x(t)$, given the observations $z(\tau)$, $t_0 \leq \tau \leq t$. Equations (6-99), (6-100), (6-102), and (6-103) were obtained earlier by Kalman and Bucy[10a] by different methods.

It can also be shown that, for $t > t_f$, the *prediction* problem [having observed $z(t)$ for $t_0 \leq \tau \leq t_f$, containing "noise" $u(t)$ and $v(t)$, find the optimal estimate of $x(t)$ for $t > t_f$] has for its solution

$$x(t) = \Phi(t,t_f)x(t_f) + \int_{t_f}^{t} \Phi(t,\tau)G(\tau)\bar{u}(\tau) \, d\tau \qquad (6\text{-}104)$$

in which $x(t_f)$ is the estimate obtained above, and $\Phi(t,\tau)$, for any $\tau \leq t$, is the $n \times n$ matrix such that

$$\Phi(\tau,\tau) = I \qquad \text{and} \qquad \frac{\partial \Phi}{\partial t}(t,\tau) = F(t)\Phi(t,\tau)$$

Estimates for the covariance matrices of the various quantities can also be obtained as the solutions of differential equations; the reader is referred to the papers cited.

Let us remark before closing that minimizing (6-87) subject to the constraints given there can be viewed as an optimum control problem. Kalman and Bucy, using different ideas, also find that the prediction problem is equivalent to the solution of a control problem, the "dual" problem. The expression they minimize is similar to (6-87), but with P_0^{-1}, R^{-1}, and Q^{-1} replaced by P_0, R, and Q.

For more recent developments in prediction theory, the reader may refer to the excellent summary by Zadeh[26a] and the references given there.

REFERENCES

1. Balakrishnan, A. V.: A General Theory of Nonlinear Estimations Problems in Control Systems, *SIAM J. on Control*, September, 1962.

2. Bartlett, M. S.: "An Introduction to Stochastic Processes," Cambridge University Press, New York, 1955.

3. Blanc-Lapierre, A., and R. Fortet: "Théorie des Fonctions Aléatoires," Masson et Cie, Paris, 1953.

3a. Bryson, A. E., and M. Frazier: Smoothing for Nonlinear Dynamic Systems, unpublished.

4. Cramer, H.: "Mathematical Methods of Statistics," Princeton University Press, Princeton, N.J., 1946.

5. ———: On the Theory of Stationary Random Processes, *Ann. Math.*, vol. 41, p. 215, 1940.

6. Davenport, W. B., Jr., and W. L. Root: "Introduction to Random Signals and Noise," McGraw-Hill Book Company, Inc., New York, 1958.

7. Doob, J. L.: "Stochastic Processes," John Wiley & Sons, Inc., New York, 1953.

8. Grenander, U., and M. Rosenblatt: "Statistical Analysis of Stationary Time Series," John Wiley & Sons, Inc., New York, 1957.

9. ———— and Gabor Szego, "Toeplitz Forms and Their Applications," University of California Press, Berkeley, Calif., 1958.

10. Helson, H., and D. Lowdenslager: Prediction Theory and Fourier Series in Several Variables, part I, *Acta Math.*, vol. 99, p. 165, 1959; part II, *Acta Math.*, vol. 106, p. 175, 1962.

10a. Kalman, R. E., and R. S. Bucy: New Results in Linear Filtering and Prediction Theory, *J. Basic Eng.*, March, 1961.

11. Karhunen, K.: Uber die Struktur Stationarer Zufalliger Funktionen, *Ark. Mat.*, vol. 1, p. 141, 1950.

12. ————: Uber Lineare Methoden in der Wahrscheinlichkeitsrechnung, *Ann. Acad. Sci. Fenn. Ser. A. I.*, no. 37, 1947.

13. Khintchine, A. Y.: Korrelationstheorie der Stationaren Stochastischen Prozesse, *Math. Ann.*, vol. 109, p. 604, 1934.

14. Kolmogorov, A. N.: Interpolation and Extrapolation of Stationary Random Sequences, *Izv. Akad. Nauk SSSR. Ser. Mat.*, vol. 5, p. 3, 1941.

15. ————: Uber die Analytischen Methoden in der Wahrscheinlichkeitsrechnung, *Math. Ann.*, vol. 104, p. 415, 1931.

16. Laning, J. H., Jr., and R. H. Battin: "Random Processes in Automatic Control," McGraw-Hill Book Company, Inc., New York, 1956.

17. Loéve, M.: "Probability Theory," 2d ed., D. Van Nostrand Company, Inc., Princeton, N.J., 1960.

18. Masani, P., and N. Wiener: "Nonlinear Prediction in Probability and Statistics," the Harold Cramer volume, John Wiley & Sons, Inc., 1959.

19. ———— and ————: Prediction Theory of Multivariate Stochastic Processes: part I, The Regularity Condition, *Acta Math.*, vol. 98, p. 111, 1957; part II, The Linear Predictor, *Acta Math.*, vol. 99, p. 93, 1958.

20. Middleton, D.: An Introduction to Statistical Communication Theory, McGraw-Hill Book Company, Inc., New York, 1960.

21. Polya, G., and G. Szego: "Aufgaben und Lehrsatze der Analysis," Springer-Verleg OHG, Berlin, 1925.

22. Rozanov, Y. A.: Spectral Theory of Multidimensional Stationary Stochastic Processes with Discrete Time, *Uspehi Mat. Nauk*, vol. 13, no. 2, p. 93, 1958; English translation in "Selected Translations in Mathematical Statistics and Probability," vol. I, American Mathematical Society, Providence, R.I., 1961.

23. Solodovnikov, V. V.: "Introduction to the Statistical Dynamics of Automatic Control Systems," Dover Publications, Inc., New York, 1960.

24. Wiener, N.: "Extrapolation, Interpolation, and Smoothing of Stationary Time Series," John Wiley & Sons, Inc., New York, 1949.

24a. ————: "Nonlinear Problems in Random Theory," John Wiley & Sons, Inc., New York, 1958.

25. Yaglom, A. M.: "Theory of Stationary Random Functions," English translation, Prentice-Hall, Inc., Englewood Cliffs, N.J., 1962.

26. Zadeh, L. A., and J. R. Ragazzini: An Extension of Wiener's Theory of Prediction, *J. Appl. Phys.*, vol. 21, p. 645, 1950.

26a. ————: Progress in Information Theory in the U.S.A., 1957–1960, part 5, *IRE Trans. on Inform. Theory*, July, 1961.

APPENDIX

The notion of a set is intuitively quite elementary, and the desirability of formalizing some notation and phraseology is indicated by the many examples throughout this book in which such notation is avoided only by use of long English phrases. A set is any aggregate or collection of things. The set of all living human beings, the set of all automobiles in the United States, the set of all complex numbers of absolute value ≤ 1, and the set of all $n \times n$ matrices with real entries are a few examples. The first two sets are examples of finite sets; the last two are infinite. (Furthermore, the first two sets are sets of tangible things; the last two are not. In mathematics, the sets considered are usually sets of abstractions, like numbers, matrices, transformations, vector spaces, or sets generally. A set of sets is often called *a class of sets, collection of sets,* or *family of sets,* for euphony.) If A and B denote sets, then $A \cup B$ (read, "A union B") is the set of things belonging to A or to B or to both. In Fig. 1, A and B are pictured as the interiors of two circles. Then $A \cup B$ is the set of all points covered by horizontal or vertical lines. Also, $A \cap B$ (read, "A intersect B") is the set of all things belonging to both A and B. In Fig. 1, $A \cap B$ is the crosshatched set of points.

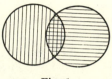

Fig. 1

If $\{A_\lambda\}$ is a family of sets, where λ is an index—e.g., λ may run through a set of integers or a set of real numbers or through any other set—then

$\cup_\lambda A_\lambda$, the *union* of the A_λ, is also a set, viz., the set of all things that belong to at least one A_λ. Similarly, the intersection $\cap_\lambda A_\lambda$ is the set of all things that belong to every A_λ. It is customary to speak of the things in a set as *points;* we say that the points of a set A *belong to* A or are *members of* A. If x is a member of A, we write $x \, \epsilon \, A$; and if x is not a member of A, we write $x \, \epsilon' \, A$.

It is also convenient to consider as a set, the set containing no members at all. This is called *the empty set* and is denoted by O. Thus $x \, \epsilon' \, O$, no matter what x is. In addition, there is in every discussion a universal set X, the set of all the things under consideration. For example, if we are talking about vectors in R_n, then we may take $X = R_n$, and every set of vectors will be a subset of R_n. More generally, if A and B denote sets, then A *is a subset of* B, written $A \subset B$, or $B \supset A$, if every member of A is a member of B. Also, A and B *are equal* (i.e., they denote the same set) if every member of A belongs to B and every member of B belongs to A; i.e., $A = B$ if $A \subset B$ and $B \subset A$. For the empty set O, we have $O \subset A$ for every A; also $A \subset X$ for every A, if X is the universal set.

Finally, the complement A' of a set is the set of all points in X that do not belong to A. We have $A \cup A' = X$, $A \cap A' = O$. If E and F are sets, then $E - F$ is, by definition, the set of all points in E and not in F; so $E - F = E \cap F'$.

If X is the universe under discussion, some points of X may enjoy a certain property P while others do not. The set of all points in X with the property P (or, the set of all points such that P) is denoted

$$\{x : x \text{ has property } P\}$$

For example, if X is the set of all real numbers, then

$$\{x : x^2 > 3 \text{ and } x > 0\}$$

is the interval $\sqrt{3} < x < \infty$. For every property, there is a subset of X consisting of exactly those points in X that have that property. Conversely, for every subset A of X, there is a property such that $x \, \epsilon \, A$ if and only if x has that property, for we can say, if nothing else, that the property is that of belonging to A.

The most important concept in mathematics is that of a mapping or transformation from a set X to a set Y. A mapping from X to Y is a rule f that assigns to each x in X a well defined point y in Y, and we write $f : X \rightarrow Y$ and read, "f sends X into Y." For example, let X be the set of integers, $X = \{1,2,3,4,5\}$, and let Y be the set of all positive integers. Then $f(1) = 10$, $f(2) = 3$, $f(3) = 8$, $f(4) = 10^{10}$, and $f(5) = 10$ define a mapping of X into Y. If X and Y each denote the set of real numbers, then the mapping f that assigns to each x in X the number $y = \sin x$ is a mapping of X into Y. If X is the set of positive integers, and Y is the set of real numbers, and $\{y_n\}$ is any sequence of real numbers, then $f(n) = y_n$ is a mapping of X into Y. (We may therefore *define* a sequence $\{y_n\}$ of real numbers as a mapping from the positive integers into the set of real numbers.) If $f : X \to Y$ and distinct x's map into distinct y's, then f is called a 1-1 (*one-to-one*) map. If every y in Y has an antecedent x in X, i.e., there is an x for which $f(x) = y$, then f is called *onto*. A mapping may be 1-1 or onto or both or neither.

The mappings of classical and modern analysis are usually, but not always, continuous. The concept of continuity depends on the concept of "nearness" of points or "neighboring points." Indeed, a function, or mapping, $f(x)$ of the real numbers into the set of real numbers is continuous at a point x_0 if, given any neighborhood $\{y : |y - f(x_0)| < \epsilon\}$, there is a corresponding neighborhood $\{x : |x - x_0| < \delta\}$ such that whenever x is in the latter neighborhood, $f(x)$ is in the former. This concept of continuity can be formulated for mappings from a set X to a set Y provided the sets X and Y are appropriately endowed with neighborhoods, i.e., if X and Y are *topological spaces*, which we shall define presently. The study of topological spaces, in addition to providing fruitful and important advances in modern mathematics, throws light on many classical proofs and provides a better understanding and unification of known results.

If X is a given set, a *topology* in X is defined by a family $\mathcal{O}$ of *neighborhoods* or *open sets* in X. The family $\mathcal{O}$ of subsets of X must satisfy the following axioms:

1. If $U_\lambda \in \mathcal{O}$ for every λ in $\wedge$, an arbitrary index set, then $\cup_\lambda U_\lambda \in \mathcal{O}$ (an arbitrary union of open sets is open).

2. If $U_1, \ldots, U_n \in \mathcal{O}$, then $\cap_{j=1}^n U_j \in \mathcal{O}$ (a *finite* intersection of open sets is open).

3. $X \in \mathcal{O}$ (the whole space is open).

4. $O \in \mathcal{O}$ (the empty set is open).

For example, let X be the set of real numbers, and call a set $U \subset X$ open if it contains with every point x in U a whole interval containing x. Then the class $\mathcal{O}$ of all such open sets satisfies axioms 1 to 4 above.

A *neighborhood* of a point x is any open set U (i.e., any U in $\mathcal{O}$) that contains x. A topological space is a set X and a family $\mathcal{O}$ of subsets of X satisfying axioms 1 to 4. Such a space is usually denoted by X, with $\mathcal{O}$ understood, in the background. A topological space X is called *a Hausdorff space* if it satisfies also:

5. For every x, y in X, with $x \neq y$, there are neighborhoods U, V of x, y, respectively, such that $U \cap V = O$.

Not every topological space is Hausdorff, but the ones most commonly encountered are. From now on, "topological space" or simply "space" will mean Hausdorff space.

Exercise Show that the space of real numbers is a Hausdorff space.

We have already said that the open sets of a topological space X are precisely those in the family $\mathcal{O}$. A subset F of X is called *closed* if its complement F' is open. An arbitrary subset of X will ordinarily be neither open nor closed.

If X is a topological space and E is any subset of X, then E is also a topological space in its own right. The open subsets of E are precisely those of the form $V = U \cap E$ where U is open in X. Of course, the open sets in E are only open *relative to E;* they will not ordinarily be open in X.

Exercise Let X be the space of real numbers and E be the closed interval $[0,1]$, or let E denote the set of all rational numbers. Describe the open subsets relative to E.

If X and Y are topological spaces and f is a mapping of X into Y, we say that f is *continuous* at a point x_0 in X if for every neighborhood V of $f(x_0)$ in Y, there is a neighborhood U of x_0 in X such that $f(U) \subset V$, i.e., every x in U is sent by f into some point in V. The mapping is called continuous if it is continuous at every point in X.

A sequence $\{x_n\}$ of points of X is called *convergent* if there is an x^* in X such that for every neighborhood U of x^* there is an n_0 such that whenever $n \geq n_0$, we have $x_n \, \epsilon \, U$. We then write $x_n \to x^*$. Note that in a Hausdorff space, a sequence $\{x_n\}$ can converge to one point at most. It is not hard to show that if $f:X \to Y$ and f is continuous at x_0, then whenever $\{x_n\}$ is a sequence converging to x_0, $\{f(x_n)\}$ converges to $f(x_0)$. But the converse is not true, in general.

A sufficient condition on X that will imply that "f is continuous at x^* if $x_n \to x^*$ implies $f(x_n) \to f(x^*)$" is that X is "first-countable" or X satisfies the first axiom of countability, viz.:

6. For each point x in X, there is a countable family $\{V_n\}$ of neighborhoods of x such that every neighborhood U of x contains some neighborhood V_n. (We may suppose that $V_1 \supset V_2 \supset \cdots \supset V_n \supset \cdots$, for if not, take $W_n = V_1 \cap \cdots \cap V_n$ and $\{W_n\}$ instead of $\{V_n\}$.)

Exercise Suppose that X is first-countable. Suppose that $f:X \to Y$ and that $x_n \to x^*$ implies $f(x_n) \to f(x^*)$. Show that f is continuous at x^*. (*Hint:* If not, then there is a neighborhood V of $f(x^*)$ such that every neighborhood U of x^* contains a point x such that $f(x) \, \epsilon' \, V$. Let $\{V_n\}$ be a decreasing sequence

$$V_1 \supset V_2 \supset \cdots \supset V_n \supset \cdots$$

satisfying axiom 6 at x^*. Choose x_n in V_n with $f(x_n) \, \epsilon' \, V$. Show that $x_n \to x^*$. Then $f(x_n) \to f(x^*)$, contradicting $f(x_n) \, \epsilon' \, V$ for all n.)

A set X is called a *metric* space if there is defined a *distance* $\rho(x,y)$ between every two points x, y in X such that:

a. $\rho(x,y)$ is a nonnegative (finite) real number.

b. $\rho(x,y) = \rho(y,x)$ for every x, y.

c. $\rho(x,z) \leq \rho(x,y) + \rho(y,z)$ for all x, y, z in X (the triangle inequality).

d. $\rho(x,y) = 0$ if and only if $x = y$.

If X is a metric space, a subset U of X is called *open* if it contains with every point x in U a whole *sphere*, $\{y : \rho(y,x) < \epsilon\}$. The class of all such open sets satisfies axioms 1 to 6 above. A metric space is a first-countable Hausdorff space. [To show axiom 6, let $V_n = \{y : \rho(y,x) < 1/n\}$.]

If X is a set and $\mathcal{O}_1$ and $\mathcal{O}_2$ are two families of subsets satisfying axioms 1 to 4, we say that $\mathcal{O}_1$ is *stronger* than $\mathcal{O}_2$ or $\mathcal{O}_2$ is *weaker* than $\mathcal{O}_1$ if $\mathcal{O}_1 \supset \mathcal{O}_2$, that is, if every set in $\mathcal{O}_2$ belongs also to $\mathcal{O}_1$; every set that is open in the $\mathcal{O}_2$ topology is also open in the $\mathcal{O}_1$ topology. If we let $(X, \mathcal{O}_1)$ denote the space X with the $\mathcal{O}_1$ topology, and correspondingly for $(X, \mathcal{O}_2)$, then $\mathcal{O}_1$ is stronger than $\mathcal{O}_2$ if and only if the identity map, $f : (X, \mathcal{O}_1) \rightarrow (X, \mathcal{O}_2)$ with $f(x) = x$ for all x, is continuous.

Exercise Prove the last statement.

Two topologies $\mathcal{O}_1$ and $\mathcal{O}_2$ on X are called *equivalent* if $\mathcal{O}_1$ is stronger than $\mathcal{O}_2$ and $\mathcal{O}_2$ is stronger than $\mathcal{O}_1$, i.e., $\mathcal{O}_1$ and $\mathcal{O}_2$ contain the same sets. Two topologies $\mathcal{O}_1$ and $\mathcal{O}_2$ are evidently equivalent if both identity maps, $f : (X, \mathcal{O}_1) \rightarrow (X, \mathcal{O}_2)$ and $g : (X, \mathcal{O}_2) \rightarrow (X, \mathcal{O}_1)$ with $f(x) = x$ and $g(x) = x$ for all x, are continuous.

Exercise Let $X = R_n$, and define the distance $\rho_p(x,y) = \left\{ \sum_{i=1}^{n} |x_i - y_i|^p \right\}^{1/p}$ for fixed p, $1 \leq p < \infty$, and let $p_\infty(x,y) = \max_{1 \leq i \leq n} |x_i - y_i|$. Each such distance gives rise to a topology as discussed above. Show that all these topologies are equivalent.

If X and Y are topological spaces, if $f : X \rightarrow Y$, and if f is 1-1, onto, and continuous, and if f^{-1} [the inverse map, $f^{-1}(y) = x$ when $f(x) = y$; this is a well defined map from Y onto X because f is 1-1 and onto] is also continuous, then f is called a *homeomorphism* of X onto Y. Two spaces X and Y are *homeomorphic* (or topologically equivalent) if there is a homeomorphism of one onto the other.

Example Let X be the set of all real numbers, and let Y be the open interval $-1 < y < 1$, both with their usual open-interval topologies. Let $x = g(y) = \tan\left[(\pi/2)y\right]$. Then g is a homeomorphism of Y onto X; X and Y are homeomorphic.

The concept of compactness is a very useful one; it is at the basis of all proofs in analysis which contain the part, "choose a sequence $\{x_n\}$ such that . . . ; a subsequence $\{x_{n_k}\}$ converges . . . ." A space X is called *compact* if for every family $\{V_\lambda\}$ of open subsets of X such that $X = \cup_\lambda V_\lambda$, there is a finite number of indices $\lambda_1, \ldots, \lambda_n$ such that $X = \cup_{j=1}^n V_{\lambda_j}$. In words, X is compact if every open cover of X contains a finite subcover.

If E is a subset of a space X, a point x^* in X is called a *limit point* of E if every neighborhood of x^* contains infinitely many points of E. If X is compact, then every sequence $\{x_n\} \subset X$ has a limit point x^*, or else $\{x_n\}$ contains only a finite number of distinct points. (The last clause would be unnecessary if we defined, for sequences $\{x_n\}$, x^* is a limit point of $\{x_n\}$ if every neighborhood of x^* contains an x_n for infinitely many n; this is the more common procedure.) The converse is not true. There is a noncompact space in which every sequence has a limit point; the space is first-countable besides. (This is the space of all ordinals less than the first uncountable ordinal; the reader is referred to standard texts for the necessary definitions.)

If X is first-countable, and E is a subset of X with a limit point x^*, then there is a sequence $\{x_n\}$ in E which converges to x^*. Consequently, if X is compact and first-countable, every infinite sequence contains a convergent subsequence.

A Hausdorff space X is called *separable* if it satisfies axioms 1 to 5 given earlier and also the *second axiom of countability*:

7. There is a countable family $\{V_n\}$ of open sets such that for every x in X and every open set U containing x, there is a V_n such that $x \in V_n \subset U$.

Evidently such a space also satisfies axiom 6; if X is separable, it is surely first-countable.

The *closure* $\bar{E}$ of a set E in X is the set whose points are either in E or are limit points of E. This is the smallest closed set containing E. If E is a subset of X and $\bar{E} = X$, then E is *dense* in X; every point in X is either in E or is a limit point of E. Of course, E is always dense in $\bar{E}$.

If X is a metric space and there is a countable set $E = \{r_n\}$ in X such that E is dense in X, then X is separable. Indeed, for each r_n in E, let $V_{n,m} = S(r_n, 2^{-m})$, the open sphere with center r_n and radius 2^{-m}, $m = 1,2,3, \ldots$. If $x \in X$ and U is an open set containing x, there is an m such that $S(x, 2^{-m}) \subset U$. Choose r_n such that the distance $\rho(x, r_n) < 2^{-m-2}$. We will then have $x \in S(r_n, 2^{-m-1}) \subset S(x, 2^{-m}) \subset U$.

Since R_n is a metric space, and the points of R_n each of whose coordinates is rational form a countable dense set in R_n, it follows that R_n is separable. The same is true for a Hilbert space H with a countable CON set $\{e_n\}$, for the set of points of the form $x = \sum_{n=1}^{N} \lambda_n e_n$ with λ_n rational is countable and dense in H. If E is a measurable subset of R_n and $1 \leq p < \infty$, then $L_p(E)$ is separable. If F is a closed bounded subset of R_n, then the space $C(F)$ of all real or complex continuous functions on F is separable.

If X is separable, X is compact if and only if every sequence $\{x_n\}$ in X has a limit point. For if $\{V_\lambda\}$ is an open cover of X, $X = \cup_\lambda V_\lambda$, choose, for each x in X and each V_λ such that $x \epsilon V_\lambda$, an open set W_n from a countable family given by axiom 7 such that $x \epsilon W_n \subset V_\lambda$. Number the W_n's, W_{n_1}, W_{n_2}, ... If $\cup_{j=1}^N W_{n_j} = X$ for some N, then also $\cup_{j=1}^N V_{\lambda_j} = X$ where $V_{\lambda_j} \supset W_{n_j}$, and the proof is complete. The other alternative, $\cup_{j=1}^N W_{n_j} \neq X$ for every N leads to a contradiction, for if we choose x_N so that $x_N \epsilon' \cup_{j=1}^N W_{n_j}$, then let x^* be a limit point of $\{x_N\}$. Since $x^* \epsilon U_\lambda$ for some λ, $x^* \epsilon W_{n_k}$ for some k. But then W_{n_k} contains x_N for infinitely many N while, at the same time, $x_N \epsilon' W_{n_k}$ for all $N \geq k$.

If F is a closed bounded subset of the real numbers, then every sequence $\{x_n\}$ in F has a limit point in F. For let $y_n = \sup_{k \geq n} x_k$. Then $\{y_n\}$ is bounded and decreasing. Take $x^* = \lim_{n \to \infty} y_n$. Then x^* is a limit point of $\{x_n\}$. (This is the same as $x^* = \overline{\lim_{n \to \infty}} x_n$.) If, now, F is a closed bounded subset of R_N and $\{x_n\}$ is a sequence in F, then the first coordinates of x_n have a limit point so that for a subsequence of $\{x_n\}$, the first coordinates converge. Similarly, for a subsequence of the latter subsequence, the second coordinates converge, etc. We get a subsequence finally, all of whose coordinates converge, so that the final subsequence converges. The limit must be in F since F is closed. It follows that every closed bounded subset of R_N is compact. This is the Heine-Borel theorem.

For infinite-dimensional vector spaces, this is no longer true, as we observed in Chap. 1. For the space $C[a,b]$, Ascoli's theorem tells us that if a subset E of $C[a,b]$ is uniformly bounded and equicontinuous, then $\bar{E}$ is compact. (Every infinite sequence in E has a limit point, but the limit point need not be in E.)

For the last several paragraphs, sequences are used fairly heavily. We can use these whenever space X is separable. The situation can be much different if X is not separable or even "first-countable." For example, let H be a Hilbert space with a countable CON set, but instead of using the topology that arises from the metric in H, consider the weak topology in H. Here the basic open sets are defined as follows: If $x_0 \epsilon H$, a *basic* neighborhood $V(x_0)$ of x_0 is obtained by specifying y_1, . . . , y_N in H and $\epsilon > 0$, and writing

$$V(x_0) = \{x : |(x - x_0, y_j)| < \epsilon, j = 1, \ldots, N\}$$

A set W is open if it is the union of sets of the form $V(x_0)$ for various x_0,

N, ϵ.   It can be shown that if $\mathfrak{O}_1$ is the topology arising from the metric in H, and $\mathfrak{O}_2$ is the weak topology just defined, $\mathfrak{O}_2$ is weaker than $\mathfrak{O}_1$.

The following is an example of a (countable) set E of vectors in H with a weak limit point $x = 0$ and such that no sequence in E converges to 0 weakly.   For every pair of positive integers m, k with $m < k$, let $x^{m,k}$ be the vector

$$x^{m,k} = (0,0, \ . \ . \ . \ ,0,1,0, \ . \ . \ . \ ,0,m,0,0, \ . \ . \ .)$$

with 1 in the mth place, m in the kth place, and zero elsewhere.   Then every weak neighborhood $V(0)$ of $x = 0$ of the form given above contains some $x^{m,k}$.   But no sequence of $\{x^{m,k}\}$ converges weakly to 0, because such a sequence would have to be bounded in norm by Exercise 1-22. This means that m is bounded, so that for some position infinitely many vectors of our subsequence would have the component 1 in that position. But if a sequence of vectors converges to 0 weakly, then every component must converge to 0.   The current standard treatise in English is "General Topology" by J. T. Kelley.

We can now introduce the generalization of the Schauder fixed-point theorem (1-10) due to Tychonoff.

A real linear topological space is a real vector space which is a topological space such that addition, i.e., $(x,y) \rightarrow x + y$, and scalar multiplication, i.e., $(\lambda,x) \rightarrow \lambda x$, are continuous for all x and y in the space and all real λ.   Such a space is called *locally convex* if every neighborhood of an arbitrary vector x contains a convex neighborhood of x.

Theorem (*Schauder-Tychonoff*)   A compact convex subset of a locally convex linear topological space has the fixed-point property (see p. 456 of Ref. 4, Chap. 1).

Note that this theorem is a generalization of Theorem 1-10, since a Banach space is a locally convex space.   We may also remark that any Banach space with the weak topology is a locally convex space.   A set U is a weak neighborhood of the vector x_0 if there is an $\epsilon > 0$ and continuous linear functionals $\varphi_1, \ . \ . \ . \ , \varphi_n$ on our space such that

$$\{x: |\varphi_i(x - x_0)| < \epsilon, i = 1, \ . \ . \ . \ , n\} \subset U$$

To apply this theorem, one essentially verifies the same points mentioned in Theorem 1-10.   To set up a locally convex space one uses seminorms.   A seminorm satisfies the same conditions as a norm except that $\|x\| = 0$ is permissible for $x \neq 0$.   An example of a seminorm for differentiable functions on the unit interval is max $|f'(x)|$.   The basic fact is that the topology of every locally convex space admits a family of seminorms in the sense that the basic open sets are those which are finite intersections of "spheres" $\{x: v(x - x_0) < \epsilon\}$ where v is a seminorm in the family.

INDEX

A CATALOGUE OF
SELECTED DOVER BOOKS
IN ALL FIELDS OF INTEREST

A CATALOGUE OF SELECTED DOVER
BOOKS IN ALL FIELDS OF INTEREST

CELESTIAL OBJECTS FOR COMMON TELESCOPES, T. W. Webb. The most used book in amateur astronomy: inestimable aid for locating and identifying nearly 4,000 celestial objects. Edited, updated by Margaret W. Mayall. 77 illustrations. Total of 645pp. 5⅜ x 8½.
20917-2, 20918-0 Pa., Two-vol. set $9.00

HISTORICAL STUDIES IN THE LANGUAGE OF CHEMISTRY, M. P. Crosland. The important part language has played in the development of chemistry from the symbolism of alchemy to the adoption of systematic nomenclature in 1892. ". . . wholeheartedly recommended,"—Science. 15 illustrations. 416pp. of text. 5⅝ x 8¼.
63702-6 Pa. $6.00

BURNHAM'S CELESTIAL HANDBOOK, Robert Burnham, Jr. Thorough, readable guide to the stars beyond our solar system. Exhaustive treatment, fully illustrated. Breakdown is alphabetical by constellation: Andromeda to Cetus in Vol. 1; Chamaeleon to Orion in Vol. 2; and Pavo to Vulpecula in Vol. 3. Hundreds of illustrations. Total of about 2000pp. 6⅛ x 9¼.
23567-X, 23568-8, 23673-0 Pa., Three-vol. set $26.85

THEORY OF WING SECTIONS: INCLUDING A SUMMARY OF AIR-FOIL DATA, Ira H. Abbott and A. E. von Doenhoff. Concise compilation of subatomic aerodynamic characteristics of modern NASA wing sections, plus description of theory. 350pp. of tables. 693pp. 5⅜ x 8½.
60586-8 Pa. $7.00

DE RE METALLICA, Georgius Agricola. Translated by Herbert C. Hoover and Lou H. Hoover. The famous Hoover translation of greatest treatise on technological chemistry, engineering, geology, mining of early modern times (1556). All 289 original woodcuts. 638pp. 6¾ x 11.
60006-8 Clothbd. $17.95

THE ORIGIN OF CONTINENTS AND OCEANS, Alfred Wegener. One of the most influential, most controversial books in science, the classic statement for continental drift. Full 1966 translation of Wegener's final (1929) version. 64 illustrations. 246pp. 5⅜ x 8½.
61708-4 Pa. $4.50

THE PRINCIPLES OF PSYCHOLOGY, William James. Famous long course complete, unabridged. Stream of thought, time perception, memory, experimental methods; great work decades ahead of its time. Still valid, useful; read in many classes. 94 figures. Total of 1391pp. 5⅜ x 8½.
20381-6, 20382-4 Pa., Two-vol. set $13.00

THE PHILOSOPHY OF HISTORY, Georg W. Hegel. Great classic of Western thought develops concept that history is not chance but a rational process, the evolution of freedom. 457pp. 5⅜ x 8½.　20112-0 Pa. $4.50

LANGUAGE, TRUTH AND LOGIC, Alfred J. Ayer. Famous, clear introduction to Vienna, Cambridge schools of Logical Positivism. Role of philosophy, elimination of metaphysics, nature of analysis, etc. 160pp. 5⅜ x 8½. (Available in U.S. only)　20010-8 Pa. $2.00

A PREFACE TO LOGIC, Morris R. Cohen. Great City College teacher in renowned, easily followed exposition of formal logic, probability, values, logic and world order and similar topics; no previous background needed. 209pp. 5⅜ x 8½.　23517-3 Pa. $3.50

REASON AND NATURE, Morris R. Cohen. Brilliant analysis of reason and its multitudinous ramifications by charismatic teacher. Interdisciplinary, synthesizing work widely praised when it first appeared in 1931. Second (1953) edition. Indexes. 496pp. 5⅜ x 8½.　23633-1 Pa. $6.50

AN ESSAY CONCERNING HUMAN UNDERSTANDING, John Locke. The only complete edition of enormously important classic, with authoritative editorial material by A. C. Fraser. Total of 1176pp. 5⅜ x 8½.
20530-4, 20531-2 Pa., Two-vol. set $14.00

HANDBOOK OF MATHEMATICAL FUNCTIONS WITH FORMULAS, GRAPHS, AND MATHEMATICAL TABLES, edited by Milton Abramowitz and Irene A. Stegun. Vast compendium: 29 sets of tables, some to as high as 20 places. 1,046pp. 8 x 10½.　61272-4 Pa. $14.95

MATHEMATICS FOR THE PHYSICAL SCIENCES, Herbert S. Wilf. Highly acclaimed work offers clear presentations of vector spaces and matrices, orthogonal functions, roots of polynomial equations, conformal mapping, calculus of variations, etc. Knowledge of theory of functions of real and complex variables is assumed. Exercises and solutions. Index. 284pp. 5⅝ x 8¼.　63635-6 Pa. $5.00

THE PRINCIPLE OF RELATIVITY, Albert Einstein et al. Eleven most important original papers on special and general theories. Seven by Einstein, two by Lorentz, one each by Minkowski and Weyl. All translated, unabridged. 216pp. 5⅜ x 8½.　60081-5 Pa. $3.00

THERMODYNAMICS, Enrico Fermi. A classic of modern science. Clear, organized treatment of systems, first and second laws, entropy, thermodynamic potentials, gaseous reactions, dilute solutions, entropy constant. No math beyond calculus required. Problems. 160pp. 5⅜ x 8½.
60361-X Pa. $3.00

ELEMENTARY MECHANICS OF FLUIDS, Hunter Rouse. Classic undergraduate text widely considered to be far better than many later books. Ranges from fluid velocity and acceleration to role of compressibility in fluid motion. Numerous examples, questions, problems. 224 illustrations. 376pp. 5⅝ x 8¼.　63699-2 Pa. $5.00

YUCATAN BEFORE AND AFTER THE CONQUEST, Diego de Landa. First English translation of basic book in Maya studies, the only significant account of Yucatan written in the early post-Conquest era. Translated by distinguished Maya scholar William Gates. Appendices, introduction, 4 maps and over 120 illustrations added by translator. 162pp. 5⅜ x 8½.
23622-6 Pa. $3.00

THE MALAY ARCHIPELAGO, Alfred R. Wallace. Spirited travel account by one of founders of modern biology. Touches on zoology, botany, ethnography, geography, and geology. 62 illustrations, maps. 515pp. 5⅜ x 8½.
20187-2 Pa. $6.95

THE DISCOVERY OF THE TOMB OF TUTANKHAMEN, Howard Carter, A. C. Mace. Accompany Carter in the thrill of discovery, as ruined passage suddenly reveals unique, untouched, fabulously rich tomb. Fascinating account, with 106 illustrations. New introduction by J. M. White. Total of 382pp. 5⅜ x 8½. (Available in U.S. only)   23500-9 Pa. $4.00

THE WORLD'S GREATEST SPEECHES, edited by Lewis Copeland and Lawrence W. Lamm. Vast collection of 278 speeches from Greeks up to present. Powerful and effective models; unique look at history. Revised to 1970. Indices. 842pp. 5⅜ x 8½.   20468-5 Pa. $8.95

THE 100 GREATEST ADVERTISEMENTS, Julian Watkins. The priceless ingredient; His master's voice; 99 44/100% pure; over 100 others. How they were written, their impact, etc. Remarkable record. 130 illustrations. 233pp. 7⅞ x 10 3/5.   20540-1 Pa. $5.00

CRUICKSHANK PRINTS FOR HAND COLORING, George Cruickshank. 18 illustrations, one side of a page, on fine-quality paper suitable for watercolors. Caricatures of people in society (c. 1820) full of trenchant wit. Very large format. 32pp. 11 x 16.   23684-6 Pa. $5.00

THIRTY-TWO COLOR POSTCARDS OF TWENTIETH-CENTURY AMERICAN ART, Whitney Museum of American Art. Reproduced in full color in postcard form are 31 art works and one shot of the museum. Calder, Hopper, Rauschenberg, others. Detachable. 16pp. 8¼ x 11.
23629-3 Pa. $2.50

MUSIC OF THE SPHERES: THE MATERIAL UNIVERSE FROM ATOM TO QUASAR SIMPLY EXPLAINED, Guy Murchie. Planets, stars, geology, atoms, radiation, relativity, quantum theory, light, antimatter, similar topics. 319 figures. 664pp. 5⅜ x 8½.
21809-0, 21810-4 Pa., Two-vol. set $10.00

EINSTEIN'S THEORY OF RELATIVITY, Max Born. Finest semi-technical account; covers Einstein, Lorentz, Minkowski, and others, with much detail, much explanation of ideas and math not readily available elsewhere on this level. For student, non-specialist. 376pp. 5⅜ x 8½.
60769-0 Pa. $4.50

THE SENSE OF BEAUTY, George Santayana. Masterfully written discussion of nature of beauty, materials of beauty, form, expression; art, literature, social sciences all involved. 168pp. 5⅜ x 8½.   20238-0 Pa. $2.50

ON THE IMPROVEMENT OF THE UNDERSTANDING, Benedict Spinoza. Also contains *Ethics, Correspondence,* all in excellent R. Elwes translation. Basic works on entry to philosophy, pantheism, exchange of ideas with great contemporaries. 402pp. 5⅜ x 8½.   20250-X Pa. $4.50

THE TRAGIC SENSE OF LIFE, Miguel de Unamuno. Acknowledged masterpiece of existential literature, one of most important books of 20th century. Introduction by Madariaga. 367pp. 5⅜ x 8½.

20257-7 Pa. $4.50

THE GUIDE FOR THE PERPLEXED, Moses Maimonides. Great classic of medieval Judaism attempts to reconcile revealed religion (Pentateuch, commentaries) with Aristotelian philosophy. Important historically, still relevant in problems. Unabridged Friedlander translation. Total of 473pp. 5⅜ x 8½.   20351-4 Pa. $6.00

THE I CHING (THE BOOK OF CHANGES), translated by James Legge. Complete translation of basic text plus appendices by Confucius, and Chinese commentary of most penetrating divination manual ever prepared. Indispensable to study of early Oriental civilizations, to modern inquiring reader. 448pp. 5⅜ x 8½.   21062-6 Pa. $4.00

THE EGYPTIAN BOOK OF THE DEAD, E. A. Wallis Budge. Complete reproduction of Ani's papyrus, finest ever found. Full hieroglyphic text, interlinear transliteration, word for word translation, smooth translation. Basic work, for Egyptology, for modern study of psychic matters. Total of 533pp. 6½ x 9¼. (Available in U.S. only)   21866-X Pa. $5.95

THE GODS OF THE EGYPTIANS, E. A. Wallis Budge. Never excelled for richness, fullness: all gods, goddesses, demons, mythical figures of Ancient Egypt; their legends, rites, incarnations, variations, powers, etc. Many hieroglyphic texts cited. Over 225 illustrations, plus 6 color plates. Total of 988pp. 6⅛ x 9¼. (Available in U.S. only)
22055-9, 22056-7 Pa., Two-vol. set $12.00

THE ENGLISH AND SCOTTISH POPULAR BALLADS, Francis J. Child. Monumental, still unsuperseded; all known variants of Child ballads, commentary on origins, literary references, Continental parallels, other features. Added: papers by G. L. Kittredge, W. M. Hart. Total of 2761pp. 6½ x 9¼.
21409-5, 21410-9, 21411-7, 21412-5,   21413-3 Pa., Five-vol. set $37.50

CORAL GARDENS AND THEIR MAGIC, Bronsilaw Malinowski. Classic study of the methods of tilling the soil and of agricultural rites in the Trobriand Islands of Melanesia. Author is one of the most important figures in the field of modern social anthropology. 143 illustrations. Indexes. Total of 911pp. of text. 5⅝ x 8¼. (Available in U.S. only)
23597-1 Pa. $12.95

THE CURVES OF LIFE, Theodore A. Cook. Examination of shells, leaves, horns, human body, art, etc., in "*the* classic reference on how the golden ratio applies to spirals and helices in nature . . . ."—Martin Gardner. 426 illustrations. Total of 512pp. 5⅜ x 8½. 23701-X Pa. $5.95

AN ILLUSTRATED FLORA OF THE NORTHERN UNITED STATES AND CANADA, Nathaniel L. Britton, Addison Brown. Encyclopedic work covers 4666 species, ferns on up. Everything. Full botanical information, illustration for each. This earlier edition is preferred ˙by many to more recent revisions. 1913 edition. Over 4000 illustrations, total of 2087pp. 6⅛ x 9¼. 22642-5, 22643-3, 22644-1 Pa., Three-vol. set $24.00

MANUAL OF THE GRASSES OF THE UNITED STATES, A. S. Hitchcock, U.S. Dept. of Agriculture. The basic study of American grasses, both indigenous and escapes, cultivated and wild. Over 1400 species. Full descriptions, information. Over 1100 maps, illustrations. Total of 1051pp. 5⅜ x 8½. 22717-0, 22718-9 Pa., Two-vol. set $15.00

THE CACTACEAE,, Nathaniel L. Britton, John N. Rose. Exhaustive, definitive. Every cactus in the world. Full botanical descriptions. Thorough statement of nomenclatures, habitat, detailed finding keys. The one book needed by every cactus enthusiast. Over 1275 illustrations. Total of 1080pp. 8 x 10¼. 21191-6, 21192-4 Clothbd., Two-vol. set $35.00

AMERICAN MEDICINAL PLANTS, Charles F. Millspaugh. Full descriptions, 180 plants covered: history; physical description; methods of preparation with all chemical constituents extracted; all claimed curative or adverse effects. 180 full-page plates. Classification table. 804pp. 6½ x 9¼. 23034-1 Pa. $10.00

A MODERN HERBAL, Margaret Grieve. Much the fullest, most exact, most useful compilation of herbal material. Gigantic alphabetical encyclopedia, from aconite to zedoary, gives botanical information, medical properties, folklore, economic uses, and much else. Indispensable to serious reader. 161 illustrations. 888pp. 6½ x 9¼. (Available in U.S. only) 22798-7, 22799-5 Pa., Two-vol. set $12.00

THE HERBAL or GENERAL HISTORY OF PLANTS, John Gerard. The 1633 edition revised and enlarged by Thomas Johnson. Containing almost 2850 plant descriptions and 2705 superb illustrations, Gerard's *Herbal* is a monumental work, the book all modern English herbals are derived from, the one herbal every serious enthusiast should have in its entirety. Original editions are worth perhaps $750. 1678pp. 8½ x 12¼. 23147-X Clothbd. $50.00

MANUAL OF THE TREES OF NORTH AMERICA, Charles S. Sargent. The basic survey of every native tree and tree-like shrub, 717 species in all. Extremely full descriptions, information on habitat, growth, locales, economics, etc. Necessary to every serious tree lover. Over 100 finding keys. 783 illustrations. Total of 986pp. 5⅜ x 8½. 20277-1, 20278-X Pa., Two-vol. set $10.00

A MAYA GRAMMAR, Alfred M. Tozzer. Practical, useful English-language grammar by the Harvard anthropologist who was one of the three greatest American scholars in the area of Maya culture. Phonetics, grammatical processes, syntax, more. 301pp. 5⅜ x 8½.   23465-7 Pa. $4.00

THE JOURNAL OF HENRY D. THOREAU, edited by Bradford Torrey, F. H. Allen. Complete reprinting of 14 volumes, 1837-61, over two million words; the sourcebooks for *Walden*, etc. Definitive. All original sketches, plus 75 photographs. Introduction by Walter Harding. Total of 1804pp. 8½ x 12¼.   20312-3, 20313-1 Clothbd., Two-vol. set $50.00

CLASSIC GHOST STORIES, Charles Dickens and others. 18 wonderful stories you've wanted to reread: "The Monkey's Paw," "The House and the Brain," "The Upper Berth," "The Signalman," "Dracula's Guest," "The Tapestried Chamber," etc. Dickens, Scott, Mary Shelley, Stoker, etc. 330pp. 5⅜ x 8½.   20735-8 Pa. $3.50

SEVEN SCIENCE FICTION NOVELS, H. G. Wells. Full novels. *First Men in the Moon, Island of Dr. Moreau, War of the Worlds, Food of the Gods, Invisible Man, Time Machine, In the Days of the Comet*. A basic science-fiction library. 1015pp. 5⅜ x 8½. (Available in U.S. only)   20264-X Clothbd. $8.95

ARMADALE, Wilkie Collins. Third great mystery novel by the author of *The Woman in White* and *The Moonstone*. Ingeniously plotted narrative shows an exceptional command of character, incident and mood. Original magazine version with 40 illustrations. 597pp. 5⅜ x 8½.   23429-0 Pa. $5.00

MASTERS OF MYSTERY, H. Douglas Thomson. The first book in English (1931) devoted to history and aesthetics of detective story. Poe, Doyle, LeFanu, Dickens, many others, up to 1930. New introduction and notes by E. F. Bleiler. 288pp. 5⅜ x 8½. (Available in U.S. only)   23606-4 Pa. $4.00

FLATLAND, E. A. Abbott. Science-fiction classic explores life of 2-D being in 3-D world. Read also as introduction to thought about hyperspace. Introduction by Banesh Hoffmann. 16 illustrations. 103pp. 5⅜ x 8½.   20001-9 Pa. $1.75

THREE SUPERNATURAL NOVELS OF THE VICTORIAN PERIOD, edited, with an introduction, by E. F. Bleiler. Reprinted complete and unabridged, three great classics of the supernatural: *The Haunted Hotel* by Wilkie Collins, *The Haunted House at Latchford* by Mrs. J. H. Riddell, and *The Lost Stradivarius* by J. Meade Falkner. 325pp. 5⅜ x 8½.   22571-2 Pa. $4.00

AYESHA: THE RETURN OF "SHE," H. Rider Haggard. Virtuoso sequel featuring the great mythic creation, Ayesha, in an adventure that is fully as good as the first book, *She*. Original magazine version, with 47 original illustrations by Maurice Greiffenhagen. 189pp. 6½ x 9¼.   23649-8 Pa. $3.50

DRAWINGS OF WILLIAM BLAKE, William Blake. 92 plates from Book of Job, *Divine Comedy, Paradise Lost,* visionary heads, mythological figures, Laocoon, etc. Selection, introduction, commentary by Sir Geoffrey Keynes. 178pp. 8⅛ x 11.   22303-5 Pa. $4.00

ENGRAVINGS OF HOGARTH, William Hogarth. 101 of Hogarth's greatest works: *Rake's Progress, Harlot's Progress, Illustrations for Hudibras, Before and After, Beer Street and Gin Lane,* many more. Full commentary. 256pp. 11 x 13¾.   22479-1 Pa. $7.95

DAUMIER: 120 GREAT LITHOGRAPHS, Honore Daumier. Wide-ranging collection of lithographs by the greatest caricaturist of the 19th century. Concentrates on eternally popular series on lawyers, on married life, on liberated women, etc. Selection, introduction, and notes on plates by Charles F. Ramus. Total of 158pp. 9⅜ x 12¼.   23512-2 Pa. $5.50

DRAWINGS OF MUCHA, Alphonse Maria Mucha. Work reveals drafts-man of highest caliber: studies for famous posters and paintings, render-ings for book illustrations and ads, etc. 70 works, 9 in color; including 6 items not drawings. Introduction. List of illustrations. 72pp. 9⅜ x 12¼. (Available in U.S. only)   23672-2 Pa. $4.00

GIOVANNI BATTISTA PIRANESI: DRAWINGS IN THE PIERPONT MORGAN LIBRARY, Giovanni Battista Piranesi. For first time ever all of Morgan Library's collection, world's largest. 167 illustrations of rare Piranesi drawings—archeological, architectural, decorative and visionary. Essay, detailed list of drawings, chronology, captions. Edited by Felice Stampfle. 144pp. 9⅜ x 12¼.   23714-1 Pa. $7.50

NEW YORK ETCHINGS (1905-1949), John Sloan. All of important American artist's N.Y. life etchings. 67 works include some of his best art; also lively historical record—Greenwich Village, tenement scenes. Edited by Sloan's widow. Introduction and captions. 79pp. 8⅜ x 11¼.
                                              23651-X Pa. $4.00

CHINESE PAINTING AND CALLIGRAPHY: A PICTORIAL SURVEY, Wan-go Weng. 69 fine examples from John M. Crawford's matchless private collection: landscapes, birds, flowers, human figures, etc., plus calligraphy. Every basic form included: hanging scrolls, handscrolls, album leaves, fans, etc. 109 illustrations. Introduction. Captions. 192pp. 8⅞ x 11¾.
                                              23707-9 Pa. $7.95

DRAWINGS OF REMBRANDT, edited by Seymour Slive. Updated Lipp-mann, Hofstede de Groot edition, with definitive scholarly apparatus. All portraits, biblical sketches, landscapes, nudes, Oriental figures, classical studies, together with selection of work by followers. 550 illustrations. Total of 630pp. 9⅛ x 12¼.   21485-0, 21486-9 Pa., Two-vol. set $15.00

THE DISASTERS OF WAR, Francisco Goya. 83 etchings record horrors of Napoleonic wars in Spain and war in general. Reprint of 1st edition, plus 3 additional plates. Introduction by Philip Hofer. 97pp. 9⅜ x 8¼.
                                              21872-4 Pa. $3.75

THE COMPLETE WOODCUTS OF ALBRECHT DURER, edited by Dr. W. Kurth. 346 in all: "Old Testament," "St. Jerome," "Passion," "Life of Virgin," Apocalypse," many others. Introduction by Campbell Dodgson. 285pp. 8½ x 12¼. 21097-9 Pa. $7.50

DRAWINGS OF ALBRECHT DURER, edited by Heinrich Wolfflin. 81 plates show development from youth to full style. Many favorites; many new. Introduction by Alfred Werner. 96pp. 8⅛ x 11. 22352-3 Pa. $5.00

THE HUMAN FIGURE, Albrecht Dürer. Experiments in various techniques—stereometric, progressive proportional, and others. Also life studies that rank among finest ever done. Complete reprinting of *Dresden Sketchbook*. 170 plates. 355pp. 8⅜ x 11¼. 21042-1 Pa. $7.95

OF THE JUST SHAPING OF LETTERS, Albrecht Dürer. Renaissance artist explains design of Roman majuscules by geometry, also Gothic lower and capitals. Grolier Club edition. 43pp. 7⅞ x 10¾ 21306-4 Pa. $8.00

TEN BOOKS ON ARCHITECTURE, Vitruvius. The most important book ever written on architecture. Early Roman aesthetics, technology, classical orders, site selection, all other aspects. Stands behind everything since. Morgan translation. 331pp. 5⅜ x 8½. 20645-9 Pa. $4.00

THE FOUR BOOKS OF ARCHITECTURE, Andrea Palladio. 16th-century classic responsible for Palladian movement and style. Covers classical architectural remains, Renaissance revivals, classical orders, etc. 1738 Ware English edition. Introduction by A. Placzek. 216 plates. 110pp. of text. 9½ x 12¾. 21308-0 Pa. $8.95

HORIZONS, Norman Bel Geddes. Great industrialist stage designer, "father of streamlining," on application of aesthetics to transportation, amusement, architecture, etc. 1932 prophetic account; function, theory, specific projects. 222 illustrations. 312pp. 7⅞ x 10¾. 23514-9 Pa. $6.95

FRANK LLOYD WRIGHT'S FALLINGWATER, Donald Hoffmann. Full, illustrated story of conception and building of Wright's masterwork at Bear Run, Pa. 100 photographs of site, construction, and details of completed structure. 112pp. 9¼ x 10. 23671-4 Pa. $5.50

THE ELEMENTS OF DRAWING, John Ruskin. Timeless classic by great Viltorian; starts with basic ideas, works through more difficult. Many practical exercises. 48 illustrations. Introduction by Lawrence Campbell. 228pp. 5⅜ x 8½. 22730-8 Pa. $2.75

GIST OF ART, John Sloan. Greatest modern American teacher, Art Students League, offers innumerable hints, instructions, guided comments to help you in painting. Not a formal course. 46 illustrations. Introduction by Helen Sloan. 200pp. 5⅜ x 8½. 23435-5 Pa. $4.00

HISTORY OF BACTERIOLOGY, William Bulloch. The only comprehensive history of bacteriology from the beginnings through the 19th century. Special emphasis is given to biography-Leeuwenhoek, etc. Brief accounts of 350 bacteriologists form a separate section. No clearer, fuller study, suitable to scientists and general readers, has yet been written. 52 illustrations. 448pp. 5⅝ x 8¼.   23761-3 Pa. $6.50

THE COMPLETE NONSENSE OF EDWARD LEAR, Edward Lear. All nonsense limericks, zany alphabets, Owl and Pussycat, songs, nonsense botany, etc., illustrated by Lear. Total of 321pp. 5⅜ x 8½. (Available in U.S. only)   20167-8 Pa. $3.00

INGENIOUS MATHEMATICAL PROBLEMS AND METHODS, Louis A. Graham. Sophisticated material from Graham *Dial*, applied and pure; stresses solution methods. Logic, number theory, networks, inversions, etc. 237pp. 5⅜ x 8½.   20545-2 Pa. $3.50

BEST MATHEMATICAL PUZZLES OF SAM LOYD, edited by Martin Gardner. Bizarre, original, whimsical puzzles by America's greatest puzzler. From fabulously rare *Cyclopedia*, including famous 14-15 puzzles, the Horse of a Different Color, 115 more. Elementary math. 150 illustrations. 167pp. 5⅜ x 8½.   20498-7 Pa. $2.75

THE BASIS OF COMBINATION IN CHESS, J. du Mont. Easy-to-follow, instructive book on elements of combination play, with chapters on each piece and every powerful combination team—two knights, bishop and knight, rook and bishop, etc. 250 diagrams. 218pp. 5⅜ x 8½. (Available in U.S. only)   23644-7 Pa. $3.50

MODERN CHESS STRATEGY, Ludek Pachman. The use of the queen, the active king, exchanges, pawn play, the center, weak squares, etc. Section on rook alone worth price of the book. Stress on the moderns. Often considered the most important book on strategy. 314pp. 5⅜ x 8½.   20290-9 Pa. $4.50

LASKER'S MANUAL OF CHESS, Dr. Emanuel Lasker. Great world champion offers very thorough coverage of all aspects of chess. Combinations, position play, openings, end game, aesthetics of chess, philosophy of struggle, much more. Filled with analyzed games. 390pp. 5⅜ x 8½.   20640-8 Pa. $5.00

500 MASTER GAMES OF CHESS, S. Tartakower, J. du Mont. Vast collection of great chess games from 1798-1938, with much material nowhere else readily available. Fully annotated, arranged by opening for easier study. 664pp. 5⅜ x 8½.   23208-5 Pa. $7.50

A GUIDE TO CHESS ENDINGS, Dr. Max Euwe, David Hooper. One of the finest modern works on chess endings. Thorough analysis of the most frequently encountered endings by former world champion. 331 examples, each with diagram. 248pp. 5⅜ x 8½.   23332-4 Pa. $3.50

SECOND PIATIGORSKY CUP, edited by Isaac Kashdan. One of the greatest tournament books ever produced in the English language. All 90 games of the 1966 tournament, annotated by players, most annotated by both players. Features Petrosian, Spassky, Fischer, Larsen, six others. 228pp. 5⅜ x 8½. 23572-6 Pa. $3.50

ENCYCLOPEDIA OF CARD TRICKS, revised and edited by Jean Hugard. How to perform over 600 card tricks, devised by the world's greatest magicians: impromptus, spelling tricks, key cards, using special packs, much, much more. Additional chapter on card technique. 66 illustrations. 402pp. 5⅜ x 8½. (Available in U.S. only) 21252-1 Pa. $3.95

MAGIC: STAGE ILLUSIONS, SPECIAL EFFECTS AND TRICK PHO-TOGRAPHY, Albert A. Hopkins, Henry R. Evans. One of the great classics; fullest, most authorative explanation of vanishing lady, levitations, scores of other great stage effects. Also small magic, automata, stunts. 446 illustrations. 556pp. 5⅜ x 8½. 23344-8 Pa. $6.95

THE SECRETS OF HOUDINI, J. C. Cannell. Classic study of Houdini's incredible magic, exposing closely-kept professional secrets and revealing, in general terms, the whole art of stage magic. 67 illustrations. 279pp. 5⅜ x 8½. 22913-0 Pa. $3.00

HOFFMANN'S MODERN MAGIC, Professor Hoffmann. One of the best, and best-known, magicians' manuals of the past century. Hundreds of tricks from card tricks and simple sleight of hand to elaborate illusions involving construction of complicated machinery. 332 illustrations. 563pp. 5⅜ x 8½. 23623-4 Pa. $6.00

MADAME PRUNIER'S FISH COOKERY BOOK, Mme. S. B. Prunier. More than 1000 recipes from world famous Prunier's of Paris and London, specially adapted here for American kitchen. Grilled tournedos with anchovy butter, Lobster a la Bordelaise, Prunier's prized desserts, more. Glossary. 340pp. 5⅜ x 8½. (Available in U.S. only) 22679-4 Pa. $3.00

FRENCH COUNTRY COOKING FOR AMERICANS, Louis Diat. 500 easy-to-make, authentic provincial recipes compiled by former head chef at New York's Fitz-Carlton Hotel: onion soup, lamb stew, potato pie, more. 309pp. 5⅜ x 8½. 23665-X Pa. $3.95

SAUCES, FRENCH AND FAMOUS, Louis Diat. Complete book gives over 200 specific recipes: bechamel, Bordelaise, hollandaise, Cumberland, apricot, etc. Author was one of this century's finest chefs, originator of vichyssoise and many other dishes. Index. 156pp. 5⅜ x 8.
23663-3 Pa. $2.50

TOLL HOUSE TRIED AND TRUE RECIPES, Ruth Graves Wakefield. Authentic recipes from the famous Mass. restaurant: popovers, veal and ham loaf, Toll House baked beans, chocolate cake crumb pudding, much more. Many helpful hints. Nearly 700 recipes. Index. 376pp. 5⅜ x 8½.
23560-2 Pa. $4.50

THE EARLY WORK OF AUBREY BEARDSLEY, Aubrey Beardsley. 157 plates, 2 in color: *Manon Lescaut, Madame Bovary, Morte Darthur, Salome,* other. Introduction by H. Marillier. 182pp. 8⅛ x 11. 21816-3 Pa. $4.50

THE LATER WORK OF AUBREY BEARDSLEY, Aubrey Beardsley. Exotic masterpieces of full maturity: *Venus and Tannhauser, Lysistrata, Rape of the Lock, Volpone,* Savoy material, etc. 174 plates, 2 in color. 186pp. 8⅛ x 11. 21817-1 Pa. $4.50

THOMAS NAST'S CHRISTMAS DRAWINGS, Thomas Nast. Almost all Christmas drawings by creator of image of Santa Claus as we know it, and one of America's foremost illustrators and political cartoonists. 66 illustrations. 3 illustrations in color on covers. 96pp. 8⅜ x 11¼. 23660-9 Pa. $3.50

THE DORÉ ILLUSTRATIONS FOR DANTE'S DIVINE COMEDY, Gustave Doré. All 135 plates from Inferno, Purgatory, Paradise; fantastic tortures, infernal landscapes, celestial wonders. Each plate with appropriate (translated) verses. 141pp. 9 x 12. 23231-X Pa. $4.50

DORÉ'S ILLUSTRATIONS FOR RABELAIS, Gustave Doré. 252 striking illustrations of *Gargantua and Pantagruel* books by foremost 19th-century illustrator. Including 60 plates, 192 delightful smaller illustrations. 153pp. 9 x 12. 23656-0 Pa. $5.00

LONDON: A PILGRIMAGE, Gustave Doré, Blanchard Jerrold. Squalor, riches, misery, beauty of mid-Victorian metropolis; 55 wonderful plates, 125 other illustrations, full social, cultural text by Jerrold. 191pp. of text. 9⅜ x 12¼. 22306-X Pa. $6.00

THE RIME OF THE ANCIENT MARINER, Gustave Doré, S. T. Coleridge. Dore's finest work, 34 plates capture moods, subtleties of poem. Full text. Introduction by Millicent Rose. 77pp. 9¼ x 12. 22305-1 Pa. $3.50

THE DORE BIBLE ILLUSTRATIONS, Gustave Doré. All wonderful, detailed plates: Adam and Eve, Flood, Babylon, Life of Jesus, etc. Brief King James text with each plate. Introduction by Millicent Rose. 241 plates. 241pp. 9 x 12. 23004-X Pa. $6.00

THE COMPLETE ENGRAVINGS, ETCHINGS AND DRYPOINTS OF ALBRECHT DURER. "Knight, Death and Devil"; "Melencolia," and more—all Dürer's known works in all three media, including 6 works formerly attributed to him. 120 plates. 235pp. 8⅜ x 11¼. 22851-7 Pa. $6.50

MAXIMILIAN'S TRIUMPHAL ARCH, Albrecht Dürer and others. Incredible monument of woodcut art: 8 foot high elaborate arch—heraldic figures, humans, battle scenes, fantastic elements—that you can assemble yourself. Printed on one side, layout for assembly. 143pp. 11 x 16. 21451-6 Pa. $5.00

UNCLE SILAS, J. Sheridan LeFanu. Victorian Gothic mystery novel, considered by many best of period, even better than Collins or Dickens. Wonderful psychological terror. Introduction by Frederick Shroyer. 436pp. 5⅜ x 8½. 21715-9 Pa. **$6.00**

JURGEN, James Branch Cabell. The great erotic fantasy of the 1920's that delighted thousands, shocked thousands more. Full final text, Lane edition with 13 plates by Frank Pape. 346pp. 5⅜ x 8½.
23507-6 Pa. **$4.50**

THE CLAVERINGS, Anthony Trollope. Major novel, chronicling aspects of British Victorian society, personalities. Reprint of Cornhill serialization, 16 plates by M. Edwards; first reprint of full text. Introduction by Norman Donaldson. 412pp. 5⅜ x 8½. 23464-9 Pa. **$5.00**

KEPT IN THE DARK, Anthony Trollope. Unusual short novel about Victorian morality and abnormal psychology by the great English author. Probably the first American publication. Frontispiece by Sir John Millais. 92pp. 6½ x 9¼. 23609-9 Pa. **$2.50**

RALPH THE HEIR, Anthony Trollope. Forgotten tale of illegitimacy, inheritance. Master novel of Trollope's later years. Victorian country estates, clubs, Parliament, fox hunting, world of fully realized characters. Reprint of 1871 edition. 12 illustrations by F. A. Faser. 434pp. of text. 5⅜ x 8½. 23642-0 Pa. **$5.00**

YEKL and THE IMPORTED BRIDEGROOM AND OTHER STORIES OF THE NEW YORK GHETTO, Abraham Cahan. Film *Hester Street* based on *Yekl* (1896). Novel, other stories among first about Jewish immigrants of N.Y.'s East Side. Highly praised by W. D. Howells—Cahan "a new star of realism." New introduction by Bernard G. Richards. 240pp. 5⅜ x 8½. 22427-9 Pa. **$3.50**

THE HIGH PLACE, James Branch Cabell. Great fantasy writer's enchanting comedy of disenchantment set in 18th-century France. Considered by some critics to be even better than his famous *Jurgen*. 10 illustrations and numerous vignettes by noted fantasy artist Frank C. Pape. 320pp. 5⅜ x 8½. 23670-6 Pa. **$4.00**

ALICE'S ADVENTURES UNDER GROUND, Lewis Carroll. Facsimile of ms. Carroll gave Alice Liddell in 1864. Different in many ways from final Alice. Handlettered, illustrated by Carroll. Introduction by Martin Gardner. 128pp. 5⅜ x 8½. 21482-6 Pa. **$2.00**

FAVORITE ANDREW LANG FAIRY TALE BOOKS IN MANY COLORS, Andrew Lang. The four Lang favorites in a boxed set—the complete *Red, Green, Yellow* and *Blue* Fairy Books. 164 stories; 439 illustrations by Lancelot Speed, Henry Ford and G. P. Jacomb Hood. Total of about 1500pp. 5⅜ x 8½. 23407-X Boxed set, Pa. **$14.95**

PRINCIPLES OF ORCHESTRATION, Nikolay Rimsky-Korsakov. Great classical orchestrator provides fundamentals of tonal resonance, progression of parts, voice and orchestra, tutti effects, much else in major document. 330pp. of musical excerpts. 489pp. 6½ x 9¼. 21266-1 Pa. $6.00

TRISTAN UND ISOLDE, Richard Wagner. Full orchestral score with complete instrumentation. Do not confuse with piano reduction. Commentary by Felix Mottl, great Wagnerian conductor and scholar. Study score. 655pp. 8⅛ x 11. 22915-7 Pa. $12.50

REQUIEM IN FULL SCORE, Giuseppe Verdi. Immensely popular with choral groups and music lovers. Republication of edition published by C. F. Peters, Leipzig, n. d. German frontmaker in English translation. Glossary. Text in Latin. Study score. 204pp. 9⅜ x 12¼.
23682-X Pa. $6.00

COMPLETE CHAMBER MUSIC FOR STRINGS, Felix Mendelssohn. All of Mendelssohn's chamber music: Octet, 2 Quintets, 6 Quartets, and Four Pieces for String Quartet. (Nothing with piano is included). Complete works edition (1874-7). Study score. 283 pp. 9⅜ x 12¼.
23679-X Pa. $6.95

POPULAR SONGS OF NINETEENTH-CENTURY AMERICA, edited by Richard Jackson. 64 most important songs: "Old Oaken Bucket," "Arkansas Traveler," "Yellow Rose of Texas," etc. Authentic original sheet music, full introduction and commentaries. 290pp. 9 x 12. 23270-0 Pa. $6.00

COLLECTED PIANO WORKS, Scott Joplin. Edited by Vera Brodsky Lawrence. Practically all of Joplin's piano works—rags, two-steps, marches, waltzes, etc., 51 works in all. Extensive introduction by Rudi Blesh. Total of 345pp. 9 x 12. 23106-2 Pa. $14.95

BASIC PRINCIPLES OF CLASSICAL BALLET, Agrippina Vaganova. Great Russian theoretician, teacher explains methods for teaching classical ballet; incorporates best from French, Italian, Russian schools. 118 illustrations. 175pp. 5⅜ x 8½. 22036-2 Pa. $2.50

CHINESE CHARACTERS, L. Wieger. Rich analysis of 2300 characters according to traditional systems into primitives. Historical-semantic analysis to phonetics (Classical Mandarin) and radicals. 820pp. 6⅛ x 9¼.
21321-8 Pa. $10.00

EGYPTIAN LANGUAGE: EASY LESSONS IN EGYPTIAN HIEROGLYPHICS, E. A. Wallis Budge. Foremost Egyptologist offers Egyptian grammar, explanation of hieroglyphics, many reading texts, dictionary of symbols. 246pp. 5 x 7½. (Available in U.S. only)
21394-3 Clothbd. $7.50

AN ETYMOLOGICAL DICTIONARY OF MODERN ENGLISH, Ernest Weekley. Richest, fullest work, by foremost British lexicographer. Detailed word histories. Inexhaustible. Do not confuse this with *Concise Etymological Dictionary*, which is abridged. Total of 856pp. 6½ x 9¼.
21873-2, 21874-0 Pa., Two-vol. set $12.00

GEOMETRY, RELATIVITY AND THE FOURTH DIMENSION, Rudolf Rucker. Exposition of fourth dimension, means of visualization, concepts of relativity as Flatland characters continue adventures. Popular, easily followed yet accurate, profound. 141 illustrations. 133pp. 5⅜ x 8½.
23400-2 Pa. $2.75

THE ORIGIN OF LIFE, A. I. Oparin. Modern classic in biochemistry, the first rigorous examination of possible evolution of life from nitrocarbon compounds. Non-technical, easily followed. Total of 295pp. 5⅜ x 8½.
60213-3 Pa. $4.00

PLANETS, STARS AND GALAXIES, A. E. Fanning. Comprehensive introductory survey: the sun, solar system, stars, galaxies, universe, cosmology; quasars, radio stars, etc. 24pp. of photographs. 189pp. 5⅜ x 8½. (Available in U.S. only)
21680-2 Pa. $3.00

THE THIRTEEN BOOKS OF EUCLID'S ELEMENTS, translated with introduction and commentary by Sir Thomas L. Heath. Definitive edition. Textual and linguistic notes, mathematical analysis, 2500 years of critical commentary. Do not confuse with abridged school editions. Total of 1414pp. 5⅜ x 8½.
60088-2, 60089-0, 60090-4 Pa., Three-vol. set $18.50

DIALOGUES CONCERNING TWO NEW SCIENCES, Galileo Galilei. Encompassing 30 years of experiment and thought, these dialogues deal with geometric demonstrations of fracture of solid bodies, cohesion, leverage, speed of light and sound, pendulums, falling bodies, accelerated motion, etc. 300pp. 5⅜ x 8½.
60099-8 Pa. $4.00

Prices subject to change without notice.

Available at your book dealer or write for free catalogue to Dept. GI, Dover Publications, Inc., 180 Varick St., N.Y., N.Y. 10014. Dover publishes more than 175 books each year on science, elementary and advanced mathematics, biology, music, art, literary history, social sciences and other areas.